DEEP-SKY WONDERS

*Scotty's asteroid citation
from Minor Planet Circular 10845*

**3031 Houston = 1984 CX
June 22, 1986**

Discovered 1984 February 8 by E. Bowell at Anderson Mesa.

Named in honor of Walter Scott Houston, American amateur astronomer well known for his column Deep-Sky Wonders in the magazine *Sky & Telescope*. Houston has specialized in the visual study of deep-sky objects and has guided countless amateurs to view and marvel at the varied objects within the grasp of small telescopes. Name proposed by the discoverer following a suggestion by P. L. Dombrowski.

DEEP-SKY WONDERS

By Walter Scott Houston

Adapted from his columns in *Sky & Telescope*

Selections and commentary
By Stephen James O'Meara

Sky Publishing Corporation
Cambridge, Massachusetts

© 1999 Sky Publishing Corporation

Published by Sky Publishing Corporation, 49 Bay State Road, Cambridge, MA 02138-1200, USA
http://SkyandTelescope.com

All rights reserved. Except for brief passages quoted in a review, no part of this book may be reproduced by any mechanical, photographic, or electronic process, nor may it be quoted in an information-retrieval system, transmitted, or otherwise copied for public or private use, without the written permission of the copyright holder. Requests for permission or further information should be addressed to Permissions, Sky Publishing Corporation, 49 Bay State Road, Cambridge, MA 02138-1200, USA.

Originally published in hardcover by Sky Publishing Corp. in 1999

Library of Congress cataloging-in-publication data

Houston, Walter Scott

Deep-Sky Wonders/Walter Scott Houston; Selections and Commentary by Stephen James O'Meara: foreword by Brian A. Skiff

       p. cm.

       Includes bibliographical references and index.
       ISBN: 1-931559-23-6 (alk. paper)
       I. Astronomy. I. O'Meara, Stephen James, 1956-
       II. Title

       QB51.H795 1999   99-047381
       523—dc21
       Printed in Canada

CONTENTS

	VII	Preface *by Stephen James O'Meara*
	XIII	Blue Postcards from Twinky *by Brian A. Skiff*
	XV	Portrait of an Icon *by Dennis di Cicco*
CHAPTER 1	1	January
CHAPTER 2	27	February
CHAPTER 3	49	March
CHAPTER 4	75	April
CHAPTER 5	99	May
CHAPTER 6	123	June
CHAPTER 7	151	July
CHAPTER 8	173	August
CHAPTER 9	195	September
CHAPTER 10	217	October
CHAPTER 11	241	November
CHAPTER 12	261	December
	283	Sources
	289	Bibliography
	293	Index

Preface

By Stephen James O'Meara

When I joined the editorial staff of *Sky & Telescope* in 1979, I found myself surrounded by astronomical icons — Charles Federer, Joseph Ashbrook, Leif Robinson, William Shawcross, Kelly Beatty, Dennis di Cicco, Roger Sinnott, and Dennis Milon; all names recognized and respected by the astronomical community.

A phone call away was yet another tier of luminaries who comprised the magazine's columnists: comet expert John Bortle, sky-lore aficionado George Lovi, and the ever-popular deep-sky wizard Walter Scott Houston, known as Scotty to his friends and Twinky to his correspondents. To work with these people (all pivotal figures in shaping our hobby into the fun and challenging field it is today) meant enjoying a life of constant education; not hard to take. I have enjoyed that privilege now for two decades, and continue to reap the benefits of knowing them and other *S&T* staffers who have since joined the magazine's ranks.

One highlight of my career occurred in the early 1990s, when Leif Robinson and Dennis di Cicco offered me the task of editing Deep-Sky Wonders, Scotty's trademark column. I immediately accepted the job and cherished my time working with Scotty until his untimely death in December 1993. I first met him in the early 1970s at an American Association of Variable Star Observers' meeting. He was an amateur's amateur, a man who could grind a mirror as well as he could estimate the magnitude of a cataclysmic variable, expound on the vagaries of atmospheric phenomena, or explain how to see dim details in a galactic nebula. He was also a groundbreaker and a ground shaker. He was generally the first to announce an observing trend and predict its promise or demise. His words reflected his love for the hobby and fueled our imaginations. And his Deep-Sky Wonders column was filled with all this observational magic.

Taking Scotty's work through our vigorous layers of editing and fact-checking was never a chore, but always an enjoyable learning experience. So why would a collection of Scotty's writings need further editing? Why not simply slap together a selection of existing columns, rub our hands in satisfaction, and say with a smile, "Well, that was easy!"? There are several reasons. First, Scotty generally divided the sky into 12 monthly strips, each two hours of right ascension wide, running between the north and south celestial poles. During mid-May, for example, for around 10:30 p.m. local time, he would discuss objects found in the strip of sky centered on the meridian between 12 and 14 hours right ascension ... but not always. If he thought readers might be inspired by Orion rising in the east after sunset, instead of waiting for it to transit the meridian, he would toss his editorial framework to the wind and wax poetic about Orion, a move that accentuated his passion for the moment.

Second, Scotty's columns were structurally complex — a mixture of colorful prose, descriptive history, helpful hints, and observational commentary, in no fixed order. Sometimes the components would be distributed in his columns over months or years. Many times he would wrap up a column on a particular subject by offering a potpourri of additional objects nearby, any one of which might become the focus of future discussion. So he did not always write about, say, deep-sky objects in Cepheus only in November, but also in June, July, September, and December.

Scotty's writings cover nearly a half century of deep-sky exploration. How does one select the columns that best contain Scotty's writings on a given object or topic? Take the Ring Nebula, for example. In the August 1953 Deep-Sky Wonders Scotty presents some historical information about the Ring, then goes on to tell novice observers how to find it. (That's important.) In his July 1984 column he details the appearance of the Ring through a variety of instruments and with a range of powers. (That's also important.) His July 1988 column focuses on the Ring's central star and its controversial sightings. (Yet another important article.) Other columns on other dates add more observational perspective and historical anecdotes.

So which article about the Ring Nebula is the most important? The fact is that they are all equally useful, for they offer information that is of interest to a wide range of readers. To exclude any one of these columns would be to leave out vital information and data. So the solution, though seemingly daunting, was simple. I decided to go through the nearly half century of Scotty's writings and begin the familiar process of cutting and pasting.

How did I decide which topics to select? First, I reviewed photocopies of the roughly 550 published columns, which were collated into their respective months. Since Scotty discussed many of the same objects over the years, it was not difficult to select about a half dozen features as representative samples for each month. I used each feature as a foundation, into which I incorporated sentences, paragraphs, and sections from other columns on the same topic. This second tier of the editing process was performed in three stages.

First I made keyword searches through each published column and created a background database for each feature. Next I reviewed the background text and weeded out repetitive sections. The challenge was to cut and paste passages so that they preserved Scotty's original sentence structures, but also flowed naturally. It was important that the "feel" of the work remained Scotty's. As an example of this cut-and-paste process, consider the following sentences, which come from three separate Deep-Sky Wonders columns about the Ring Nebula:

August 1953: It seems to have been discovered by Darquier in 1779, and he and Messier saw it in their imperfect instruments as a pale planetary disk. By Herschel's time its "smoke-ring" form was known and justly admired.

August 1976: It was discovered by the French astronomer Antoine Darquier in 1779 while comet hunting with a 3-inch refractor. He described it as "a very dull nebula, but perfectly outlined; as large as Jupiter and looks like a fading planet."

August 1980: It was discovered by the French amateur Antoine Darquier in 1779, who described it as about the size of Jupiter, but dim and looking like a "faded planet." William Herschel thought it to be a ring of stars just beyond the resolution of his telescopes. His son John first called attention to the fainter nebulosity which fills the interior of the ring, likening it to gauze stretched over a hoop.

Notice that though the above passages are very similar, they differ in the amount of detail Scotty revealed about Darquier's discovery. After gleaning the vital information from each of these passages, I created the following final version:

It was discovered by the French astronomer Antoine Darquier in 1779 while comet hunting with a 3-inch refractor. He described it as "a very dull nebula, but perfectly outlined; as large as Jupiter and looking like a fading planet." By Herschel's time its "smoke-ring" form was known and justly admired. William Herschel thought it to be a ring of stars just beyond the resolution of his telescopes. His son John first called attention to the fainter nebulosity which fills the interior of the ring, likening it to gauze stretched over a hoop.

Although few of the essays in this book will duplicate any single Deep-Sky Wonders column, 99 percent of the words in this book are Scotty's. They begin in classic Scotty fashion, with an eloquent and colorful description of the season or the month, or perhaps some topic of astronomy that happens to be on his mind. He then launches into an informative discussion of a particular object of interest.

The text follows a general pattern; namely, it first presents any historical information, followed by Scotty's advice on how to find the object with the naked eye or binoculars, then the object's appearance through small to large telescopes, and finally a wrap-up discussion including observing mysteries and challenges. I have not replaced any of his sentences with my own. I did not add any sentences or thoughts. Occasionally I needed to add a word or a phrase for a smooth transition. To make Scotty's references to books and people appear in a logical fashion in each chapter, we spell out a book or person's full name if it is the first reference, and let subsequent references remain as Scotty originally wrote them. Where something might seem ambiguous or obscure, footnotes have been added to clarify the material for the reader.

Occasionally Scotty referred observers to a star chart that is now out of print, such as the Skalnate Pleso *Atlas of the Heavens*. At other times, Scotty used a work's earlier edition, such as *Norton's Star Atlas*. I did not alter Scotty's words to reflect the existence of newer atlases. Scotty was not alive when, say, the *Millennium Star Atlas* came out, so replacing the *Atlas of the Heavens* with it would be anachronistic. But Scotty was aware of the original *Sky Atlas 2000.0*, *Uranometria 2000.0*, and the *Deep-Sky Field Guide*, and these atlases do appear here by name.

All chapters and sections open with a brief introduction in which I set the tone for what follows. The chapters all end with tables of objects that Scotty discussed and are meant to guide the reader who wants to go out and take up Scotty's observing challenges. The book concludes with a bibliography, a list by chapter of the issues of *Sky & Telescope* in which the original text appeared, and an index.

That Scotty's columns span several decades also means they include variations in nomenclature and editorial style. For example, in the 1950s, Scotty might write, "the great nebula in Andromeda," while in the 1980s he would write "the Great Galaxy in Andromeda"; so the text has been edited for consistency of style. The sizes, magnitudes, dimensions, and distances of deep-sky objects also show marked variation over the years. For consistency (and accuracy), I selected the most recent data used by Scotty in his columns.

It is important to remember that this book is not a field guide but a collection of works by a man who kept his hand on the pulse of amateur astronomy as it underwent sometimes dramatic revolutions of thought.

This anthology of Scotty's writings is intended to encapsulate the man's vast knowledge of the night sky, which he accumulated during a long and fruitful career as a writer and an observer. It captures the full spirit of Scotty. He was a man who did not fear going against the grain of traditional thinking. He often "questioned authority;" at times his words were quixotic, especially when he encouraged us to pursue those elusive windmills of the deep sky — dim galaxies at the limit of vision, painfully faint stars, and phantom nebulae. If there was a visual challenge, he was willing to face it. That flavor and style is embodied in this book. And Scotty was not all talk. He was a good listener and a generous supporter of amateur claims; he would at least give us the benefit of the doubt. He aimed to define, and redefine, the limits of observing and thus the ever-changing truth, and so he supported us and our views in his column. And that is how deep-sky observing progressed, in fits and starts, in Scotty's verbal arena.

Working on this book, I found myself growing even fonder of this man who gave so much to the amateur astronomical community. His words instilled in us a feeling of dignity and hope. "So it's with a great sense of pride," Scotty said, "that we, as amateurs, go outside and enjoy the night sky."

Acknowledgements

No anthology is the work of a sole author and editor, and this one is no exception. At least a dozen people formed the "Scotty Production Team," and the project couldn't have been done without them.

Sky & Telescope associate editor Dennis di Cicco worked with Scotty for nearly 20 years. Most of the original editing of the columns, then, was done by him. A world-renowned imager of the night sky, Dennis also supplied several deep-sky shots and photographs of Scotty. He also authored and updated the Scotty retrospective on page xv, first published in *Sky & Telescope* to celebrate the occasion of Scotty's 75th birthday.

One of Scotty's best friends, a favorite deep-sky observer and frequent correspondent, Brian A. Skiff of Lowell Observatory, graciously agreed to write the book's introduction.

Canadian amateur astronomer James Lucyk, an avid Deep-Sky Wonders reader, keyboarded in all the Scotty columns as a labor of love. Sky Publishing's special projects assistant E. Talmadge Mentall compiled the tables and lists of sources and painstakingly checked all of the measurements, figures, and citations.

Sky Publishing's image archivist Imelda B. Joson did daily penance in the "image crypt," finding just the right images to illustrate Scotty's work. Assisting her in this task was Jacqueline S. Mitchell. Craig Michael Utter's expertise in digital imaging was essential in preparing the book's many photographs. Associate editor and acclaimed atlas wizard Roger W. Sinnott carved away valuable time from his *Sky & Telescope* editorial tasks to create three charts especially for the book.

Our proofreaders were John Woodruff, who also supplied some hard-to-find references and background information, and Nina Barron who checked the book's front and back matter. John Woodruff also created the index.

Publications manager Sally MacGillivray got the right printer lined up and coordinated all the production details. Designer Lynn Sternbergh created the elegant look and feel of the text and jacket.

Richard Tresch Fienberg, president and publisher of Sky Publishing Corporation, recognized that Deep-Sky Wonders is an astronomical treasure and gave me the opportunity and honor to work on putting Scotty's words together.

Carolyn Collins Petersen, Sky's books and products editor, has been *my* editor, guardian angel, and sounding board throughout the project.

Finally, Miriam and Margaret Houston patiently supported and awaited the publication of this book. In lieu of a dedication, I'm sure Scotty would have said that he loved and cherished you both.

Blue Postcards from Twinky

By Brian A. Skiff

In my letters file is a treasured folder with notes from Walter Scott Houston. Our correspondence ran from 1979 through 1990, after which we saw each other often enough to keep up. Usually Scotty's responses came in the form of blue postcards featuring a pot-bellied cartoon stargazer on the front.

As often as not the reverse contained a compressed, personal Deep-Sky Wonders column signed "Twinky." Since I have ready access to the true dark skies of Arizona, Scotty often wrote me to ask about the visibility of objects at the "edge," and occasionally the results would be distilled into a few lines of a column many months later. On one occasion, with sky-suppressing nebula filters newly available, we surprised each other by reporting — in notes that crossed in the mail — the naked-eye visibility of the California Nebula using them. It was typical of his exploration to examine what was possible to observe without extraordinary means, and of my being inspired by his words to push the envelope.

In reading through these chapters, it's clear to me that Scotty was well ahead of the curve in challenging observers to see as much as they could. Within these pages, we find accounts of attempts to view not only well-known objects, such as the Horsehead and Veil nebulae, but also such low-surface-brightness denizens as Barnard's Loop, the Sculptor dwarf galaxy, the "nonexistent" NGC open clusters, and objects that were simply off the beaten track. Who else could direct observers with such ease to a faint clump of IC galaxies near Castor or to a string of overlooked objects in the southern constellation Grus?

The column was driven largely by the responses of readers, whose results Scotty quoted with abandon in order to first follow up on objects and techniques already discussed and then to steer the discussion in new directions. Thus we read of observations by folks now well known in the field, including John Bortle, Brenda Branchett, Steve Gottlieb, Phil Harrington, David Riddle, and Barbara Wilson. The column was often a "moderated newsgroup" for active observers, focused through Scotty's keen eye for where the vanguard was going.

Scotty's articles ranged far and wide over the whole subject of visual astronomical observation. His topics included the idiosyncracies of data in catalogs and the psychological effects of such unpredictability on an observer's attempts to see a supposedly tough object; the seasonal changes in weather; seeing and transparency; aperture and magnification in the telescope; the great observers of the 19th century; and the effect on viewing planetary nebulae after the lens in Scotty's own cataract-clouded eye was replaced with a UV-transparent lens implant.

Scotty had a light touch and avoided being distracted by technical details. You don't find any invidious comparisons of different telescope or eyepiece brands in his writing or much about the nitty-gritty of equipment at all, because Scotty knew that the most important piece of equipment was the eye, and its training the most important activity; all else was trivial in comparison. Time wasted arguing the virtues of one eyepiece over another was time not spent honing your observing skills. Similarly the articles are quietly but solidly grounded in the up-to-date research of the time. Although it is clear that he did his homework to prepare each column by looking up details in professional journals and catalogs, we get only the gist of it — he was not out to impress you with his library.

Beyond any specific targets, Scotty saw his job as being the person to cajole, coerce, and otherwise inspire others to go out and observe, no matter what the target.

David Levy tells the story of meeting Scotty at a Deep-Sky Wonder Night in northern Vermont in late August 1966. He had just begun comet hunting some months earlier. In the middle of the night, David took a break and began telling Scotty of his hopes to discover a comet someday. Puffing slowly on his pipe, Scotty asked David what the sky was like outside. He answered that it was pretty clear, dark, and moonless. Scotty then asked if David's telescope was out there, to which the answer was "yes." Scotty took another puff on his pipe, looked up quizzically and said, "Well, David, you sure aren't going to find a comet as long as we're inside talking about it!"

That encounter gave David the dedication to persevere until he'd bagged his first comet. Despite his tenure behind the eyepiece, Scotty never managed to spy a comet of his own. But in 1986, at the suggestion of amateur Phil Dombrowski, Lowell Observatory astronomer Ted Bowell named an asteroid — 3031 Houston — in Scotty's honor. Readers will recognize that this asteroid number is the same as the NGC number of one of the brightest handful of galaxies in the sky, M81 — a coincidence not lost on Scotty. In a letter to Bowell, he wrote that he had given a theme talk at Stellafane on the occasion of an asteroid being presented to the meteorite expert Oscar Monnig [2780 Monnig]. "At the time I fantasized about getting one myself, but gave up the idea as an impossible job. Now look how bright the Sun shines."

Like many of the devoted followers of the column and readers of this collection, I became a deep-sky observer 30 years ago because of Scotty Houston. The next generation will find here the reasons why far bigger telescopes, more sophisticated sky-charting software, and other modern accouterments aren't the most important things in the hobby. Sit and read one of the masters, and then put the book down and go look at the sky!

Portrait of an Icon

By Dennis di Cicco

The wonderful areas of Cygnus are overhead.
Sweep along the Milky Way
with binoculars or a rich-field telescope . . .
The dark nebulae are plainly visible.

Anyone familiar with Walter Scott Houston's writing would not be the least bit surprised to learn that the quote above came from his monthly Deep-Sky Wonders column in *Sky & Telescope*. What might be surprising to contemporary readers, however, would be the date — September 1946 — since this was well before many of them were born, if we are to believe the magazine's demographic surveys. The quote is from Houston's first "column," which was a mere 12 sentences long and served mainly as a caption to a one-page, black-and-white star chart. During the ensuing decades Deep-Sky Wonders grew in size and reputation and became one of the most widely read and enjoyed sections of *Sky & Telescope*. When Houston succumbed after a brief illness while traveling in Mexico just before Christmas 1993, he was riding a seemingly perpetual wave of popularity that crested in the mid-1980s and accompanied the large-aperture Dobsonian telescope revolution. He had been a continuous contributor to the magazine for more than 47 years, a record that still stands.

On May 30, 1987, Houston celebrated his 75th birthday. The milestone served as an excuse to profile one of the magazine's most-recognized personalities. At the time, Twinky, as he was known to many friends and correspondents, was still an avid skywatcher with a special interest in variable stars and, of course, deep-sky objects. (His nickname, by the way, which was prominently displayed on the license plate of his automobile, was conferred upon him by his three daughters for nonastronomical reasons.)

Born in Tippecanoe, Wisconsin, in 1912, Scotty (another of his countless aliases that include names like Three-Toed Pete and Hu) acquired an interest in optics as a preteenager when he read about the subject in a physics book. His first telescope was a 1-inch 40× refractor made from scavenged lenses and a paper-and-paste tube. He transformed other lenses into an 80× compound microscope, and, until he headed off to college, Scotty knew far more about microscopes than telescopes.

In 1930 he ground a 6-inch telescope mirror that was a failure. The 10-inch that followed, however, gave excellent images. Made from thin plate glass, the mirror is now on display in the R. W. Porter Museum of Amateur Telescope Making, located in the underground chambers of Hartness House in Springfield, Vermont.

While attending the University of Wisconsin, Scotty met Joseph Meek, who encouraged him to begin observing variable stars. He joined the American Association of Variable Star Observers in 1931 and ultimately submitted more than 12,500 observations. He always maintained a major interest in variables.

With degrees in English, Scotty taught at universities and public schools in Wisconsin, Alabama, Ohio, Kansas, Missouri, and Connecticut. During World War II he was an instructor of advanced pilots at the Army-Air Force's Navigation School at Selman Field in Monroe, Louisiana. In 1960 he moved to Connecticut and became an editor for American Education Publications, where he worked until his retirement in 1974.

Scotty was an avid traveler. Many times he and his wife, Miriam, took to the highways of the United States and Central America pulling a small trailer behind their automobile. On many occasions the copy for his monthly column arrived in an envelope with the return address simply rubber-stamped, "Walter Scott Houston on the Road." The postmarks were from Wisconsin, the American Southwest, and Mexico. There was usually a small note attached to the copy that apologized for the condition of the manuscript due to this or that key sticking on his portable typewriter.

Scotty was particularly active in Kansas during the late 1950s. Along with Cliff Simpson, he used various radio techniques to monitor meteor activity, record the passage of artificial satellites, and map the sky at 108 MHz. These experiments later led him to an interest in detecting solar flares with a sudden-atmospheric-enhancement (SAE) recorder. He identified his first meteorite while in Kansas, and went on to locate at least a dozen more.

In cooperation with the Smithsonian Astrophysical Observatory, Scotty established a Moonwatch station in his backyard near Manhattan, Kansas. In early 1958 it was the first station to catch sight of Explorer 1, the United States' first satellite.

During his Kansas years Scotty was editor of the freewheeling amateur newsletter *The Great Plains Observer,* which had more than a thousand subscribers. Typical of its style was the opening line from a story chronicling hurried plans to observe a satellite. It read, "Manhattan Moonwatch got caught with its rompers well down around the varicose veins."

Although it was but a small part of his wide-ranging astronomical activities, the Deep-Sky Wonders column gave Scotty his greatest public exposure. He did much of his early observing with a 10-inch scope under the clear, dark Kansas skies. Those observations were often recounted in his columns. Deep-Sky Wonders evolved from little more than a stark listing of the brightest objects visible during a given month to a free-flowing journal that was alive with Scotty's own observing experiences as well as those of others who wrote to him. These were mixed with observing tips, asides on double or variable stars, and a dash of constellation lore.

It was at the annual Stellafane gathering of telescope makers outside Springfield, Vermont, that Scotty seemed most in his element. Surrounded by hundreds of friends and admirers, he continually found someone stopping to greet him or ask an observing question.

The appellation "raconteur" was well suited to Scotty, as he was never hesitant to launch into a story. Some listeners would raise an eyebrow as the tale unfolded. But if anyone questioned the reality of the story, Scotty would likely reply with a broad grin and a mischievous twinkle in his eye. Some interpreted this as "listener beware," but it was not always clear if that meant "beware if you do believe it" or "beware if you don't." As a case in point, in 1959 he wrote an April Fools' story about the two moons of Mars being artificial satellites. It became more than a passing curiosity when a Russian scientist put forth the same "theory" soon after.

Scotty remained forever active. In early 1986 I received a note on one of his famous blue postcards that was signed "Deerslayer." When I inquired about the new nickname, I learned that Scotty, then 74, had collided with a deer at 3:30 a.m. while on the way to show a group of schoolchildren Halley's Comet. Neither the deer nor his automobile survived, but he did keep his appointment.

In the summer of 1980 he underwent cataract surgery. Although some might consider this unlucky for a person who spent much of his life looking at the sky, for Scotty it opened up a new world of observing. With the eye's lens removed, a flood of ultraviolet light could reach his retina, and some faint blue stars, like the central star in the Ring Nebula, appeared brighter to him in small telescopes.

In 1986 he had a hip-replacement operation, and at that year's Stellafane, supported by crutches, he delivered his annual "Shadowgram" talk to an enthusiastic crowd, which had just been told that asteroid 3031 was named Houston in his honor. In response to a standing ovation he waved the crutches high in the air as if the applause alone had cured him.

My first encounter with Scotty, albeit an indirect one, was in the early 1960s. I had already explored the Moon and Jupiter and "discovered" the rings of Saturn with a 60-mm refractor when I came across his column in a library copy of *Sky & Telescope*. "A glory of the northern sky is M13," he had written. One look at this cluster and I was hooked on deep-sky observing. Little did I know then that some dozen years later I would become Scotty's editor. I always saved his copy so it would be the last thing I'd work on each month — it was my editorial "dessert," and I savored every word of it as much as I savored our working relationship. We were friends as well as colleagues, and we were always on the lookout for ways to amuse ourselves at each other's expense. Unable to locate Tippecanoe on a Wisconsin map, I dropped him a tongue-in-cheek note asking if he really knew where he was born. Return mail brought a photocopy of a map with the Milwaukee suburb indicated with a deliberately suspicious-looking handmade label, leaving me more in doubt than ever.

In February 1987 I visited him at his Connecticut home. The intent was an interview, but we both knew beforehand that the conventional question-and-answer session would never work — it would be too easy for us to drift from the topic at hand. As we sat in the den that served as his office, the noontime sun streamed through a south-facing window and illuminated the clouds of smoke that billowed from his pipe — one of his many trademarks.

In his backyard, surrounded by the snow of an unusually hard winter, was his latest astronomical project: a new observatory. He had completed much of its cement-block foundation soon after his hip operation. The walls had been hastily covered for winter. Nevertheless, a temporary roof with hinged panels ensured that observing would continue.

The following "interview" was pieced together from our shared afternoon. It gives some insight into one of the 20th century's best-known astronomical writers.

S&T: How did you get your start in deep-sky observing?

Scotty: Through writing the column. At first the writing was easy since I had viewed all the Messier objects when I was in the 6th grade, and the early copy was mainly about bright objects. The column first appeared in November 1942, and was done by Leland Copeland when I took it over in 1946. I didn't want to write about anything I hadn't seen myself, so as the column grew to include fainter objects, I would go out in the early morning sky to look at things I would be writing about for evening observers later in the year.

In those days there weren't many observing guides. There was *Norton's Star Atlas*, which was utterly useless for any instrument larger than a pair of binoculars. I had access to a library copy of Admiral Smyth's *The Bedford Catalogue* and Reverend Webb's observing handbook. They were good for reference and for making sure I didn't overlook any bright object.

Of course in those days there was no one to argue with me, since readers didn't know any more than I did. Today it's a different story. There are a lot of people who know more than me, especially about specific types of objects. There are people who specialize in planetaries or diffuse nebulae. As better catalogs come along, we have observers who will go after unusual objects. With the recent publication of Vehrenberg's *Atlas of Galactic Nebulae*, I suspect we will find people looking at many of these objects, especially the ones that haven't made it onto popular star charts in the past.

S&T: Do observers in general seem more skilled today?

Scotty: Oh my, yes. Years ago I would bait the column with some difficult observation, like the dark lanes in M13. There would rarely be a reply, so I would wait a few more years and try again. Lately, I get responses to even the most difficult observations. And it's not due just to the large-aperture telescopes in the hands of amateurs today. Many of the current sightings come from people using 12-inch and smaller instruments. While they weren't as common a decade or two ago, there were still plenty of large scopes being used by amateurs.

Certainly some of the credit can be given to modern light-pollution filters. They have been the basis for some wonderful observations, but some of the results must also be due to a positive attitude on the part of the observer. Before, people were too willing to assume that a particular object was beyond the reach of their telescopes without really trying to look for it.

I get a lot of pleasure in throwing plugs out into the stream and seeing who will pick them up. Often something interesting will come of it. I have no interest in making a discovery myself, except perhaps to find a supernova in our own Milky Way galaxy.

S&T: Interesting you should say that on the heels of the discovery of the supernova in the Large Magellanic Cloud. Have you ever had a good look at the southern sky?

Scotty: The farthest south I've been is about 15° north latitude. I've never been far enough to see the Magellanic Clouds, but I have seen the Coalsack and much of the bright southern Milky Way. I was completely unimpressed by the Southern Cross. I do like the bright stars Alpha and Beta Centauri to the east of it, however. The Eta Carinae Nebula is an easy naked-eye object, even when low on the horizon. The first time I saw it I thought it was a cloud.

S&T: What do you think the future holds for deep-sky observers?

Scotty: I don't really know. Every time I think I know the answer, something different happens. It's one of the pleasures of astronomy — you can't always predict the future, and so much has happened in the past few years.

Take astrophotography. Every now and then some knowledgeable person comes along and says amateur astrophotography is dead; it has progressed as far as it can. The next thing you know there's a new film or hypersensitizing technique, and amateurs are taking the best pictures ever.

Every so often someone proclaims the death of amateur telescope making. Who would have thought the Dobsonian revolution would have caught on the way it did? There are probably more people building scopes now than ever before.

No, I don't think I'd like to try to predict what the next stage of deep-sky observing will be.

JANUARY

CHAPTER 1

The Glory of a Thousand Stars in a Thousand Hues

Walter Scott Houston poured his heart and soul into his writing. Reading his words, you often felt as if someone you love had wrapped an arm around your shoulder on a cold winter night and, pointing a finger toward the heavens, spoke with passion and eloquence about the night sky. It was as if he were teaching his daughter, or son, or best friend about the sky. In this column, adapted from Deep-Sky Wonders for January 1991 and written just three years before his death, Scotty recalls a time when, as a boy with his telescope under the moonless sky of Wisconsin, he felt "lucky to walk in the January night under the glory of Orion."

I learned my constellations in Tippecanoe, Wisconsin, a town that long ago vanished into the urban sprawl of Milwaukee. Back then Tippecanoe was a rather treeless tract of farmland bounded by the great clay bluffs of western Lake Michigan. The sky ran right down to the horizon and, with an almost irresistible force, called for you to look at it. In January 1926, after a midnight walk home from ice-skating, I wrote:

> Snow crystals sparkle like blue diamonds, but with a dreamy gentle radiance totally unlike the harsh gem. A rail fence as black as Pluto himself runs along the road. The forest is black in the distance. The landscape is a masterpiece in ultramarine and sable.
>
> As if in contrast, the heavens above blaze with a thousand tints. Incomparable Orion leads the hosts with blue Rigel, ruby Betelgeuse, and bright Bellatrix. His silver belt and sword flash like burnished stellar steel. And more advanced is dark and somber Aldebaran, so heavy and gloomy. In fitting contrast are the delicate Pleiades, who sparkle "like a swarm of fireflies tangled in a silver braid."*

---

* *Scotty was referring to the poem* Locksley Hall *by Alfred, Lord Tennyson. One stanza in particular is a favorite of stargazers:*
   "Many a night I saw the Pleiads, rising through the mellow shade,
   Glitter like a swarm of fireflies tangled in a silver braid."

Figure 1.1 Orion the Hunter dominates the winter sky. The constellation is a veritable treasure-trove of sights for stargazers. (All images are oriented so north is up, unless otherwise indicated.)

How can a person ever forget the scene, the glory of a thousand stars in a thousand hues, the radiant heavens and the peaceful Earth? There is nothing else like it. It may well be beauty in its purest form.

On a cosmic time scale, we are lucky to walk in the January night under the glory of Orion. The slow but constant wobble of Earth's poles, which we call precession, made things very different in the distant past. In 10,000 B.C. Orion was a summer constellation barely visible from the Northern Hemisphere, as it will be again in A.D. 16,000. But for the present, no matter on which side of the equator we live, Orion rides high in the January evening sky.

I know of no constellation brighter than Orion **(Figure 1.1)**. It contains two 1st-magnitude stars and five of 2nd magnitude. I have long felt Orion is the first constellation amateurs learn, though mine was Scorpius because I began with the summer sky. I wonder if this gut feeling is borne out by actual numbers?

The Nebulous Wonder of Orion

As night falls in January, Orion hoists his weight over the eastern horizon, bringing into view the "incomparable M42, the great Orion Nebula, about which words fail," as Scotty penned in 1955. "No amount of intensive gazing ever encompasses all its vivid splendor." To Scotty, the Orion Nebula was the most dramatic nebula visible from *both* hemispheres. Although the Eta Carinae Nebula is larger, it is visible to observers only from more southerly latitudes. But the Orion Nebula is more than an awe-inspiring sight: it is a historical wonder, an inspiration to astrophotographers, and an observer's paradise. It can fill the field of a telescope, regardless of size, with an array of visual splendors.

At this season, when most of the United States is covered by the cold and transparent blanket of the polar continental air mass, the nebula observer can do no better than to direct a telescope to the greatest of all spectacles — the Orion Nebula, middle "star" in the mighty Hunter's Sword. Against the intense, sharp blackness of these skies, with the stars actually resembling their prototype diamonds, there is no other sight so well calculated to stir the observer's feelings of awe and wonder. Even the naked eye perceives M42 as having a real diameter.

The Orion Nebula (**Figure 1.2**) has inspired more adjectives than any other deep-sky object. None, however, do real justice to this great mass of swirling, pale green, chaotic gas. Even "overpowering" is a most inadequate word when the nebula is seen in a really dark sky. Intertwined with the Sword on the Hunter's Belt, M42 requires no charts or setting circles. There's no need for a finder either, since you can simply sight along the edge of the telescope tube to bring the nebula into view.

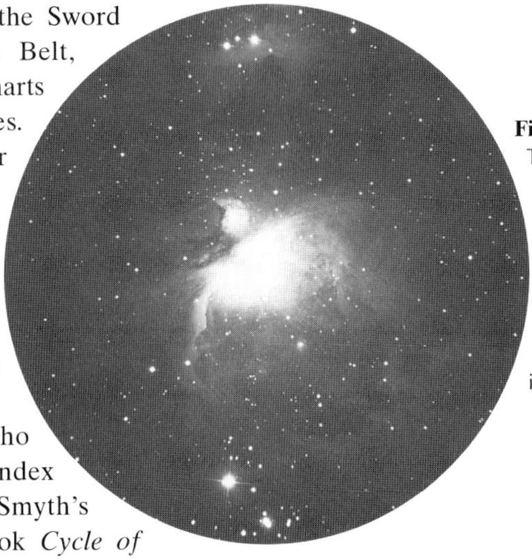

The reader who looks to the index in William H. Smyth's observing handbook *Cycle of Celestial Objects* for the great Orion Nebula will fail to find it listed. This is not an incredible omission born of some careless printing, but rather a reflection of the prevailing amateur interests of the early 19th century. This most impressive nebula in the entire heavens is listed by Smyth as a mere addition to the entry on the double star Theta (θ) Orionis.

Figure 1.2
The great Orion Nebula (M42) is a cloud of gas and dust glowing from the light of hot young stars embedded in it.

Most (but not all) of the great deep-sky objects have been handed down to us by a succession of famous past observers. The Orion Nebula, for example, was first discovered by the French naturalist Nicolas Peiresc in 1611, two years after Galileo began turning the newly invented telescope skyward. Reputedly missed by Galileo, the Orion Nebula was independently discovered by the Swiss Jesuit priest Johann Cysat in 1618 and observed by Christiaan Huygens in 1656. Legend has it as the first object viewed by Sir William Herschel in 1774 with his first successful telescope (after two hundred obvious failures at speculum making); a few years later, it was the final object to be looked at with the 40-foot reflector before its great glass was laid aside forever. It was the first nebula to be photographed (by Henry Draper), and its spectrum (observed by British astronomer William Huggins) sounded the death knell for the Herschelian theory that all nebulae were composed of stars and therefore ultimately resolvable.

Today, it is the first object the stargazer turns to on these crisp winter evenings. Every amateur maker of a Schmidt camera finds that the Orion Nebula is the only proven test of workmanship. The variable-star observer includes it while following the perplexing nebular variable, T Orionis, immersed in the misty light. Whether viewed with opera glasses or with a 16-inch reflector, no spectacle is better devised to open the year's observing, no object brings more gasps from casual visitors, and no other nebula has such continuing power to stimulate the veteran lover of the stars.

Most published descriptions of M42 dwell on the aesthetic shock experienced by the observer rather than an actual description of the nebulosity. William Herschel was so taken by the view that nebulae became a lifelong interest. His son John later compared the nebula to a "surface strewn with flocks of wool — or like the breaking up of a mackerel sky when the clouds of which it consists begin to assume a cirrus appearance."

The 19th-century French astronomical artist Leopold Trouvelot wrote that he saw M42 to advantage only once with the 15-inch refractor at Harvard Observatory in Cambridge, Massachusetts. That time he was able to see the cirrus pattern mentioned by Herschel. This detail is a real observing challenge.

Garrett P. Serviss, the author who taught my generation and the one before it to appreciate the heavens, referred to stars in the Orion Nebula as "shining like gems just dropped from the hand of the polisher."*

Drawings of the Orion Nebula made before the influence of photography raise more questions than they answer. Only superficially do the sketches bear any resemblance to one another. The bright section of the nebula drawn by Bindon Stoney using Lord Rosse's 3-foot reflector in Ireland doesn't begin to match what I saw in 1935 with the 36-inch reflector at Steward Observatory in Arizona. Trouvelot's 1882 lithograph based on observations with the Harvard 15-inch is a reasonable match to my view through a 3-inch. On the other hand, John Mallas's drawing in the *Messier Album*, made in the 1960s with a 4-inch telescope, shows features that most observers need a 10-inch to see.

As for details in the brightest portion of the nebula, try using very high magnifications. This may bring out the cirrus effect that Herschel and Trouvelot mentioned.

---

* From Serviss's 1901 book Pleasures of the Telescope. *Serviss was a well-known science popularizer and early science-fiction writer.*

The nebula stands up well under all sizes of telescope and all powers. With high magnifications, the intricate curdlings of its luminous masses rival and resemble frost paintings. Its chaotic form gives a strong impression of twisting and turbulent motions that are too slow to follow. It is so bright that its green tint is obvious to most. Some have even glimpsed reds and pinks in the curdling clouds. With low powers and a field wide enough to include the whole nebula, it becomes an object compelling enough to draw exclamations of delight from even the most disinterested bystander. Observers should use averted vision until they can trace the luminous beauty of the Orion Nebula far beyond the bounds normally assigned to this entrancing object. The telescopic view of M42 overspilling the widest eyepiece field more than compensates for cold feet and numbed fingers.

The Quest for Barnard's Loop and the Horsehead

"Orion contains three deep-sky wonders that are unmatched elsewhere in the sky," Scotty once exclaimed. The brightest and best of these is obvious: the great Orion Nebula. But the other two objects — Barnard's Loop and the Horsehead Nebula — are hardly sights for untrained eyes. Some experienced observers have spent decades trying to view them with a variety of telescopes, but without luck. Obviously, the photographic appearance of these objects is what impressed Scotty. Still, he believed that what could be recorded on film might also be within the realm of possibility for the observer, and his columns quickly became the place to turn to for new challenges. Indeed, it was Scotty's curiosity about the chances of seeing such "impossible" wonders that spurred generations of amateurs to put their skills to the test and rewrite the annals of visual astronomy.

One observing frustration in my younger days was that I could not see Barnard's Loop — a huge, dim arc of glowing gas believed to be a remnant of an ancient unrecorded supernova. It is superficially similar to, but larger than, the Veil Nebula in Cygnus. The eastern half is quite well defined, while the western portion, if it actually exists, is seen only as a few nebulous patches. If it were a complete circle, it would be about 10° in diameter and roughly centered on Orion's Sword.

Although the name implies that astronomer Edward Emerson Barnard discovered this nebulosity, he did not. After finding it on his photographs made in October 1894, Barnard remembered that the Harvard astronomer William H. Pickering had written about the arc more than four years earlier. Pickering had discovered it on photographs made at Mount Wilson (then called Wilson's Peak) in 1889. Why the loop became associated with Barnard is a bit of a mystery, especially since Barnard's original description of the nebula quoted Pickering's work extensively.

For years Barnard's Loop (**Figure 1.3**) was a goal only for astrophotographers. No one I knew had ever seen it visually. I had searched with my homemade 10-

Figure. 1.3 Barnard's Loop is a three-million-year-old emission nebula. The eastern portion is shown here at left. While it shows up often in photographs, the Loop presents a visual challenge for observers.

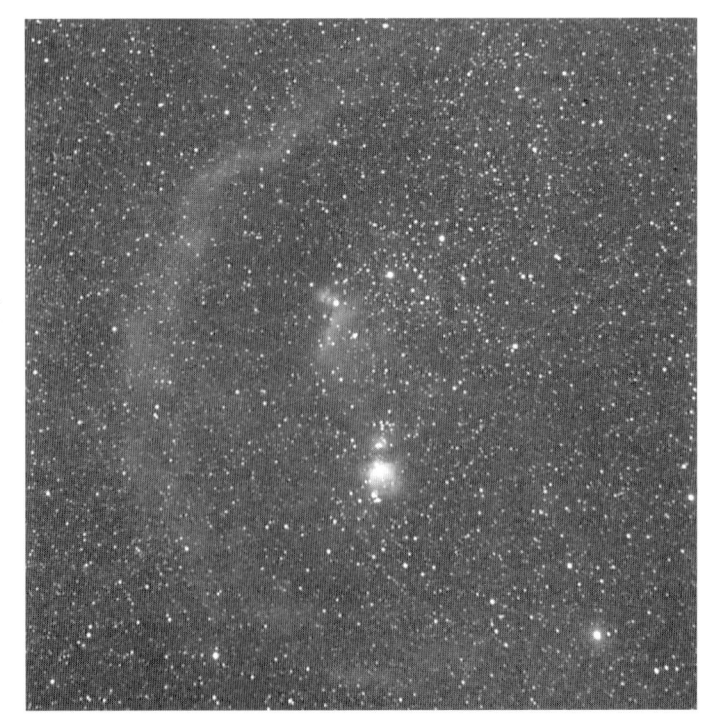

inch f/8.6 reflector as well as 6- and 15-inch Clark refractors* to no avail. The object defied all my bizarre optical creations until I saw it with, of all things, the naked eye. That, however, was because Lumicon began manufacturing O III nebula filters. I was absent-mindedly comparing naked-eye views of the sky with and without the filter under the mediocre observing conditions here in Connecticut when I chanced upon the loop. My wife says I jumped clear over the observatory (it's a small building). Since then, I have duplicated the feat at dark sites in Texas, Southern California, and Mexico, my only accessories being a Lumicon ultrahigh-contrast (UHC) filter and a dark cloth over my head.

Nevada observer Richard Johnson and California amateur John Bartels have been doing the same thing. By contrast, attempts to catch sight of even its edges with a 4-inch 20× rich-field telescope, under good conditions but without the filter, have so far met with despair. Many astrophotographers consider it quite a feat to capture the loop on film; seeing it visually is even more remarkable.

The Horsehead

The Horsehead Nebula (Barnard 33) is one of the most photographed but least observed objects in the sky (**Figure 1.4**). It is incredibly challenging for visu-

---

* *Renowned optical firm Alvan Clark & Sons of Cambridge, Massachusetts, supplied finely crafted telescopes to both amateur and professional astronomers from 1850 to 1958.*

al observers, and until rather recently it was a prize worth crowing about even among photographers. The blob of darkness lies halfway along the streamer of faint nebulosity that runs for 1° south from Zeta (ζ) Orionis, the easternmost belt star. The streamer IC 434 (the abbreviation IC stands for *Index Catalogue*, a supplement to the *New General Catalogue of Nebulae and Clusters of Stars* by J. L. E. Dreyer), is a bit brighter than the Veil Nebula in Cygnus, and no great feat to see. But recognizing the dark blotch B33 is another matter.

The Horsehead was first photographed about 1900, but was believed to be only a void in IC 434. Barnard appears to have been the first person to suggest that it was actually an obscuring cloud of material seen in silhouette. When I first suggested that amateurs look for the Horsehead in 1969, veteran deep-sky observers disagreed on its visibility in amateur telescopes. I then queried several active nebula hunters, and none reported seeing this object. Some thought that IC 434 was itself the target, while others mistook the rift in NGC 2024, northeast of Zeta, for the Horsehead.

The Horsehead is harder to see by far than the nebula around R Monocerotis (Hubble's Variable Nebula). It is much harder than the Veil in Cygnus or spiral galaxy M33 in Triangulum. Possibly it is more difficult than the Merope Nebula in the Pleiades. In any case, the observer should wait for special weather and use a photograph as a detailed finding chart.

Figure 1.4 The Horsehead is the dark nebula B33 seen in silhouette against IC 434, which is much larger and very bright. NGC 2024 is at lower left. North is to the left.

Leslie Peltier, the well-known comet hunter of Delphos, Ohio, mentioned in a letter that he once saw it long ago. His telescope is a 6-inch refractor of about f/10, a type more common in Europe than this country, and he uses low-power oculars.

IC 434, but no Horsehead, was seen with a 2.4-inch refractor by Larry Krumenaker in New Jersey, with a 12-inch f/6.5 reflector by Ted Komorowski in North Carolina, and with 6-inch f/8 telescopes by Stephen Barnhart in Ohio and Mark Grunwald in Indiana. However, other observers were more successful.

California amateurs made quite a few sightings of B33, mostly from mountain locations. Harold Simmons of Broderick took three trips to the 2,500-foot site of Colfax Observatory before he found the elusive object in a 16-inch at 200×. Simmons attributed his success to an exceptionally clear, windless sky and superior optics.

On Onyx Peak, California, Leonard P. Farrar had difficulty in finding the Horsehead with a 10-inch and averted vision. But once it was located he could show B33 to several other people. His method was to identify five nearby faint stars on a photograph of the region and use these as a guide to the exact spot. "The nebula appeared to me as a slightly fuzzy jelly bean," Mr. Farrar writes. "The color was a dark smoke gray."

From De Funiak Springs, Florida, Wayne Wooten reports: "I have observed the Horsehead several times with my 10-inch at 50×, mostly on a cold night following a rainstorm." He has been able to see it several times with his 6-inch rich-field telescope.

From Connecticut my 4-inch refractor failed to reveal the Horsehead, but my notebook indicates that it was visible from Kansas with a 10-inch reflector. I have since fished it out using a 4-inch Clark refractor, a 4-inch off-axis Newtonian telescope made by Margaret Snow, and a 5-inch Moonwatch Apogee telescope under the same circumstances as Mr. Wooten, immediately after the passage of a cold front. Scattered light from 2nd-magnitude Zeta foils many attempts to find the Horsehead, since the two are separated by only 1/2°.

Another reason that many searches fail is that observers are looking for the wrong-sized object. When I have seen it with telescopes between 10 and 16 inches in aperture, my first reaction has always been how tiny it is! Knowing just where to look for it is half the battle. The Horsehead is only about 5′ across. Amateurs accustomed to seeing it on large-scale photographs made with professional telescopes end up looking for an object that is much too big. I was fortunate in my original quest since the only photograph I had to go by was from the *Atlas of the Northern Milky Way,* compiled in 1934 by Frank Ross and Mary Calvert. The three stars in Orion's Belt spanned only 2 inches on these small-scale charts, and the image of the Horsehead looked pretty much the way it did in my 10-inch reflector. Having found it, however, I could easily translate the view to other telescopes.

John Bortle of Stormville, New York, sees it with his 12-inch f/5.5 reflector, using 40× and averted vision. In picking out faint objects near a bright star, it might help to use an ocular fitted with an iris diaphragm in the focal plane. This would be used wide open in setting on the field, and then, centered on (say) the Horsehead, it would be closed down to shut off all disturbing bright light.

Nebula filters are the main reason the Horsehead has fallen from the list of test objects. Anthony Nutley of Western Australia could see no trace of the Horsehead with his 10-inch reflector until he added an H-beta filter that made it obvious. This filter, sold by Lumicon, is tuned to transmit nebular light from the hydrogen-beta (Hβ) emission line. Darian Rachal of Louisiana had a similar experience with an H-beta filter and his 17½-inch reflector.

Without a filter the Horsehead is still a challenge. If you want to take it on, then here's your competition: Peltier saw it with a 6-inch refractor, I've glimpsed it with the 4-inch Clark, and a group of four amateurs sighted it with Snow's 4-inch off-axis reflector.

Also, there is a question as to how easily it can be photographed. The well-known French observer Jean Texereau recorded the Horsehead with an f/5 lens of 1.6-inch aperture. Can you beat that?

The Great Triangulum Spiral

In his July 1949 Deep-Sky Wonders column, Scotty wrote, "A number of readers have written of their inability to locate M33, the large spiral in Triangulum. This column will discuss the matter and the overall problem of the visibility of nebulous objects in a future issue." Then, in January 1955, he introduced one of his famous visual challenges: to spot M33 without a telescope. Scotty continually pushed his vision to the limit and challenged the limits of others. He never considered himself the "final word" on any observation. Scotty made observing a sport and enjoyed creating friendly competitions, through which he would ultimately discover the truth behind the power of the eye — not *his* eye, but "the" eye.

The great spiral in Triangulum (**Figure 1.5**), also known as M33 and NGC 598 (**Figure 1.6**), is a favorite of photographers and the despair of many visual

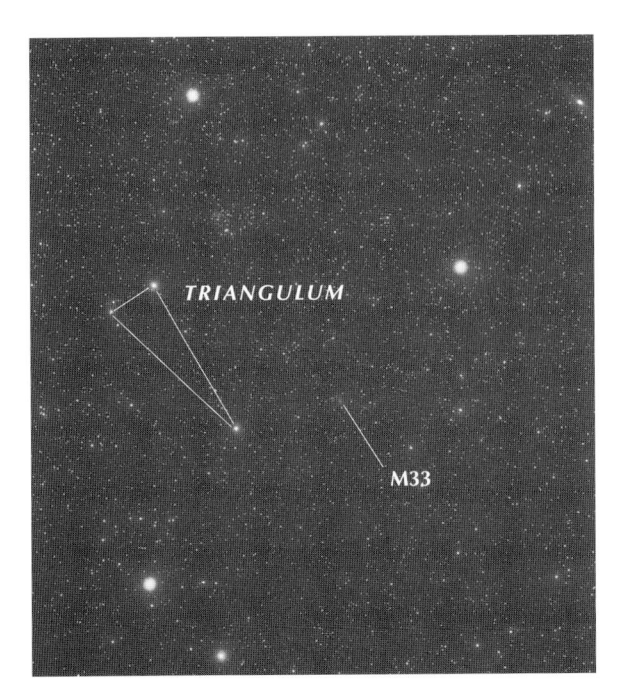

Figure 1.5
The constellation Triangulum may appear small and unimportant, but it hosts the beautiful spiral galaxy M33.

observers. Although W. H. Smyth, in his *Cycle of Celestial Objects*, called it "large and distinct" in his 6-inch refractor, this galaxy is much harder to find in a telescope than a beginner might expect. With a diameter of about 1°, the 7th-magnitude spiral more than fills the field of view in high-power binoculars and presents an almost featureless glow that is easily missed. Therefore, very low powers or even small binoculars give the best view.

Amateurs have often searched telescopically without avail for this apparently elusive object, while others report it visible in finders or binoculars. Until recently no one had reported a naked-eye observation to me, although the rather discordant measurements of its integrated visual magnitude, 5.8 to 6.8, suggest that M33 might be seen in a really clear sky. There is a common misconception that the faintest stars the human eye can see are 6th magnitude. But with a pure dark sky and good dark adaptation, the threshold of vision is actually considerably fainter. In fact, stars as faint as 8.5 have been detected by very keen-eyed persons when special precautions were taken.

Figure 1.6. M33 lies about 2.4 million light-years from Earth and is part of the Local Group that contains the Andromeda and Milky Way galaxies.

Only one published reference to such naked-eye visibility has been found. Knut Lundmark, discussing the limits of non-telescopic vision in the *Handbuch der Astrophysik*, Vol. VI, Part 1, page 354, states, "The present writer is able to see with his unaided eye the object M33 having a total magnitude of 6.8."

In September 1954, while observing meteors eight miles east of Manhattan, Kansas, I stumbled on the nebula accidentally. It was so obvious that at first I took it for a bit of cloud, but binoculars corrected the identification. To the unaided eye, stars well below magnitude 7.0 were easily visible that evening, and the nebula itself was a conspicuous luminous patch, very slightly brighter at the center, and apparently larger than its published major diameter of 60'. I made it easily 90', but somewhat less in binoculars.

Postcards were sent to other amateurs, whose attempts to verify the naked-eye visibility of M33 are reported here. Harold Peterson, after several failures in

Durango, Colorado, finally saw "a faint diffuse patch of light with a suggestion of a core," which he was not able to hold steadily in view. This was in the Vallecito River canyon. A. D. Peterson saw M33 clearly from the mountains 40 miles west of Grand Junction, Colorado, and his son made it out from the city itself. S. R. B. Cooke writes that while it was invisible from Minneapolis, he could see it clearly with the unaided eye at 6,000 feet in the Montana Rockies. At Neosho, Missouri, R. Adams detected it as a faint haze, best seen by averted vision.

Five other observers wrote me of unsuccessful attempts to see M33. One of these, James Corn of Phoenix, Arizona, also remarked, "Dr. Carpenter showed me M33 through the 36-inch reflector at Steward Observatory, and I noticed a condensation in one of the spiral arms. When I got home I tried my 12-inch and found the condensation easily: it is visible whenever the seeing is good."

Since then the writer has seen M33 at House Springs, only 30 miles from St. Louis, Missouri, but I have never been able to see the galaxy from the city itself. Apparently M33 is visible to the naked eye, but only when skies are quite clear and city smog is entirely absent. In August 1961 I told the Stellafane convention in Vermont that I doubted that anyone there could detect M33 without optical aid. But shortly after midnight the sky grew so transparent that this galaxy became visible, first by glimpses and later steadily, even without use of averted vision.

Telescopically, the texture of M33 is usually smooth, but one night I saw the whole surface surprisingly mottled, with the southeast part considerably brighter than the northeast. Dreyer's *New General Catalogue* notes the galaxy as very gradually brighter toward the center, and Edwin Hubble suggests that the nucleus resembles a huge globular cluster. Some of the misty knots in the spiral arms are conspicuous enough to have been cataloged in the *NGC* as separate nebulae. Most observers settle for locating NGC 604, a bright knot in one arm 9.1' east and 7.6' north of the galaxy's nucleus.

One night in an 8-inch, a congested mass of bright patches was seen superimposed on an overall spiral pattern. Once before I had a similar view in a 36-inch. Mr. Bartek (a fellow observer) pointed out a patch some distance from the main whirl that made M33 resemble M51. The patch was faint, but I held it steadily without averted vision.

The Mystery of "Nonexistent" Star Clusters

In his December 1975 column, Scotty noted a seeming paradox concerning large, loose star clusters that was originally reported by William and John Herschel. Observers Pat Brennan and Dave Ambrosi, he said, found that many of these objects, designated "nonexistent" in the *Revised New General Catalogue of Nonstellar Astronomical Objects*, are nevertheless visible in medium-sized telescopes. The authors of the *RNGC*, Jack Sulentic and William Tifft, called an object "nonexistent" if it could not be recognized on prints of the *Palomar Sky*

Survey.* But, as Scotty pointed out, the survey's large scale and deep limiting magnitude tend to conceal sparse open clusters. By the way, you still will not find these objects plotted on most star atlases, including the *Millennium Star Atlas*. Since Scotty's original column appeared, however, Brent Archinal (U.S. Naval Observatory) compiled the *Non-Existent Star Clusters of the RNGC*, published by the Webb Society in 1993, which includes the very objects Scotty introduced in his columns. We have plotted three of them on special charts for this book.

How well do descriptions in catalogs match what amateurs see in their telescopes? With his 6-inch f/7 Newtonian and its finder, Pat Brennan of Regina, Saskatchewan, has been looking at open star clusters, especially William Herschel's Class VIII, the big scattered groups. All of these objects are listed in Dreyer's classic *New General Catalogue*, together with descriptions based on visual observations. But often these characterizations do not match what Mr. Brennan sees with his instruments. One example he cites is NGC 1662 in northwestern Orion. The *NGC* description suggests a poor cluster, as does the fact that it is not plotted in either Norton's *Star Atlas and Reference Handbook* or the Skalnate Pleso *Atlas of the Heavens.*** But when he looked for this object, he was surprised to spot it easily in the 10 × 42 finder.

Mr. Brennan describes NGC 1662 as "a bright coarse splash of stars, the brightest being of 9th magnitude. Few of the 20 or so apparent members were fainter than magnitude 12. The cluster stands out well and is elongated approximately northwest to southeast, its maximum extent being about 12'. Near the center is a very small but distinct triangle of 12th-magnitude stars, the northernmost of which has a companion of equal brightness 1' to the west. While not a fine cluster, NGC 1662 can be seen easily in small apertures and is worth a look."

This description fits quite well with what I see in my 4-inch refractor, but others may get a different impression.

Another loose object is NGC 2251 in Monoceros. It is plotted in the Skalnate Pleso *Atlas*, and its companion *Atlas Catalogue* indicates that it is 10' in diameter with 35 stars. A 12-inch telescope, however, reveals about twice that number, with a bright star in the center. In the 4-inch the group is smeared out like a long narrow arrowhead, but larger apertures round out the shape.

Also in the Skalnate Pleso *Atlas* is NGC 2169 in Orion, a cluster much less impressive than the omitted NGC 1662. Using a small telescope, look for a scatter of half a dozen stars. Larger apertures reveal as many as 30 within a 5' circle. Were it not already in the *NGC*, I would never have recognized it as a cluster.

---

* *The National Geographic Society-Palomar Observatory* Sky Survey *is an atlas of 1,830 photographs taken with the 48-inch Schmidt telescope on Palomar Mountain.*

** *Now out of print, this atlas, compiled by Antonín Bečvár and his colleagues at the Skalnate Pleso Observatory (in the former Czechoslovakia), was first published in 1948 under the Latin name* Atlas Coeli 1950.0. *Sky Publishing released it with an English title in 1962. Its successor is* Sky Atlas 2000.0 *(now in its second edition) by Wil Tirion and Roger W. Sinnott.*

This object was originally discovered by William Herschel, who placed it in his class VIII.

An especially interesting task is to search for clusters that are designated "nonexistent" in the *Revised New General Catalogue*. Then, by direct observation, try to determine why they were included in the *NGC*. Mr. Brennan writes, "Have you ever encountered a 'nonexistent' *RNGC* cluster alive and well, so to speak? I've found several, albeit the vast majority were coarse and less than impressive."

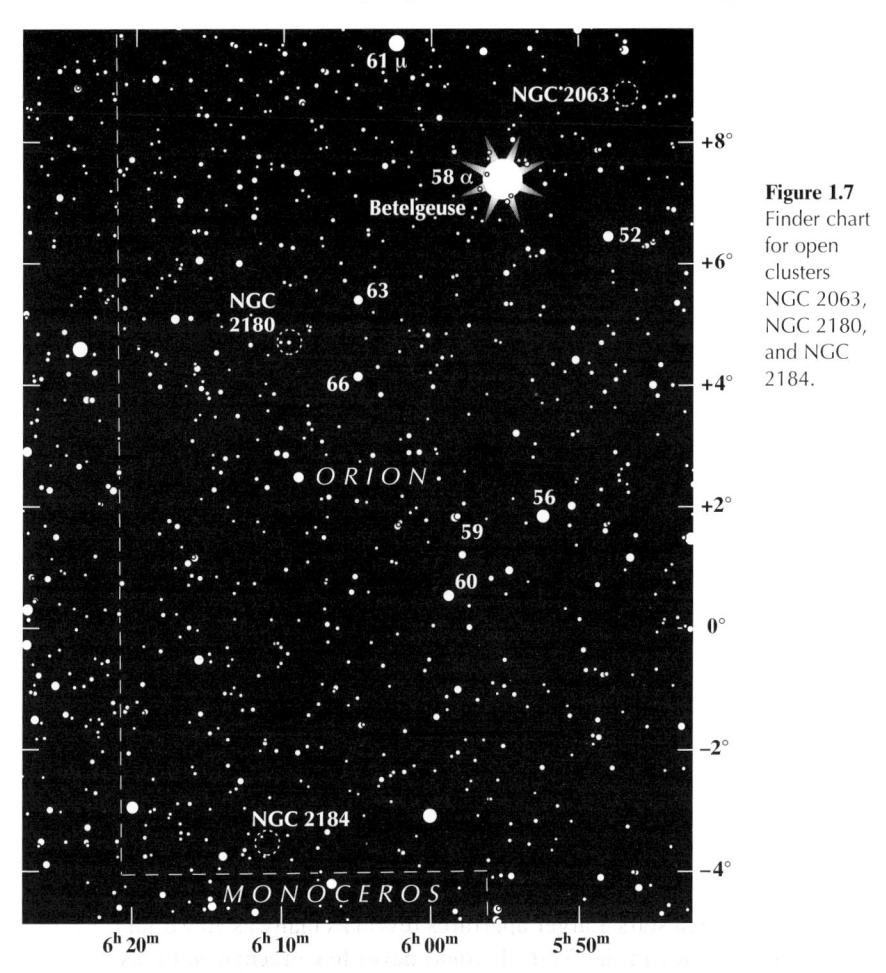

Figure 1.7 Finder chart for open clusters NGC 2063, NGC 2180, and NGC 2184.

NGC 2184 in Orion is such an object (**Figure 1.7**). Mr. Brennan saw it in his 6-inch and I did with my 4-inch. The *NGC* description is "cluster, large, very little compressed." Mr. Brennan's account is more detailed. "At 67× a very pretty sight. It contains 30 to 40 stars, the brightest of 9th magnitude with others down to 12th. They appear evenly arranged in a roundish grouping 12' to 15' in diameter. Lovely view at 38×. I would certainly rank NGC 2184 as one of the top clusters in Orion and better than most of the 25 I have observed in Monoceros."

This apparent discrepancy results from the procedure used by Jack W. Sulentic to check the reality of each object. For this he used prints of the *Palomar Sky Survey*, which shows galaxies and nebulae beautifully. But the large scale and deep limiting magnitude tend to conceal sparse open clusters.

I also checked a Palomar print and found a vague indication of the cluster, but it was so spread out that some imagination was needed to discern it. However, on the much smaller scale of the *Atlas of the Northern Milky Way*, NGC 2184 is clearly visible, and on a still smaller-scale picture by E. E. Barnard it is even more distinct.

NGC 2180 is also in Orion (**Figure 1.7**). "A small but pretty cluster," Brennan writes, "containing more than 20 stars, mostly 12th magnitude, in a 6' area. They are coarsely distributed, with a bunching of very faint stars around a 9th-magnitude one on the cluster's following (eastern) edge. Quite nice with averted vision." I have seen this cluster distinctly in my 4-inch. William Herschel found it and gave it the designation VIII 6.

NGC 2063 is only 2° northwest of Betelgeuse. Brennan says of it: "A coarse cluster of 10 to 15 stars, 10th to 12th magnitude, scattered across a 10' area, with other faint stars strewn to a diameter of 20'. The larger aggregation stands out better than the smaller." This cluster was discovered by William Herschel, who designated it VIII 2.

NGC 1708 is another "nonexistent" cluster (**Figure 1.8**). Brennan writes: "South following 8 Camelopardalis and near the Auriga border, this cluster stands out well, and is pleasing with averted vision. It is triangular in shape, elon-

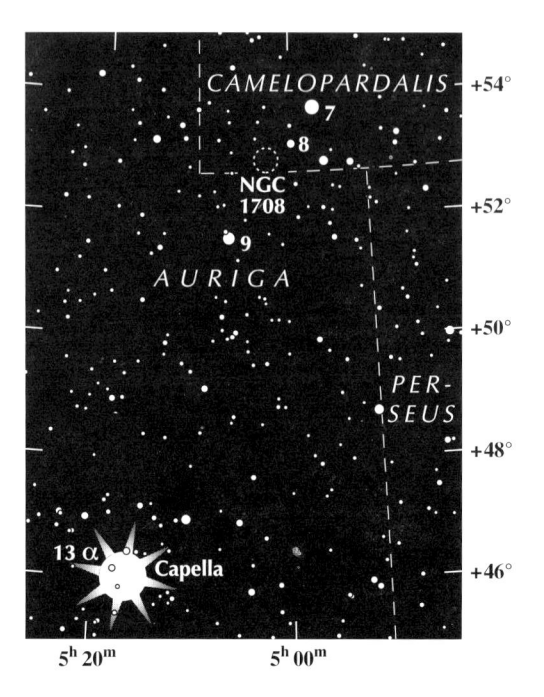

Figure 1.8
Finder chart for open cluster NGC 1708.

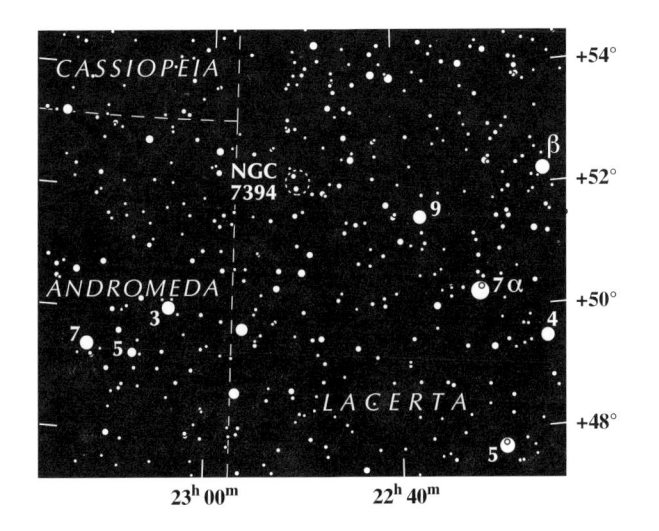

Figure 1.9
Finder chart for open cluster NGC 7394.

gated north-south, and 20 × 15 minutes of arc in size. It contains about 10 stars of magnitudes 10 and 11, with a score of fainter ones."

John Herschel, who discovered NGC 1708 in the year 1831, noted it as "very loose; pretty rich; fills field; the largest star 10th-mag.; mixed magnitudes." The writer was unable to locate NGC 1708 with his 4-inch refractor on a poor night in Connecticut.

NGC 7394 is in Lacerta (**Figure 1.9**). The writer's impression agrees with the Brennan description: "A coarse grouping, 10′ by 3′ in extent, elongated northwest to southeast. In addition to the bright star at its southeast end, there are about 10 fainter ones. Since the field is not particularly rich, the cluster can be easily recognized."

When John Herschel swept up this grouping in 1829, he noted in his record book that the measured position referred to "a double star, the last of a poor cluster of about a dozen stars."

"Coarsely scattered clusters of stars" was William Herschel's definition of Class VIII, to which he assigned 88 of the objects he discovered. No fewer than 32 of these have been rejected by the *RNGC* and, as Mr. Brennan comments, it is unlikely that such a capable deep-sky observer could have made so many erroneous observations. Hence this Canadian amateur is conducting a systematic check of all of William Herschel's "nonexistent" Class VIII clusters.

Jewel of the Night

Every constellation has its grand masters — celestial objects that have become perennial favorites of amateurs around the world. Perseus is loaded with them. Scotty said in December 1988, "This constellation . . . is filled with deep-sky wonders, no matter what your taste may be." Of course, his favorite "jewel" in Perseus was the famous Double Cluster. Known since antiquity as a fuzzy star in the Milky Way, this cluster pair has long fueled the curiosity of skywatchers. Today we can enjoy its twin multitudes of dazzling suns through our backyard telescopes. It is, as C. E. Barns wrote in 1927 in his *1001 Celestial Wonders*, "a field shot with diamond dust." Interested in astronomical history, Scotty spent much time viewing the Double Cluster with a replica of Galileo's 20× refractor. "Although I cannot fit both clusters into the same small field of this instrument," he wrote in November 1990, "I wanted to see them just as Galileo first did. It's too bad the Italian astronomer did not have a poet for company that night, for if he did, I'm sure literature would have been richer for it."

The Double Cluster, NGC 869 and 884 (**Figure 1.10**), also known as h and Chi (χ) Persei, is a jewel. To the naked eye it shines with a steady glow, while telescopically its majesty is so compelling that it causes many observers to neglect fainter offerings in the area. These rich conglomerations of stars were known to Hipparchus in the second century B.C. But we can wonder how many Mongol shepherd youths before him knew about these fuzzy twin lights in the winter sky. It was about as faint a nonstellar object as Ptolemy mentioned. Although it is conspicuous to modern stargazers, we find it but briefly mentioned in the *Alfonsine Tables** and some other old works. Messier did not put the Double Cluster in his catalog. It is sometimes suggested that he felt it was too well known to need mention, but he did list the even more famous Pleiades and

Figure 1.10
To find the beautiful Double Cluster in Perseus, first locate the W in Cassiopeia. Starting at Gamma (γ) Cas, hop to Delta (δ) and then twice that distance again in the same direction. North is to the right.

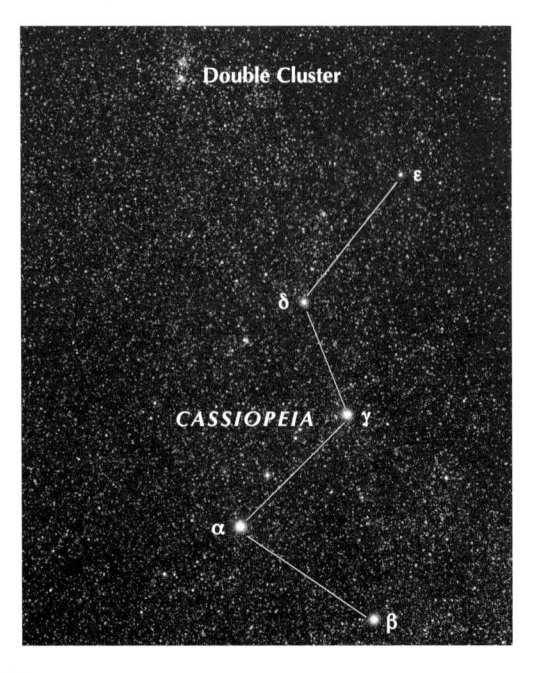

---

* *The* Alfonsine Tables *comprise some of the earliest-known lists of planetary positions for various dates, based on Ptolemy's models of planetary motions. They were named after the 13th-century King Alfonso X of Leon and Castile, a devoted patron of astronomy.*

Praesepe as M45 and M44, respectively. Perhaps he felt that there was no chance of its being mistaken for a comet.

To behold its beauty (**Figure 1.11**), first locate the famous W of Cassiopeia and the constellation Perseus. (Novices who cannot easily identify these star groups are not ready to hunt deep-sky objects.) The easternmost of the W's five stars is Epsilon (ε) Cassiopeiae. Northeast of a pont midway between Delta and Epsilon Cassiopeiae is the Double Cluster.

It should appear as a 4th-magnitude blur to the naked eye. The brightness of an open cluster is often determined by adding the magnitudes of its individual stars. Since the outer limit of a cluster is not always well defined, as more stars are included in the tally, the reported brightness increases. I have seen magnitudes of 3.5 and 3.6 quoted for NGC 869 and 884, respectively. On the other hand, Brian Skiff of Flagstaff, Arizona, considered just the stars in the cluster's cores and obtained visual magnitudes of 5.3 and 6.1 for the pair.

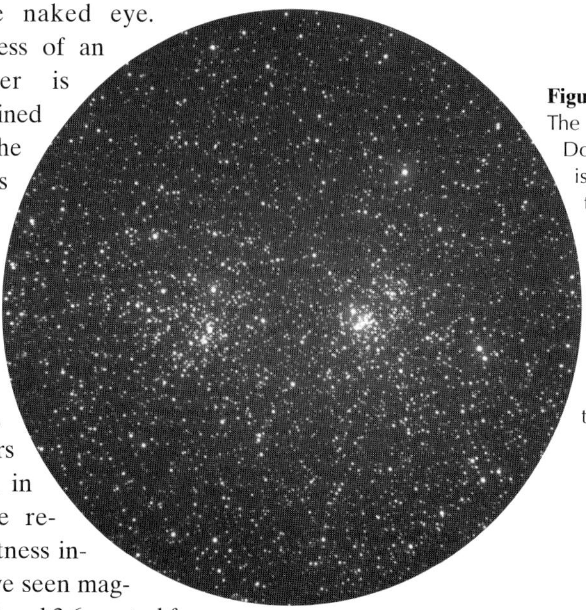

Figure 1.11 The Perseus Double Cluster is visible to the naked eye under dark-sky conditions and makes a beautiful sight through any small telescope.

Setting the telescope on them is no problem, for all you need do is sight along the tube to get the cluster in the finder. But the following operations give a good idea of how to work in less favorable skies or to locate more difficult objects. Get 2nd-magnitude Alpha (α) Persei in the finder. Then move the telescope north and west until the triangle formed by Gamma (γ), Tau (τ), and Eta (η) is in the finder field. Gamma, brightest of the three, and Tau are at the southern corners. Move slightly farther in the same direction until Eta lies just south of the center of the field. Then swing the telescope westward until the jumbled stars of the cluster come into view. If you get lost when moving from one star to another, start over again at the beginning. Try to form the bright signpost stars into simple geometrical patterns. Even after years of practice mistakes occur. If you cannot find it, it's time to join your local light-pollution committee.

Under a dark sky, a simple opera glass, the optical standby of a century ago, will reveal the two star clusters. Perhaps the most widely read observing guide around the turn of the century was Serviss's *Astronomy with an Opera-Glass,* published in 1888. This book was far more popular than William Tyler Olcott's *Field Book of the Stars,* which was published nearly 20 years later. Serviss wrote of the Double Cluster, "With a telescope of medium power, it is one of the most marvelously beautiful objects in the sky — a double swarm of stars, bright enough to be clearly distinguished from one another, and yet so numerous as to dazzle the eye with their lively beams."

The well-known 19th-century observers W. H. Smyth and Thomas W. Webb both commented on the Double Cluster in their own observing handbooks. Smyth noted that the clusters were quite distinct and that their outer stars were intermingled. He called the pair "one of the most brilliant telescopic objects in the Heavens," and Webb agreed.

While reading these early authors, I discovered that nowhere in the old descriptions is the pair referred to as the Double Cluster (though Serviss called it a double swarm). I had always assumed that the name was old simply because it is so natural. The first mention of it I found, however, is in Kelvin McKready's *A Beginner's Star-Book,* published in 1923. But when I asked if anyone knew the source of the name "Double Cluster," letters arrived from four continents. Smyth did not use the term in his classic *Bedford Catalogue* (the second volume of his *Cycle of Celestial Objects*) published in the mid-1800s. By the end of the century it was used as an uncapitalized descriptive term, not as a proper name.

J. P. Liddell of East Woodburn, England, wrote to me saying that the American astronomer Simon Newcomb noted "it may be considered as a double cluster" in his 1901 book *The Stars.* Liddell adds, "The fact that Newcomb does not call it the Double Cluster indicates that the term was not in general use at the turn of the century." But the earliest mention discovered to date was found by *Sky & Telescope*'s Dennis di Cicco in George F. Chamber's 1867 edition of *Descriptive Astronomy,* which calls the group "the magnificent double cluster in the sword-handle of Perseus." He also found the term in several popular observing guides published during the 1890s by authors on both sides of the Atlantic.

As early as the 1930s astronomers had some doubt as to the clusters being neighbors in space. *Sky Catalogue 2000.0* lists NGC 869 (the western member of the pair) at a distance of 7,200 light-years, while NGC 884 is 7,500. But the uncertainties in making such measurements could easily put 2,000 more light-years between them. With an estimated age of 5.6 million years, NGC 869 is nearly $2^1/_2$ million years older than NGC 884. Both, however, are very young, having formed shortly before humans first appeared on Earth.

But let's forget the astrophysics and simply enjoy the spectacle. In the telescope this tremendous blaze of scintillating suns makes a commanding entrance into the eyepiece field. One can look for a long time at the many doubles, the colors, the winding patterns, as the dense cores of the cluster thin out slowly to merge finally in the star-rich background of the galaxy itself. Gazing at these clusters produces a succession of feelings too subtle and too complex to be captured by words alone.

Each of these two open clusters would stand well on its own, but they are even more spectacular because, less than a degree apart, they are visible in the same low-power field. I see h Persei (NGC 869) being the slightly brighter and more concentrated of the two. Bečvár's *Atlas Catalogue** gives the star count in NGC 869 as 250. Just $1/2°$ east, χ Persei is said to contain some 300 stars. However, anyone who looks at these clusters with a 10-inch telescope will certainly consider the catalog values to be conservative.

While Robert Burnham, Jr., in his *Celestial Handbook,* suggests that the pair is best seen with a rich-field telescope, I prefer the view in a long-focus instrument that produces a darker background and more contrast. My most spectacular view of the Double Cluster came in a 6-inch refractor equipped with a unique wide-field eyepiece of 4-inch focal length made by Art Leonard.

You can also use the Double Cluster to gauge your nightly sky conditions. Begin with a photograph of the group and select an easily identified field near the edge of one cluster. Draw a map of the area with a fairly large scale of, say, 1° to 8 inches.

Plot the star brightnesses as carefully as possible. Compare the chart to an actual low-power eyepiece field and make whatever brightness corrections are needed. Next, choose a small area on the chart and include all the faint stars visible in the low-power field. Then, switching to a higher magnification, add in the faintest stars your telescope will show. This chart can be compared with the view on other nights. Although you will not have an actual value for the magnitude limit of your telescope, you will be able to tell which nights are better than others by checking the chart against the telescopic view. It would be best always to use the same eyepiece when making this test.

By the way, while in the area, be sure to search for the dark nebula Barnard 201, about a degree due west of h Persei. Try sweeping the area around the Double Cluster; there are several other dark nebulae nearby. The 12-inch f/17 Porter turret telescope at Stellafane showed several of them quite well.

Winter's Furnace

Sometimes the obvious escapes us. For instance, many Southern Hemisphere celestial wonders *are* visible from mid-northern latitudes. A few decades ago, most Northern Hemisphere amateurs would have considered seeking them out a worthless pursuit. But Scotty knew better. In his column he continually proclaimed

---

* *This catalog is the companion to Bečvár's* Atlas of the Heavens.

the possibilities of "going south." He loved the temptation to tip his telescope to extreme southern declinations and scout out the "unachievable." The Fornax Group of galaxies is a good example of the treasures he knew existed down south. In January 1957 he wrote: "Most people do not realize that this amazing group is within the reach of an observer in latitude 40° north, if he has a fine, clear sky near the southern horizon. It is worth watching for a perfect night so that this tightly clustered group may be viewed at its best on the meridian, early these January evenings. Were this array in northern declinations, it would be one of the great shows of the sky." With these words, Scotty opened the door to this southern "hot" zone of galaxies and other celestial wonders for Northern Hemisphere observers.

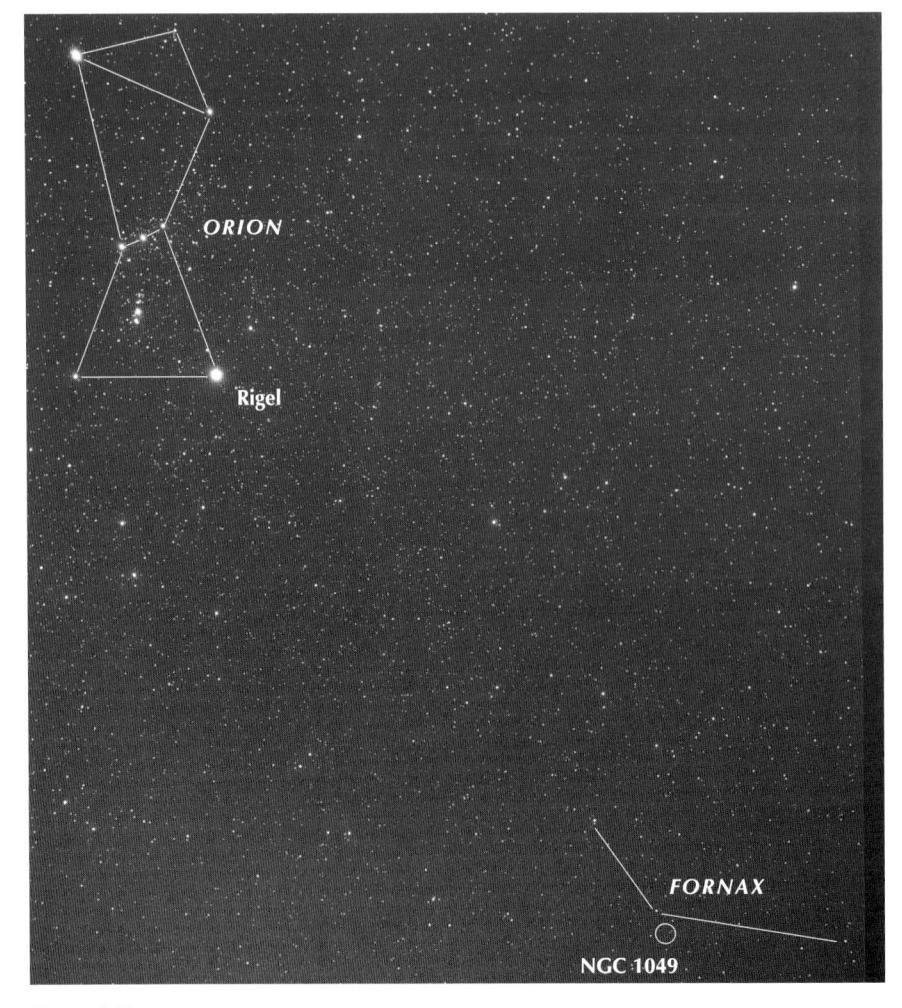

Figure 1.12
Although low on the southern horizon from mid-northern latitudes, the constellation Fornax rewards deep-sky observers hunting for galaxies. Look for it some 25° southwest of bright Rigel in Orion.

January evenings are perfect for exploring the dim expanse of the constellation Fornax, the Furnace (**Figure 1.12**). While many see Fornax as vacant space, it contains some remarkable deep-sky objects. Its lucida* is only 4th magnitude, however, so when working this area it helps to have a finder of 2-inch aperture or larger on your telescope. Hunting objects in this part of the sky is easier than some may think. Although bright stars are rather sparse, even the smallest telescope finder will show an ample selection near many of the objects of interest.

If the Stinger stars in the tail of Scorpius clear your southern horizon during the summer, then you have access to all of Fornax during the winter. A good place to begin is Alpha (α) Fornacis, a neat, 4th-magnitude double star. The 7th-magnitude companion, which is suspected of being variable, takes about 300 years to complete an orbit. Of course, readers who have enjoyed Harlow Shapley's *Galaxies* will recall his account of the Fornax Group (**Figure 1.13**), and those who have looked up the region in the Skalnate Pleso *Atlas* find a dense swarm of spirals. But even so, a clear sky will show so many galaxies fainter than the 13th-magnitude limit that identifications are almost impossible, and amateurs must satisfy themselves with the spectacle of unnamed wonders. Half a

Figure 1.13 Galaxies in Fornax show a wide variety of shapes — from ellipticals to spirals. Some appear edge on. The barred spiral in the lower right corner is NGC 1365.

---

* A term for the brightest star in a constellation.

JANUARY

dozen of these galaxies are brighter than magnitude 11.5 and hence visible in small telescopes; they are comparable with the fainter Messier objects. The brightest is 10th-magnitude NGC 1316, easy in a 3-inch. John Herschel described it in 1837 as "very bright, very large, 4' in diameter, brightening first gradually and then suddenly toward the middle to a starlike nucleus." His description still holds for my 10-inch reflector in Kansas.

Just north of NGC 1316 lies 12th-magnitude NGC 1317. A large telescope and a fine night will bring out faint NGC 1318 between the two; a 10-inch shows the

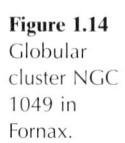

Figure 1.14
Globular cluster NGC 1049 in Fornax.

last with difficulty. But the highlight of the Fornax array is NGC 1365, a large 11th-magnitude barred spiral fully 8' long, which can be observed even with a 3-inch. With its obviously lenticular form and bright central part, it reminds one of binocular views of the Andromeda Galaxy, M31.

There are too many galaxies in Fornax to detail them all. But one last one I'll mention stands out because of its astronomical interest. Simply called the Fornax System, it is the largest of a relatively rare class of galaxies known as dwarf spheroidals — the smallest and intrinsically faintest galaxies yet to be observed. We see this object only because it is close at hand. Estimates place it 600,000 light-years from the Milky Way — more than three times farther than the Magellanic Clouds, but only a quarter the distance of M31. I know of no visual observations of the Fornax System, but other seemingly impossible observing challenges have been conquered by amateurs in recent years.

Some astronomers liken dwarf spheroidals to overgrown globular clusters. Thus it is somewhat ironic that several true globulars have been found associated with the Fornax System. NGC 1049 (**Figure 1.14**) is one such globular. It was bright enough to be picked up by John Herschel and is within reach of modern amateur telescopes. It will be a test of the observer's skill, however, as its 13th-magnitude disk is only ½' across.

A Fantastic Fornax Planetary

Down in Potter Valley, California, Todd Hansen avidly tracks every planetary nebula within reach at his 39° latitude. He is quite capable of identifying a 1″ planetary nebula that looks for all the world like a star at first glance. So when he ran across NGC 1360 in Fred Klein's selected list of 500 observing objects, he promptly turned his telescope to Fornax. There, to his surprise, he found not the tiny planetary he expected, but a "significant object not in the *Atlas of the Heavens.*" In his notebook he wrote: "Medium brightness, large, oval, unconcentrated. Quite a glowing cloud, must be 7th- or 8th magnitude overall, and a noteworthy object." But why hadn't he run across it before?

The history of this object is strange. It was found in 1857 by Lewis Swift and later by August Winnecke, both well-known comet hunters of the time. How John Herschel missed it in his sweeps, or why observer James Dunlop overlooked it in his catalog, is most curious.

The planetary was known to a few, however, because amateur astronomer Ernst J. Hartung comments on it in his *Astronomical Objects for Southern Telescopes.* Hartung's description is close to Hansen's. A short notice on this object was included in Deep-Sky Wonders for December 1972, and it surprises me now. I wrote that NGC 1360 was not seen with a 4-inch refractor but glimpsed with a fast 5-inch refractor — a sad testimony to the murk of my Connecticut skies that evening. In 1985, from Joshua Tree National Monument in the California desert, I saw it clean and bright with a little Tuthill Star Trap on a tiny Dobsonian mounting. The sky that night had a naked-eye magnitude limit of 7.5. Hansen's description seems to be accurate.

Again, do not let the Fornax label put you off from searching for it. Perhaps NGC 1360 is overlooked because it is in a nondescript constellation that U. S. observers subconsciously class as "too far south." The planetary, however, is no farther south than the globular cluster M4 near Antares, which even beginning observers quickly hunt down in the summertime sky. It should be visible easily from deep into the Canadian provinces. In their *Revue des constellations,* R. Sagot and J. Texereau state that NGC 1360 has been seen in a 2.2-inch refractor.

University of Illinois astronomer James B. Kaler points out several interesting facts concerning NGC 1360. The nebula is one of the few known examples of a large, high-excitation planetary. Its 11th-magnitude central star has a surface temperature of at least 85,000° Kelvin. It is some 14 times hotter than the Sun and at least 540 times more luminous, but has only about one-tenth the Sun's diameter. The nebula, with an estimated diameter of 2 light-years, is perhaps twice the size of the Ring Nebula in Lyra, and about half the size of the largest planetaries known.

After I asked for notes from people who had seen NGC 1360, over three dozen letters arrived from observers who were delighted with the view. This object should be added to every winter star party observing list.

JANUARY OBJECTS

Name	Type	Const.	R. A. h m	Dec. ° '	Millennium Star Atlas	Uranometria 2000.0	Sky Atlas 2000.0
Alpha (α) Fornacis	*	For	03 12.0	−29 00	380	311, 355	18
Andromeda Galaxy, M31, NGC 224	Gx	And	00 42.7	+41 16	104, 105	60	4, 9
B201	DN	Per	02 13.0	+57 05	—	—	—
Barnard's Loop, Sh2-276	BN	Ori	05 56.0	−02 00	252, 254, 229, 230, 276, 277	226	11, B2
Double Cluster (east), NGC 884	OC	Per	02 22.4	+57 07	46, 62	37	1
Double Cluster (west), NGC 869	OC	Per	02 19.0	+57 09	46, 47, 62	37	1
Fornax System, ESO 356-4	Gx	For	02 39.0	−34 30	403	354	18
Horsehead Nebula, B33	DN	Ori	05 41.0	−02 24	253, 277	225, 226	11, B2
Hubble's Variable Nebula, NGC 2261	BN	Mon	06 39.2	+08 44	202, 203, 226, 227	182, 183	11, 12
IC 434	BN	Ori	05 41.0	−02 24	253, 254, 277, 278	225, 226	11, B2
M33, NGC 598	Gx	Tri	01 33.9	+30 39	146	91	4
NGC 604	BN	Tri	01 34.4	+30 47	146	91	—
NGC 1049	GC	For	02 39.7	−34 17	403	354	18
NGC 1316	Gx	For	03 22.7	−37 12	401	355	18
NGC 1317	Gx	For	03 22.8	−37 06	401	355	18
NGC 1318	Gx	For	03 22.8	−37 07	—	—	—
NGC 1360	PN	For	03 33.3	−25 51	356	312	—
NGC 1365	Gx	For	03 33.6	−36 08	401	355	18
NGC 1662	OC	Ori	04 48.5	+10 56	208	179	11
NGC 1708	OC	Cam	05 02.6	+52 53	—	—	—
NGC 2024	BN	Ori	05 41.9	−01 51	253	225, 226	11, B2
NGC 2063	OC	Ori	05 46.8	+08 48	—	—	—
NGC 2169	OC	Ori	06 08.4	+13 57	204	181, 182	11, 12
NGC 2180	OC	Ori	06 09.6	+04 43	—	—	—
NGC 2184	OC	Ori	06 10.9	−03 31	—	—	—
NGC 2251	OC	Mon	06 34.7	+08 22	203, 227	182	11, 12

Ast = Asterism; BN = Bright Nebula; CGx = Cluster of Galaxies; DN = Dark Nebula; GC = Globular Cluster; Gx = Galaxy; OC = Open Cluster; PN = Planetary Nebula; * = Star; ** = Double/Multiple Star; Var = Variable Star

JANUARY OBJECTS (CONTINUED)

Name	Type	Const.	R. A. h m	Dec. ° ′	Millennium Star Atlas	Uranometria 2000.0	Sky Atlas 2000.0
NGC 7394	OC	Lac	22 50.6	+52 10	1085	—	—
Orion Nebula, M42, NGC 1976	BN	Ori	05 35.4	–05 27	278	225, 226, 270, 271	11, B2
Pleiades, M45	OC	Tau	03 47.0	+24 07	163	132	4, A2
T Orionis	Var	Ori	05 35.8	–05 29	278	225, 226, 270, 271	—
Trapezium, Theta (θ) Orionis	∗∗	Ori	05 35.3	–05 23	278	225, 226, 270, 271	11, B2

Ast = Asterism; BN = Bright Nebula; CGx = Cluster of Galaxies; DN = Dark Nebula; GC = Globular Cluster; Gx = Galaxy; OC = Open Cluster; PN = Planetary Nebula; ∗ = Star; ∗∗ = Double/Multiple Star; Var = Variable Star

FEBRUARY

CHAPTER 2

Wonders in the Void

"Most would agree that the area between the Big Dipper and Cassiopeia is as close to a stellar void as there is in the heavens," Scotty wrote in March 1985. "If asked, I could not point out which 4th-magnitude stars are Alpha and Beta Camelopardalis, but I can turn a telescope to the star field containing 13th-magnitude Z Cam without hesitation." Scotty knew an underdog constellation even when he couldn't see it. Perhaps that's why he devoted so many words to describing the hidden treasures in this dim corridor of stars in the northern sky. Beyond its familiarity to variable-star observers, Camelopardalis was largely avoided by skywatchers of his generation. That did not keep Scotty from grabbing his telescope and visually roaming this *stella incognita*. And he did not go it alone. He used his column to invite others to explore this new frontier. "Camelopardalis is a wonderful training arena," he wrote, "for those who want to sharpen their skills at finding inconspicuous targets."

"Camelopardalis is the absence of a constellation," wisecracked Dalmiro F. Brocchi, a well-known chart maker for the American Association of Variable Star Observers. Joseph Meeks, the man who taught me to observe variable stars, considered the Celestial Giraffe "the black hole of all the constellations." In my early years I thought of venturing into this part of the sky as something akin to exploring the den of Beowulf's dragon. But times have changed, and many of today's deep-sky observers enthusiastically probe the star-poor regions avoided by earlier generations.

These February nights the broad expanse of Camelopardalis rides high in the sky between Auriga and Ursa Minor. Although it may look bleak to the naked eye, it offers a variety of telescopic deep-sky wonders. Some objects are easy, and some are real tests. The most numerous deep-sky objects in Camelopardalis are galaxies, and the most impressive of these is NGC 2403 (**Figure 2.1**).

It's too bad that Messier missed this spiral while hunting comets. If it had been included in his list, it would certainly be one of the better-known galaxies in the northern sky. *Sky Catalogue 2000.0* lists NGC 2403 as about $1/4°$ in diameter and shining with the total light of an 8.4-magnitude star — values similar to the

Figure 2.1
NGC 2403 is a spiral galaxy with open arms, a hint of a bar, and a small central bulge. The spiral arms are bursting with star-forming regions and young supergiant stars.

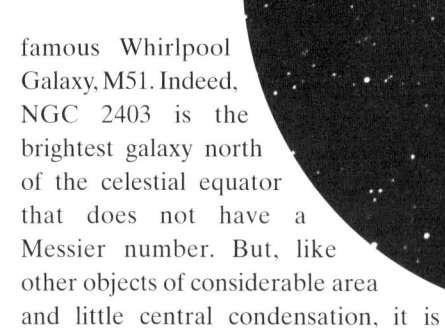

famous Whirlpool Galaxy, M51. Indeed, NGC 2403 is the brightest galaxy north of the celestial equator that does not have a Messier number. But, like other objects of considerable area and little central condensation, it is more difficult than its listed magnitude might suggest. Because of its low surface brightness, this oblong patch is easily missed if you look for too small an object. NGC 2403 lies 1° west of 6th-magnitude 51 Camelopardalis. Curiously, the galaxy is not covered in many of the old observing handbooks, but Hans Vehrenberg includes it in his *Atlas of Deep-Sky Splendors*.

In his *Celestial Handbook* Robert Burnham Jr. notes that NGC 2403 is frequently swept up by comet hunters. I suspect that those who mistake its nature are new to the comet-hunting game, since all the veterans I know agree that comets just do not look like galaxies. The very successful 19th-century New York comet hunter William R. Brooks made a similar comment. And Leslie C. Peltier once wrote, "To me there is a definite substantial appearance to the light emanating from a comet [unlike] the evanescent fugitive shimmer that finally reaches us from the immensely distant cluster or nebula."

(Interestingly, Peltier included NGC 2403 in a list of galaxies for testing the suitability of telescopes for comet hunting. This is a topic worth systematic experiments by amateur astronomers. It would be particularly helpful if the experiences of a number of observers with varied equipment could be collected, relative to the same list of galaxies.)

NGC 2403 is visible in large binoculars and presents an ever-changing vista as the aperture used to view it increases. My 4-inch Clark refractor shows it as a lovely gem. I logged it as "an ocean of turbulence and detail" as seen with a 10-inch reflector under dark Kansas skies in the 1950s. In 1992 I saw it with a 20-inch telescope from the Florida Keys — a view that transformed it into a hurricane of cosmic chaos. Much of this detail is visible because the galaxy is relatively close. NGC 2403 is about 8 million light-years away and possibly an outlying member of the group of galaxies that includes M81 and M82 in Ursa Major.

Another beautiful object in Camelopardalis is the huge 12th-magnitude spiral galaxy IC 342. The galaxy was discovered by the English amateur William F. Denning as late as the 1890s, and it may be a member of the Local Group, which includes the Milky Way, M31, M33, NGC 6822, and about a score of dwarf galaxies. Because IC 342 is one of the nearest galaxies, it appears to span almost half the Moon's diameter, and on photographs it has been traced across some 40' of sky. This large size, however, causes observing problems akin to those encountered with M33, since the galaxy's light is spread out and the actual surface brightness is low. When hunting for IC 342, use low powers and averted vision, and try wiggling the telescope tube, since very diffuse objects are sometimes better seen if the field is moving slightly.

Because it lies only 10° from the plane of the Milky Way, IC 342 is well baptized by the clouds of dust and gas in our own galaxy. We can only wonder what it would look like if the view were unobstructed. I have no record of seeing it with any of several 4-inch refractors, but some years ago using a 10-inch reflector I noted it as "easy and even stands 100× once located." I received a letter from Missouri amateur Douglas Brown who observes with a homemade 10-inch reflector. He hunted down the spiral, saying that under average skies he saw only a faint glow. From a dark site the galaxy appeared bright enough to be picked up while sweeping.

Those who enjoy hunting down more challenging objects can try for the 12th-magnitude irregular galaxy NGC 1569. It is a little over 2' long, only half as wide, and well within the range of an 8-inch instrument, but I wouldn't be surprised if it were picked up by a skilled observer using a 4-inch.

Over the years, several people have written to me about NGC 1501, a planetary nebula in southwestern Camelopardalis (**Figure 2.2**). Although the nebula and its central star are each cataloged as magnitude 13, the object is easier to find than these numbers suggest. William H. Smyth called it "bright;" the 19th-century observer Heinrich d'Arrest thought it conspicuous in a $4^1/_2$-inch refractor; and James Corn of Phoenix, Arizona, found it "easy." To me it is easy and bright even in small telescopes.

Finally, look for one more planetary: IC 3568. Although it is actually in Camelopardalis, IC 3568 lies near the center of the circular arc of the Little Dipper's handle. Because of its high declination, IC 3568 is observable from the United States every night in the year. Unfamiliar to most amateurs, it is some 18" across and near magnitude 11.5. As seen with a 4-inch, this planetary has a brighter middle, while a 10-inch brings out well the tiny 12th-magnitude central star. In a hasty sweep with low magnification, it is easily passed over as just another faintish star, but once spotted it resembles a ghostly green Mars at an average opposition distance.

Figure 2.2
The planetary nebula NGC 1501 lies nearly 4,000 light-years away. This remnant of a dying star is sometimes called the Oyster Nebula.

Kemble's Cascade and Pazmino's Cluster

"The sky is always turning a new page," wrote Henry David Thoreau in his journal for November 17, 1837. Although he was referring to the daytime sky, with its perpetual random splendor of colors and clouds, one could argue that the same applies to the night sky. In a terrestrial sense it appears static, save for the nightly changes in perspective. Of course, there are new and marvelous sights that come to our attention, like a voyaging comet or an erupting nova or supernova. But deep-sky observers surprise us from time to time by finding some new asterism or shape in a grouping of stars. Such "discoveries" do little to advance astronomy as a science, just as seeing a pretty cloud adds little to weather studies. But they do add to the enjoyment of our hobby, and Scotty often opened our minds to the possibilities of such discoveries.

Despite more than a half century of peering into nooks and crannies and looking where the guide books were silent, I missed one of the sky's more beautiful asterisms. In 1980 a letter from Lucian J. Kemble, who lives under the clean skies of Alberta, Canada, told of a fine grouping he had run across. While sweeping with 7 × 35 binoculars in Camelopardalis, Kemble found "a beautiful cascade of faint stars tumbling from the northwest down to the open cluster NGC 1502." I called the asterism Kemble's Cascade when writing about it in this column. The name has stuck. Although Kemble found the long star chain very remarkable, none of my references makes any mention of it. It was daylight when I read his letter, so all I could do was to look in *Norton's Star Atlas*, but it showed no trace of the group. Other atlases proved blank as well. In fact, it wasn't until the publication of the *Uranometria 2000.0* star atlas in 1987 that Kemble's group could be found on a map.

Figure 2.3 Kemble's Cascade in Camelopardalis is a long chain of stars near the open star cluster NGC 1502.

The asterism rides high in the northern sky during February evenings. I was shocked the night I first turned to the area with my binoculars, for there was a totally unexpected gem (**Figure 2.3**). In this pallid corner of Camelopardalis, tumbling steeply southeastward, was a celestial waterfall of dozens of 9th- and 10th-magnitude stars. Down it went, over 2½° before splashing into NGC 1502.

Although seldom mentioned in observing guides, NGC 1502 stands out nicely from its relatively starless backdrop. Were it in a brighter part of the Milky Way, the cluster might be a problem, but many observers note that it stands out clearly. It has about 45 stars that shine with a total brightness of magnitude 5.7. They are pretty evenly spread across an 8′ diameter and their magnitudes appear uniform. Binoculars or finders pick the cluster up easily. My 5-inch 20× spotting scope does an excellent job showing a score of 7th- to 10th-magnitude suns. Pat Brennan, an accomplished Canadian observer, draws the group as an obvious triangular mass, and other observers agree. It is often the case, however, that such shapes depend on the size and magnification of the telescope used for the observation. What do you see?

The cluster also includes two interesting multiple stars: Struve 484 and 485.* The former is a trio of 10th-magnitude stars with the companions separated from their primary by 5½″ and 22½″. The latter, however, is a small celestial fireworks display involving nine components, seven of which are between 7th and 13th magnitude and within reach of a good 4-inch telescope. Even the two 13.6- and 14.1-magnitude components are accessible with an 8-inch telescope. Moreover, the system's B component (brightest companion) is the eclipsing variable star SZ Camelopardalis, which dips 0.3 magnitude every 2.7 days.

---

* *Struve numbers refer to a list of double stars first compiled by the 19th-century astronomers Wilhelm Struve and his son Otto, from observations made at Dorpat and Pulkovo observatories in Russia.*

Figure 2.4
The open cluster Berkeley 10 is a 13th-magnitude cluster near Kemble's Cascade.

If you have a 16-inch or larger telescope try to see if you can detect the small open cluster Berkeley 10 (**Figure 2.4**). It is 3° north-northwest of the northern end of Kemble's Cascade and is about 12' in diameter on the *Uranometria 2000.0* charts. Some sources give it 50 stars or more, but none of the observing guides in my collection mentions this faint cluster whose brightest star is about 13th or 14th magnitude. I thought I could see it once with a friend's 17-inch reflector, but the night was a smog brew that ruined the observation.

Another amateur discovery that came to my attention is Stock 23, on the border with Cassiopeia (**Figure 2.5**). It is a coarse collection of about 25 stars spread across 10' of sky, roughly magnitude 6.5, and visible in binoculars. What is particularly surprising is that the group carries no Messier, *NGC*, or *IC* designation. Sometimes it is referred to as Pazmino's Cluster — after the New York amateur who wrote about his independent discovery of it in *Sky & Telescope* for March 1978. Stock 23 was virtually unknown in the late 1970s when John Pazmino accidentally stumbled across it while observing with a 4.3-inch refractor. I found an interesting comment about Stock 23 in the *Deep-Sky Field Guide* (the companion volume to the *Uranometria 2000.0* star atlas). The guide mentions that Stock 23 is involved with nebulosity. I've not heard of this before, and none is visible from Connecticut in my 4-inch Clark refractor. Perhaps someone with a larger telescope and better skies can check this out.

Figure 2.5 Look near the southeastern corner of Cassiopeia, just over the border in Camelopardalis, for the open cluster Stock 23. It is also known as Pazmino's Cluster.

By the way, Kemble uses a trick for making drawings at the telescope that I have not heard mentioned before. He racks the eyepiece out of focus until only the brightest stars are visible, and then plots their relative positions. Once this "skeleton" is made, he refocuses and adds the fainter stars.

Going to California

In December 1981 Scotty pondered whether the California Nebula (NGC 1499) in Perseus could be seen with the naked eye, and he asked readers for their comments. "This large, faint, diffuse nebula," he wrote, "is plotted as an inviting green blob over 2° long on the new color edition of Wil Tirion's *Sky Atlas 2000.0*. However, I have never looked for it and do not have any letters in my file mentioning visual sightings of it. . . . I suspect that large binoculars, perhaps equipped with nebula filters, will have the best chance of revealing it." Of course, the positive responses poured in. Scotty had begun what was fast becoming the hottest new visual challenge for telescopic and naked-eye observers.

At the 1982 Texas Star Party, I was asked what was the best new challenging deep-sky object after the large-aperture Dobsonian revolution had dispatched most of the test objects from the 1950s and '60s. I suggested the California Nebula, not knowing that a piece of modern technology would soon remove it from the list of challenges: a skyglow-piercing nebula filter. In fact, I remember saying that it is the ultimate test object for visual observers. So much for that wisdom, for little did I realize when I made the comment that before I returned to Connecticut I would see the nebula with my naked eye through an O III filter. In the winter of 1992, in Mexico, the same filter showed the California Nebula as "bright."

Figure 2.6
The California Nebula (NGC 1499) lies about 12° north of the Pleiades. This emission nebula is about 2,000 light-years away in the constellation Perseus.

The California Nebula (**Figure 2.6**) is a vast cloud of glowing hydrogen (plus trace amounts of other gases) about $2\frac{1}{2}°$ long and $\frac{3}{4}°$ wide. It was discovered visually by E. E. Barnard on November 3, 1885, while he was using the 6-inch refractor at Vanderbilt University in Nashville, Tennessee. It is known as the California Nebula because its shape, recorded on long-exposure photographs, is similar to that of the western state. NGC 1499 has a surface brightness which is less than that of M33. So conventional wisdom has long held that the surface brightness of the California Nebula is too low to be seen against the background sky visually in any telescope. Of course, the fact that Barnard discovered the nebula visually seems to have been long forgotten.

Today the term *low surface brightness* (LSB) is used to describe large, diffuse objects such as the California Nebula. Their surface brightnesses are so close to that of the sky background that the observer can run his or her eye across the edge of the object without realizing that the view has changed from sky to nebulosity.

These objects are seen only when the sky is dark and transparent. A low magnification should be used so that the field of view shows plenty of sky to contrast with the object. The telescope's optics should be well collimated and free from dust and dirt that would scatter light and reduce the image contrast. The eyepiece also should be clean, and all air-to-glass surfaces antireflection-coated. While a number of things affect the visibility of LSB objects, I suspect that seeing them depends more on observer experience and eye training than on specific telescope f/ratios and magnifications.

Despite its difficulty, NGC 1499 has been viewed by a fair number of amateurs. John Bortle of Stormville, New York, did not have the best sky conditions when he observed the nebula. He writes: "NGC 1499 is visible, although extremely faint, in both 10 × 50 binoculars and 20 × 80 binoculars; the northern side of the nebula appears rather sharply defined even though the object's entire outline is generally quite vague. The presence of Xi (ξ) Persei (4th magnitude and about 1° south of the nebula) is an annoyance and hinders observation. It is decidedly fainter in surface brightness than the nearby Merope Nebula in the Pleiades and much fainter than the North America Nebula. I imagine its visibility is limited to binoculars for want of contrast in narrow-field telescopes."

Alister Ling of Montreal, Canada, also furnished a report on NGC 1499. He was observing from a small boat in the middle of a lake by David Levy's cottage in northern Quebec. While this may seem rather unusual, the water acts as a thermal stabilizer and often provides better seeing conditions than on the nearby land. Ling comments: "I made a monocular from my 400-mm telephoto lens by attaching a 28-mm orthoscopic eyepiece to it. This gives a magnification of about 14× and a field several degrees in diameter. No sooner had I located Xi Persei than the extended nebula was quite obvious. It was about $1^1/_2$° long with two fairly bright stars embedded near its edge. Roughly at its midpoint there is an obvious kink in the nebulosity. It appears more like a mass of unresolved stars than a gas cloud — very much as the Milky Way appears to the naked eye. Later, a crescent Moon rose in Gemini and rendered the California Nebula invisible."

During a trip to the West Coast I spoke with several amateurs who had seen NGC 1499 with telescopes as well as with binoculars. Their technique was to locate the edge of the cloud and follow it around. Here in Connecticut I have glimpsed NGC 1499 with 6 × 30 binoculars and can see it nicely with a 5-inch Apogee scope at 20×. However, my 4-inch Clark refractor will not reveal the faint glow.

As a final note, in response to my query about visual observations of the California Nebula, Lucian Kemble writes that he has seen parts of it with a Celestron 8. In September 1981, he notes: "I had a perfectly dark sky and decided to try my luck. With a Celestron 8 and orthoscopic eyepiece giving 74×, I was able to cover the entire object without detecting any variations in its uniform, very pale surface. As it spans an area much greater than my eyepiece field, the nebula had to be viewed by sweeping. Its outline appeared as a very slightly lighter patch of sky against the background blackness. It was much like tracing the far wispy reaches of M31 or the edges of the Helix or Rosette Nebula."

Thus, it appears that NGC 1499 is accessible to a wide range of instruments. The most important requirements for locating it are an experienced eye and clear, dark skies. If your observing conditions are good and you cannot detect the nebula, try practicing on the galaxy M33 in Triangulum, which has a considerably higher surface brightness. Careful cleaning of the optics, especially the eyepiece, can help improve the image contrast of a telescope. Another suggestion is to apply gentle heat to the optics, including the eyepiece, to remove subtle condensation. This rather unconventional practice has worked very well for me with both reflectors and refractors.

The Little Dumbbell

Scotty often talked about the Big Four of Perseus, a diverse assortment of deep-sky objects ranging from the easy (the Double Cluster) to the difficult (the California Nebula). In the middle he listed the pleasing open cluster M34 and the intriguing planetary nebula M76, the Little Dumbbell. M76 is often touted as being a difficult object to spot, but Scotty knew otherwise. He suspected that this belief was based more on perception than on reality — that M76 appears to be a difficult-to-find object because many published magnitudes suggest it *is* one. As Scotty explained, this simply is not the case. M76 is indeed a treasure worth the hunt.

In the clear, cold nights of winter, the dazzling constellation Perseus stretches its silvery fishhook high in the northern sky. The Milky Way narrows considerably in Perseus, being partly veiled by interstellar dust, and we are looking well away from the center of the galaxy, in Sagittarius. Star charts show that the open star clusters which abound in Cassiopeia and Auriga are noticeably fewer in Perseus. But the constellation does offer many objects that will reward the observer who braves the cold weather to observe them.

One object that everyone should enjoy is M76, an unusual planetary nebula on the edge of a dense part of the Milky Way (**Figure 2.7**). Often called the Little Dumbbell because it resembles the more-familiar Dumbbell Nebula (M27) in Vulpecula, to me it always looks more like a dog biscuit. The nebula has a reputation of being hard to find, so here's how best to find it. Start with Phi (ϕ) Persei. This star and a dimmer one just to the south form a pointer, with Phi at the head, that directs the observer to a diamond of faint stars, within which M76 is dimly perceptible.

Do not be misled by the photographic magnitude of 12.2 given in some catalogs, which makes it the faintest object in the Messier catalog. I believe that it is much brighter. If I try to defocus a 12th-magnitude star until the image spans 1′ or 2′ (about the size of M76), the star disappears. Yet I can see the planetary in the same telescope, indicating that the nebula is brighter. *Sky Catalogue 2000.0* lists M76 as magnitude 11.5, nearly a magnitude brighter than older listings. My own estimate has it brighter still, perhaps 11.0, and a nebula filter helps to make it stand out in a less-than-perfect sky.

With a small aperture or in indifferent sky conditions, M76 shows only a dim irregular oval with ragged edges. But one night, with an 8-inch reflector in the hills north of the Golden Gate in San Francisco, M76 was a most exciting object. It appeared more than 2' by 1' (large for a planetary) and high magnifications brought out an intricate network of turbulent celestial clouds. At Stellafane in Springfield, Vermont, M76 appeared as a marvelous object in George Scotten's 12-inch f/5.7 Dobsonian reflector. The nebula seemed to float between us and the starry background, its edges appearing even more frayed than when smaller

Figure 2.7 The Little Dumbbell (M76) lies more than 3,000 light-years away. It is sometimes referred to as the Cork Nebula or the Butterfly Nebula.

telescopes are used. Its curled twists and streamers seemed to show the whole mass in turmoil. At the 1992 Winter Star Party in the Florida Keys I had a chance to view M76 through the 36-inch Dobsonian reflector built by Tom and Jeannie Clark. To reach the eyepiece required climbing a stepladder half as high as the surrounding palm trees, but the view was worth it. It made anything I had ever seen in my old 10-inch reflector just a dusty memory.

Indeed, M76 has the oddest shape for a planetary nebula. Instead of the common smoke ring or circular disk, this nebula looks more like a brick about 2' long and half as wide. William Herschel thought it was two separate objects in contact; therefore its disk has two *NGC* numbers — 650 and 651, the latter applying to the northeastern part. Ireland's Lord Rosse[*] thought he saw hints

---

[*] *A mid-19th-century telescope maker and astronomer, Lord Rosse (William Parsons, 3rd Earl of Rosse) made numerous observations and discoveries with his fantastic reflectors, which he erected at Birr Castle. His largest telescope, a 72-inch, was so enormous that it was nicknamed the Leviathan of Parsonstown.*

of a curious spiral structure. Since Rosse discovered the spiral structure of galaxies (known only as "nebulae" at the time), it's easy to say that his interests influenced his view of M76. However, deep photographs do show circular arcs in the outer nebulosity surrounding M76, and these could have given the object a spiral appearance in Rosse's huge reflector. And W. H. Smyth used this planetary as a visibility test during a lunar eclipse on October 13, 1837. It was well seen during totality with his 6-inch, but gradually faded as the Moon came out of the Earth's shadow. Some amateurs might like to repeat his experiment.

M76 is an emission nebula, and as such should be a good test object for light-pollution filters that are designed to transmit light from emission nebulae while blocking the skyglow from natural and artificial sources. A good plan would be to observe M76 on as many nights as possible over the next few months. Record the sky conditions on each night, and the nebula's appearance without the filter and then with it. It is important to look through the filter for several minutes to allow the eye time to adapt.

Probing the Depths of Perseus

The night sky is full of undiscovered entertainers. Scotty reminded us time and again that there is more on the celestial stage than its primary players and often encouraged readers to take a good look at the sky's supporting cast. Often he would do this by tossing out an occasional "by-the-way." For example, in this section Scotty voices his concern about how an amateur is supposed to identify galaxies fainter than magnitude 13.5 — those not plotted on traditional star charts. Had he lived a few years longer, Scotty would have had a chance to congratulate Texas amateur Larry Mitchell, who compiled and plotted the positions of more than 27,350 anonymous galaxies. These galaxies now appear in *Mitchell's Anonymous Catalogue*, which is incorporated into the deep-sky database of the *Megastar* software package.

M34 in Perseus is an open cluster that receives relatively little amateur attention, because it is overshadowed by the more spectacular Double Cluster. While not as rich, M34 (**Figure 2.8**) is one of the finest sights in wide-field telescopes that can be found. It lies about halfway between the stars Algol and Gamma (γ) Andromedae, with a total brightness equal to that of a 5.2-magnitude star. Messier discovered the cluster in 1764. Ten years later astronomer Johann Bode called it "a star cluster, visible to the naked eye." W. H. Smyth found M34 a "scattered but elegant group." Thomas W. Webb was more enthusiastic, exclaiming, "one of the finest objects of its class."

According to *Sky Catalogue 2000.0,* nearly one-third of the Messier objects are brighter than the 6.5-magnitude limit often cited for the dark-adapted eye. However, brightness alone is not enough to determine the visibility of a given object — size also plays a very important role. For instance, I had a good view of M34 with

Figure 2.8 The bright open cluster M34 is a fine sight in binoculars and small telescopes.

the Milwaukee Astronomical Society's 13-inch reflector, when it appeared somewhat less than 30' in diameter, but this estimate was uncertain. Consider that Åke Wallenquist notes there are about 80 members brighter than 12th magnitude within an area 42' across. Why not try determining the size yourself?

Webb justly called attention to M34 as a grand, low-power, rich-field object. I feel that 15 × 65 binoculars give the best impression; the low power allows plenty of dark sky surrounding the cluster to enhance contrast. The sky background is sprinkled with faint stars, so it may be easier to decide where the cluster edge lies by using a small rather than a large telescope. More magnification merely spreads out the few bright stars that the binoculars show perfectly well.

Many observers see its stars arranged into distinct curved lanes that diverge from the cluster's center. I see three noteworthy curved rays of stars running out from the center which are very evident in my 4-inch Clark refractor at 40×; indeed, they even show in binoculars. Many of the stars also form pairs, as mentioned by both Webb and Smyth. Near the center of the swarm lies the double star Otto Struve 44 (OΣ 44), which my 4-inch refractor splits nicely at 100×, especially when the heater is turned on to remove any trace of dew from the objective. The primary star is of magnitude 8.5 and the 9.2-magnitude companion is 1.4" distant at position angle 55° (toward the northeast).

Digging Deeper

Until the late 1970s an unwritten rule prohibited amateur charts, observing handbooks, and columns like Deep-Sky Wonders from including objects fainter than about 13th magnitude. The reason was that big telescope mirrors were hard to come by because everyone followed the rule that a mirror's thickness had to be one-sixth its diameter. Such mirrors larger than 12 inches cost a fortune. They are heavy, and a heavy mirror requires a heavy mounting with big castings and ball bearings.

There were exceptions. In 1932 I made a 10-inch reflector from $^1/_2$-inch plate glass. The mirror had to be carefully supported or else it made every star in the field appear double — pretty, but hardly suitable for astronomy.

In the early 1930s the Milwaukee Astronomical Society (MAS) received a 13-inch mirror from Cornell University. For a while it was the largest in the country solely dedicated to serious amateur observing. Ed Halbach made the mounting from welded steel plates and used brass bearings instead of expensive ball bearings. That mounting still serves the same telescope in the same MAS observatory, but at the time few telescope makers were eager to switch to the unconventional techniques that were used in its construction.

Go to any good-sized star party today and you'll find telescopes with an aperture of 20 inches or more. It's now possible to break the 13th-magnitude limit. If you're interested in trying, there's a good starting point in Perseus. About $2^1/_2°$ west-northwest of the famous variable star Algol lie three little galaxies that all fit in the same eyepiece field. They form a nice north-south chain about 10′ long. The northernmost galaxy, NGC 1130, at magnitude 13.0, is also the brightest. In the middle is NGC 1129, at magnitude 14.5, while at the southern end of the chain is 15.5-magnitude NGC 1131. They should all show in a 10-inch scope under good skies.

None of the three is plotted on Tirion's *Sky Atlas 2000.0*, but they are on the *Uranometria 2000.0* charts. If you have a finder that shows 9th-magnitude stars, you can easily star-hop to the field by starting from Algol.

Let's assume for the moment that you've hunted this group down with a 20-inch reflector. The scope should make easy work of the galactic trio, but what of the other galaxies in the field? There are a number of them here — ones not plotted on the *Uranometria 2000.0* charts. Now what do you do, and for that matter are you still sure about the identifications of the first three? The answers are not simple. A *Palomar Sky Survey* print would certainly help in confirming what you see, but it would take some time at a major astronomy library to identify all the objects. Such are the problems confronting amateurs who push farther into the deep sky.

If this type of madness appeals to you, then try swinging to the unusual galaxy NGC 1275 (**Figure 2.9**), just over 2° east-northeast of Algol. The Herschels missed it, and we owe its discovery to the keen eyesight of H. L. d'Arrest. It was not considered anything special until astronomers in this century discovered it to be a strong radio source. Indeed, during the 1950's Cliff Simpson and I operated an amateur 200-foot radio interferometer at Zeandale, Kansas. Although this "radio telescope" could not pinpoint objects in the sky, we did see a change in the signal as the general area of NGC 1275 passed across the beam of our

Figure 2.9 NGC 1275 (center) and galaxy cluster Abell 426. The core of NGC 1275 harbors the Perseus A radio source.

antenna. It was a much weaker signal than we recorded for the radio sources in Cygnus and Cassiopeia.

A look at the *Uranometria 2000.0* chart for NGC 1275 shows the area swarming with galaxies. Identifying the individual members at the eyepiece would be near impossible even with this detailed chart. By good fortune, however, this is part of the cluster of galaxies known as Abell 426 (the Perseus Cluster) and included in Vol. 5 (Clusters of Galaxies) of the Webb Society *Deep-Sky Observer's Handbook*. This work contains detailed finder charts for several dozen members of the cluster. How many you can see will depend on your telescope, observing skill, and the quality of your night sky. NGC 1275 itself has been reported in telescopes as small as 5-inch aperture. For large instruments, the Webb handbook notes that "the Perseus cluster presents probably the most spectacular concentrated galaxy field in the northern winter sky."

Navigating the Celestial River

Meandering across 1,138 square degrees of sky, Eridanus, the celestial River, is the sixth-largest constellation. Its northernmost stars stream out from brilliant Rigel in Orion's knee. From there, it brushes the celestial equator (nearly paralleling it at first) before dipping gradually southward to about –25°. From there the River plunges deep south in a double meander toward –57°, landing at 1st-magnitude Achernar. "This star is out of reach for all observers in the United States," Scotty wrote in January 1972, "except in Florida, the Gulf Coast, southern Texas, and Hawaii." In reality, many of the celestial wonders in Eridanus *are* within range of U. S. amateur telescopes — as long as the observer has a good southern horizon and a dark sky free of haze and pollutants.

As darkness settles on the February landscape, the mighty Hunter Orion stands high over the southern horizon. Now is a fine time, however, for observers living in northern temperate latitudes to explore the backwaters and eddies of the River Eridanus cascading westward from brilliant, blue-white Rigel. Eridanus meanders in graceful loops and bends before disappearing below the southern horizon, where it ends at Achernar deep in the southern sky, at declination –57°. Eridanus is an old constellation, but Achernar is a Johnny-come-lately addition. The ancients who first outlined Eridanus could not see anything that far south. They originally considered the River's end to be the star

Figure 2.10
The bright star Achernar in the constellation Eridanus.

Acamar, Theta (θ) Eridani, some 16° north of Achernar. It wasn't until explorers headed south that mappers extended Eridanus to Achernar, picking up (and corrupting) Acamar's name.

Achernar is the ninth-brightest star in the sky (**Figure 2.10**). I once observed it from the slopes of a live volcano in southern Guatemala with a 4-inch rich-field reflector. Achernar was 10° high in a limpid sky and appeared as a terrific burst of sapphire light in the little telescope. I followed my usual practice after viewing an outstanding object of scouring several degrees of adjacent sky just to see what might be lost in the limelight of the showpiece. I found nothing. But back home in Connecticut, a glance at *Uranometria 2000.0* revealed NGC 782 about 2½° east-southeast of Achernar. This is a 12th-magnitude spiral galaxy about 2′ in diameter. I suspect it is fair game for 4-inch scopes and larger. It's probably too low to be seen at the Winter Star Party, but considering some of the huge-aperture Dobsonian reflectors seen there in recent years, who knows?

Eridanus offers no star clusters — either galactic or globular — for average amateur telescopes. Galaxies, however, swarm through it in abundance. So my first target was a hard-to-see pair of galaxies in an easy-to-find field, about 1° northwest of Upsilon[4] (υ[4]) Eridani. The brighter member is 11th-magnitude NGC 1532. At 20×, several minutes were required to glimpse this 5' × 1' spiral, but then it was held steadily. Its companion is NGC 1531, about 2' to the northwest. Since this object is at least a magnitude fainter and only about 1' in diameter, 60× was required.

About 2° north of Tau[4] (τ[4]) Eridani is NGC 1300 (**Figure 2.11**). With a total light equivalent to a star of visual magnitude 10.3, NGC 1300 is a nice barred spiral galaxy seen almost face on. It is within reach of a 4-inch, and I have seen it easily even with a 3½-inch Questar telescope. Though photographs of NGC 1300 with large telescopes reveal a central bar with two thin but tightly wound spiral arms, small amateur instruments show only a blurred spindle. A 4-inch f/12 off-axis

Figure 2.11
The barred spiral galaxy NGC 1300.

reflector suggested some detail in the glow but fell short of showing any spiral structure. A 10-inch or larger telescope will give a more distinct image, about 6' × 3', and may even reveal the faint companion to the north, NGC 1297, which is 1' in diameter and magnitude 13. The smallest telescope in which I have sighted this neighbor is an 8-inch. Of course, a 17-inch, under the clear skies of the western United States, clearly showed NGC 1300 as a barred spiral full of wisps of nebulosity to explore.

Northwest of Tau[4] Eridani and forming a right angle with it and Tau[3] (τ[3]) is another gem — NGC 1232, which lies about 3° southwest of NGC 1300. Though its total visual magnitude is about 10, it appears much less conspicuous than a star of that brightness, as its light is spread over an area about 7' in diameter. The *New General Catalogue* describes this galaxy as "pretty bright, rather large, round, gradually brighter in middle, mottled." It is a bit brighter and larger than NGC 1300, appearing nearly circular and about 7' in diameter, but I do not know of anyone seeing detailed structure with telescopes smaller than 12 inches. In my 4-inch refractor it seems to be better seen with a 150× eyepiece than with a 50× used in combination with a 3× Barlow lens. This is curious, for usually a Barlow and a long-focus eyepiece give a view superior to an eyepiece of shorter focus that is used alone. Photographs show NGC 1232 as an open spiral.

If you're interested in taking on a challenge, there is an interesting group around the 10th-magnitude elliptical NGC 1332. NGC 1332 is about as bright as NGC 1300 and also visible in a 4-inch scope. It is a rather featureless elliptical galaxy with an oval disk about 4' long. Although there is no pronounced central brightening, the edges are distinctly fuzzy at 200×. Located less than 1° to the southwest is NGC 1325, an 11th-magnitude spiral about 4' long and less than half as wide. A 13th-magnitude companion galaxy, NGC 1325A, is located just to the northeast. It is at the same declination as NGC 1332, $1/3°$ to its east. Several other galaxies in the immediate vicinity are all too faint for most amateur telescopes. Arizona amateur Ronald Morales did a fine job describing the NGC 1332 group in the February 1984 issue of *Sky & Telescope*.

Another spiral galaxy within the reach of a 4- or 6-inch telescope is NGC 1637. My 4-inch Clark refractor shows the galaxy at 34×, and though more difficult, the fixed 20× of my 5-inch Apogee scope also reveals it. This lone galaxy, which lies about 7' southeast of a 9th-magnitude star, is an 11th-magnitude glow about 2' or 3' in diameter. It has been seen with a 3-inch by an observer who knew just where to look for it, but an 8-inch or larger is preferred. On photographs, NGC 1637 is revealed as an open Sc-type spiral.

Finally, here's something to ponder. One of the weakest aspects of the *NGC* is the coded description for an object's visual appearance. As an example, it describes the galaxy NGC 1440 in Eridanus as "pB, pS, R, smbM * 13," which translates into "pretty bright, pretty small, round, smaller bright middle like a star of 13th-magnitude." All this would be fine if we knew what telescope the observer used for the description. In some cases we know but in others it remains a mystery.

Averted Vision and the Celestial Jellyfish

Averted vision is one trick of the amateur astronomer's trade that can benefit all skywatchers. Shifting your gaze slightly away from the object you're viewing makes use of peripheral vision by allowing you to peer deeper into the night, to see fainter objects, and to glimpse dim details that would otherwise go unnoticed with a direct gaze. Veteran amateur astronomers have learned how to position their eyes for optimum light-gathering power. It's all related to the eye's physiology, as Scotty explains below. He told beginning observers to practice first using averted vision without a telescope. For example, he advised those in a very dark sky to try to detect the faint outlying parts of the Milky Way. "Can you follow this glow to M31, the Andromeda Galaxy?" he asked. "Does the Milky Way fade undefinedly into the sky background, or does it have a more or less distinct edge?" Scotty understood that observing skills, like skills in any sport, are acquired through training. He used his columns to coach observers on improving their visual acuity, to eke out finer details in all they observed.

As most observers are aware, away from the center of vision the retina is more sensitive to faint light. Thus, when one looks directly at a nebula, for example, its image falls near the center of the retina (fovea) — a region of high resolving power but rather low sensitivity. By looking a bit to one side, the image falls away from the center and on a usually more sensitive region. Thus, fainter images can be seen. The skilled observer always uses "averted vision" with faint objects. But the visibility of faint objects is very sensitive to the quality of the atmosphere, the individual's eye, and experience. An observer must test each object because this method may aid some persons, while others do not find it particularly helpful.

The observer can experiment with various magnifications, averted vision, or a dark cloth to block stray light from entering the eye, to find which combinations work best for observing faint objects. Under conditions of low illumination, such as are frequently encountered in visual deep-sky observing, individuals differ greatly in retinal sensitivity. Some persons have large threshold gains, but others have very little. Also, the gain in red light is generally more than in blue.

In experiments at the Naval Research Laboratory in the late 1950s, one subject actually saw less as the image approached the edge of his retina. However, one exceptional individual's sensitivity increased steadily in both colors; the gain in red light was three magnitudes in a direction 40° from the fovea. These experiments, by L. J. Boardman, were done with scotopic (dark-adapted) vision. Amateurs can test their own eyes (at least for point sources) by using an AAVSO chart to determine the limiting magnitude of stars visible with direct vision and with averted vision. The average of a dozen or so tries might tell you a good deal about your eyes.

Until the tyro observer acquires the skills needed to ferret out fainter deep-sky targets, there is often a period of frustration at the eyepiece. Every year I receive a number of letters from persons who are facing this difficulty. My suggestion is to start with objects that are close to easily identified naked-eye stars. The telescope can be carefully moved from the star to the area to be searched. Then, confident that the nebula or cluster is within the field of view, the observer can experiment with averted vision and different magnifications to obtain the best view. In this hobby, a little experience goes a long way. You will quickly learn what techniques work best for hunting down the fainter objects.

As far as extended sources are concerned, planetary nebulae make good test objects. NGC 1535 is a good place to start (**Figure 2.12**). First, identify the field of NGC 1535, which swims through the darkness like a celestial jellyfish. Its 9th-magnitude oval disk is about ⅓′ long and surrounds a 12th-magnitude central star. However, a lack of bright guide stars in the area will make this dimly glowing patch of gas more of a challenge than its 20″ diameter and magnitude suggest. The nearest guide star is 5th-magnitude 39 Eridani, about 2° due north.

But note that a few degrees west of NGC 1535 is conspicuous 3rd-magnitude Gamma (γ) Eridani, which has a slightly curved chain of three 6th-magnitude stars north of it. The planetary is to the east at a distance about equal to the length of the chain. Since the finder's field of view should be large enough to include all three stars of the chain, it is possible to point the telescope very near the nebula.

The planetary lies in a space about 1° in diameter containing no stars brighter than 9th magnitude. Look for that dark space with the finder, and center it. Then, through the main telescope, a low-power eyepiece (1° field) should show 8th- and 9th-magnitude stars only around the edge. In the middle there will appear a dim glow about 20″ in diameter. A 6-inch or larger telescope may show some fainter stars near the planetary. Once it is located, try higher powers; they often help with planetaries. This is when averted vision is particularly useful. Let your eye roam around the margin of the field but keep your attention on the center. Your eyes should be thoroughly dark adapted, and the more transparent the sky the better for these faint extended subjects.

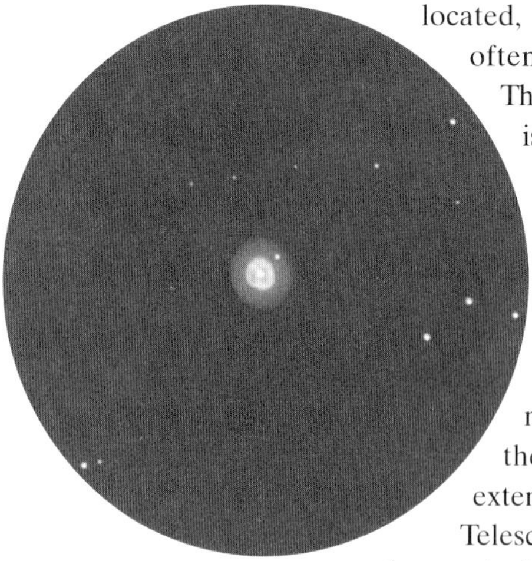

Figure 2.12
The planetary nebula NGC 1535 in Eridanus.

Telescopically, NGC 1535 has a pale, evenly illuminated disk, which W. H. Smyth saw only like a star out of focus. His contemporary William Lassell, with a very large reflector, called it "most extraordinary." Today, James Corn of Arizona reaches it without trouble. In the early days of spectroscopy William Huggins was perplexed by its "not gaseous" spectrum, but today we know that it and a few other planetaries have practically continuous spectra.

Two other planetary nebulae make good test objects. The first is PK 171–25.1 in Taurus. It is very difficult, being about 14th magnitude and only 40″ in diameter. My 4-inch refractor failed to show it, but it was seen in a 12-inch. I had to use averted vision, which adds one or two magnitudes to my threshold.

The other, NGC 246, lies to the south in Cetus. Though its total light equals that of an 8th- or 9th-magnitude star, the surface brightness of this planetary is low because the light is spread out over an area about 4′ across. Yet with averted vision I could see it in my 4-inch.

FEBRUARY OBJECTS

Name	Type	Const.	R. A. h m	Dec. ° '	Millennium Star Atlas	Uranometria 2000.0	Sky Atlas 2000.0
Achernar, Alpha (α) Eridani	*	Eri	01 37.7	−57 14	463, 478	418	24
Berkeley 10	OC	Cam	03 39.0	+66 31	31	18	—
California Nebula, NGC 1499	BN	Per	04 00.7	+36 37	117, 118	95	4, 5
IC 342	Gx	Cam	03 46.8	+68 06	31	18	1
IC 3568	PN	Cam	12 32.9	+82 33	520	9	2
Kemble's Cascade	Ast	Cam	03 56.0	+63 00	30, 31, 43, 44	18	1
Little Dumbbell, M76, NGC 650–651	PN	Per	01 42.3	+51 34	63, 82	37	1, 4
M34, NGC 1039	OC	Per	02 42.0	+42 47	100	62	4
NGC 246	PN	Cet	00 47.0	−11 53	316	261, 262	10
NGC 782	Gx	Eri	01 57.8	−57 46	477	418	24
NGC 1129	Gx	Per	02 54.5	+41 35	99	63	—
NGC 1130	Gx	Per	02 54.4	+41 37	99	63	—
NGC 1131	Gx	Per	02 54.6	+41 34	99	63	—
NGC 1232	Gx	Eri	03 09.8	−20 35	333, 357	311	18
NGC 1275	Gx	Per	03 19.8	+41 31	98	63	4, 5
NGC 1297	Gx	Eri	03 19.2	−19 06	332, 333	311	—
NGC 1300	Gx	Eri	03 19.7	−19 25	332, 333	311	10, 18
NGC 1325	Gx	Eri	03 24.4	−21 33	332, 356	311, 312	18
NGC 1325A	Gx	Eri	03 24.8	−21 20	332, 356	311, 312	—
NGC 1332	Gx	Eri	03 26.3	−21 20	332, 356	311, 312	18
NGC 1440	Gx	Eri	03 45.0	−18 16	331	312	10, 11, 18
NGC 1501	PN	Cam	04 07.0	+60 55	43	18, 39	1
NGC 1502	OC	Cam	04 07.7	+62 20	30, 43	18, 19	1
NGC 1531	Gx	Eri	04 12.0	−32 51	377, 399	356	18, 19
NGC 1532	Gx	Eri	04 12.1	−32 52	377, 399	356	18, 19
NGC 1535	PN	Eri	04 14.2	−12 44	306	268	11
NGC 1569	Gx	Cam	04 30.8	+64 51	30	19	1
NGC 1637	Gx	Eri	04 41.5	−02 51	256, 280	224	11
NGC 2403	Gx	Cam	07 36.9	+65 36	25	21	1

Ast = Asterism; BN = Bright Nebula; CGx = Cluster of Galaxies; DN = Dark Nebula; GC = Globular Cluster; Gx = Galaxy; OC = Open Cluster; PN = Planetary Nebula; ∗ = Star; ∗∗ = Double/Multiple Star; Var = Variable Star

FEBRUARY OBJECTS (CONTINUED)

Name	Type	Const.	R. A. h m	Dec. ° '	Millennium Star Atlas	Uranometria 2000.0	Sky Atlas 2000.0
OΣ 44	**	Per	02 42.2	+42 42	—	—	—
Pazmino's Cluster, Stock 23	OC	Cam	03 17.0	+60 00	45	18, 38	1
Perseus Cluster, Abell 426	CGx	Per	03 19.0	+41 30	98	63	—
PK 171–25.1	PN	Tau	03 53.5	+19 28	187	132	4, 10, 11
Struve 484	**	Cam	04 07.4	+62 23	30, 43	18, 19	1
Struve 485	**	Cam	04 07.9	+62 20	30, 43	18, 19	1
Whirlpool Galaxy, M51, NGC 5194	Gx	CVn	13 29.9	+47 12	589	76	7

Ast = Asterism; BN = Bright Nebula; CGx = Cluster of Galaxies; DN = Dark Nebula; GC = Globular Cluster; Gx = Galaxy; OC = Open Cluster; PN = Planetary Nebula; * = Star; ** = Double/Multiple Star; Var = Variable Star

MARCH

CHAPTER 3

The Elusive Winter Wreath

Over the four decades that Deep-Sky Wonders appeared in *Sky & Telescope*, Scotty watched visual astronomy evolve from its tentative and reserved roots to a burgeoning sport whose amateurs continually extend the possibilities of night vision. Advances in telescope technology and observing techniques certainly helped to revolutionize visual astronomy. Scotty often reflected on such changes, saying once that "several of winter's showpieces were considered 'test objects' not so long ago. But experience has shown that yesterday's challenges have become the stock-in-trade of today's public star parties." Through his column, Scotty almost singlehandedly advanced the rate of this progess by continually challenging observers. If one test object fell to the power of vision, Scotty would introduce another.

On one memorable trip to northern Mexico I was camped at about 6,500 feet — a rather low elevation for this region — when a front passed around midnight and brought a view of diamond-like stars. Instead of setting up a telescope, I began to explore the sky with nothing more than a set of Lumicon nebula filters and my naked eye.

The first target was an old favorite of mine, the Rosette Nebula (NGC 2237-39) to the east of Orion in the stellar wilderness we call Monoceros. Without a filter only a tiny glimmer of light was visible, but with an ultrahigh-contrast (UHC) filter the nebula burst forth in spectacular fashion. I know of no other object in the sky where flicking a filter back and forth in front of a naked eye produces such a wonderful effect.

The Rosette (**Figure 3.1**) is one of the few deep-sky objects better seen with a telescope's finder than with the main instrument. With large binoculars and good observing conditions, the Rosette may appear as a formless aura of soft light encircling a cluster. Most telescopes reveal only a neat but unassuming open star cluster. William Herschel, for example, discovered the cluster, called NGC 2244, but missed the nebulosity around it during his great visual surveys in the 18th and 19th centuries. The nebulosity was discovered piecemeal by several observers. Albert Marth was the first person to report nebulosity associated with the stars.

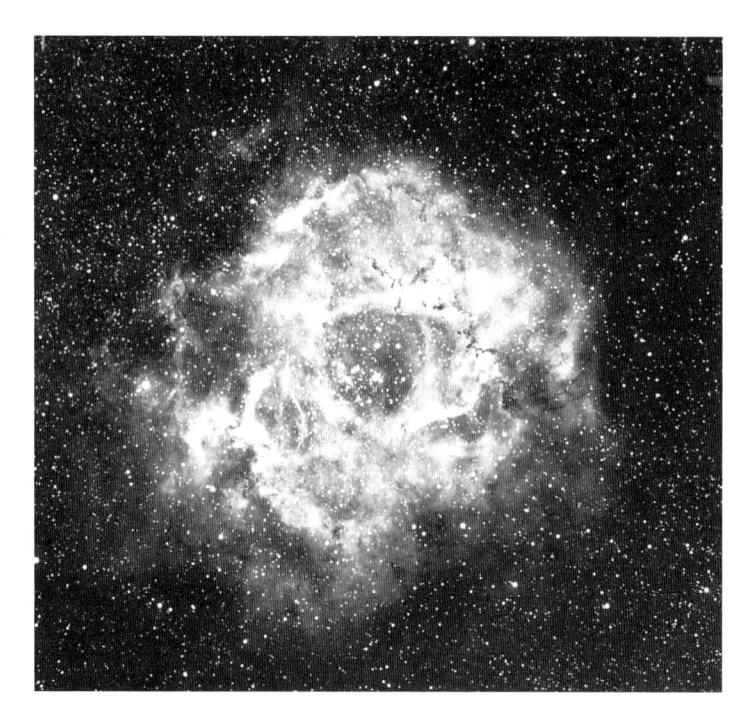

Figure 3.1
NGC 2237-39, the Rosette Nebula in Monoceros, surrounds the open star cluster NGC 2244 like a wreath.

In the early 1860s he used the 48-inch speculum-metal reflector that William Lassell installed on the Mediterranean island of Malta and found NGC 2238, a "small star in nebulosity." Herschel's son John found one section, and the skilled American comet hunter Lewis Swift added two more in the 1880s (one of them he even suspected of being the same object observed by Marth). Collectively these nebulosities form the Rosette Nebula, an object with far better name recognition among today's amateurs than NGC 2244, the cluster it surrounds, even though the latter was the only object mentioned in observing guides until recently.

NGC 2244 was itself once a test for the naked eye. The brightest star here is 6th-magnitude 12 Monocerotis, but it's quite likely in the foreground and not an actual member of the cluster. *Sky Catalogue 2000.0* lists NGC 2244 as having about a hundred stars and a total brightness equal to that of a 4.8-magnitude star. Despite these impressive credentials, the cluster was missed by Messier and his fellow comet hunters in the 1700s. William H. Smyth, who observed with a 5.9-inch refractor, rightly called it "a brilliant gathering," but he also failed to detect the faint Rosette Nebula. As I mentioned, it wasn't until William Herschel began his comprehensive sky survey toward the end of that century that the cluster was discovered. With an estimated age of 3 million years, NGC 2244 is very young; in fact, the Rosette Nebula is the cloud of gas and dust that gave the cluster its birth.

I learned of the Rosette while in grade school. During the 1940s when I began writing this column, I mentioned only the star cluster, thinking that the nebulosity could not be seen in amateur telescopes. After all, the great 19th-century English observers W. H. Smyth and T. W. Webb never mentioned the nebula in their classic observing guides. In the 1970s I persuaded Canadian amateurs Fred Lossing and Rolf Meier in Ottawa to look for it. Even using a 16-inch reflector, they could barely see the brighter parts.

It amazes me how quickly a hitherto ignored nebula took on the mantle of a test object. Mostly through word of mouth, observers were encouraged to search for the Rosette. Everyone looked but most failed to see it. By the mid-1970s letters were filling my mailbox asking for help with seeing the elusive wreathlike clouds of gas around NGC 2244. So I asked readers of this column if they would look for the Rosette, and I did the usual experimentation with the telescopes I had at hand. My 4-inch Clark refractor showed nothing, while perhaps a trace was seen with my low-power 5-inch Apogee telescope. Believable wisps of the Rosette were visible in a pair of 5-inch Japanese naval binoculars. But what really surprised me was that sections could be seen with a 4-inch off-axis reflector made by Margaret Snow. I suspect that the high-contrast images provided by the off-axis design were a big factor in improving the nebula's visibility. Any telescope that scatters light from the cluster's relatively bright stars will have a difficult time with the nebula.

As the 1970s drew to a close, the Rosette had established itself with observers. Those working with 12-inch and larger telescopes had little trouble seeing parts of the nebula. With smaller telescopes there were varying degrees of success, probably much of which depended on sky conditions. Then came the nebula-filter revolution of the 1980s. Intrigued by a report by John F. Bartels of Travis Air Force Base in California, that with such a filter he could see the Rosette Nebula in Monoceros, I borrowed a Lumicon UHC filter, which is designed primarily to pass the green emission lines of doubly ionized oxygen near 5000 Å wavelength. This is the nebula light to which the eye is most sensitive. In the early morning Connecticut sky I too was able to glimpse the Rosette. I found that the filter must be kept square to the eye; if it is tilted, the transmission characteristics change. However, this effect can be used to advantage. Rocking the filter slightly causes the nebula to flicker, which sometimes helps in seeing the object.

Equipped with a proper filter, just about any telescope will show the Rosette. Even my replica of Galileo's refractor showed the faint haze. Indeed, about 1983 both Brian Skiff at Flagstaff, Arizona, and I independently viewed the Rosette with the naked eye and a nebula filter. By the way, a 10-inch telescope equipped with a UHC filter will show a lane of black droplets that crosses the nebula's northwest corner. As would be expected, the field stars around the nebula appear about half a magnitude fainter with the filter than without it.

During the last two decades amateurs have demolished other traditional "test" objects like the Horsehead and turned them over to the rosters at star parties. Of course, the list of test objects has been updated with new, more difficult targets. But at one time the Horsehead was a prize worth claiming.

As with most new optical devices, UHC filters take time to get used to, but naked-eye exploration of the sky through them should be interesting. The eye is a very sensitive instrument, and such things as the zodiacal band and gegenschein are more easily seen than photographed. Thus, my guess is that the Milky Way will yield new vistas and unexpected delights when carefully surveyed with the naked eye and a suitable filter.

The Gem of Gemini

In his March 1964 column, Scotty reveals that he first learned the constellations from Garrett P. Serviss's old book, *Astronomy with an Opera-Glass*. "I did not have an opera glass," Scotty wrote, "but made a 1-inch telescope from my mother's reading glasses and a hand magnifier, in a tube of rolled and glued newspapers. Through this juvenile instrument I first saw M35, and still remember the impressive sight." Serviss's work is indeed old. The copy I own is a 1923 edition of the 1888 original. Unlike Scotty, I did not learn the constellations with it, yet the book holds a prominent position on my bookshelves. To read it — like reading Deep-Sky Wonders — is to experience the romance of astronomy. Take Serviss's description of M35. To the naked eye, he wrote, "It is a nebulous speck." He went on to say that with optical aid, "No one can gaze upon this marvelous phenomenon, even with the comparatively low powers of an opera-glass, and reflect that all these swarming dots of light are really suns, without a stunning sense of the immensity of the material universe." After reading Serviss's prose, it's easy to see where Scotty found much of his inspiration.

M35 is my personal favorite open cluster (**Figure 3.2**). Located about $2\frac{1}{2}°$ northwest of Eta (η) Geminorum, it is an impressive frame of bright stars with a softly flaming background of fainter ones, seemingly containing hundreds of members. William Herschel did not include the cluster in his general catalog of deep-sky objects. It was his way of honoring Messier as the man who, through his earlier catalog of about a hundred deep-sky objects, had inspired him to conduct his own sky survey. Messier, however, was not the first to call attention to M35. That honor appears to go to the Swiss astronomer Philippe de Chéseaux, who in 1746 called it "a star cluster above the northern feet of Gemini."

Amateurs often test themselves by trying to view the cluster with the naked eye. This puts it in the same class as the galaxy M33 and the zodiacal light. Brian Skiff in Arizona, counting 434 stars ranging from magnitude 8.2 to 15.3, has concluded that M35's total magnitude is 5.1. Thus, Skiff's estimate puts the cluster within the grasp of the naked eye, but the bright background of the Milky Way may be a handicap. Nevertheless, one clear morning in September 1984, while waiting for Comet Austin to rise, *Sky & Telescope* associate editor Dennis di Cicco chanced upon the cluster with his naked eye. Have others seen M35 without optical aid? I've glimpsed it that way with the help of a nebula filter from my home in Connecticut. Can anyone distinguish individual stars in the cluster without optical aid?

More than just being a feat worth bragging about, the naked-eye visibility of M35 is an excellent measure of the atmosphere's transparency. Readers have sent me many observations in which they note the transparency on a scale of 1 to 10. But invariably this turns out to be little more than a wild guess. While not perfect, it would be more meaningful to note whether M35 was seen with the naked eye.

Figure 3.2
To the naked eye, Scotty's favorite open cluster, M35, appears as a fuzzy knot in the Milky Way. It is flanked to the southwest by "tiny" NGC 2158, a rich galactic cluster about 16,000 light-years away. M35 itself is about 2,800 light-years away.

The standard measure of the night's transparency is the faintest star visible to the naked eye. But there is some question as to whether looking at a point source of light is a good test of sky conditions. For example, one night I watched as changing sky conditions transformed the telescopic appearance of Halley's Comet from an 8th-magnitude, starlike point with no trace of a coma or tail to a proper 6th-magnitude comet with a distinct tail. All the while my naked-eye judgment of the sky transparency had not changed. Really serious observers can make their own lists of naked-eye nebulae and clusters and arrange them in a series of steps to determine sky clarity.

In finders M35 resembles G. P. Serviss's description in *Astronomy With an Opera-Glass:* "a piece of frosted silver over which a twinkling light is playing." Small binoculars usually resolve a few stars in M35 but, for the most part, the whole group will show as a soft glow. In today's large-aperture binoculars this cluster is a delight. Every increase in optical power produces more detail. A telescope, whether a 4-, 6-, or 10-inch, will bring out the glorious profusion of stars. At magnifications between 40× and 60×, curved lines of them appear like exploding fireworks. Indeed, Smyth likened the view of M35 to the "bursting of a sky-rocket." Using a 4-inch refractor in Kansas, I obtained my best views with a Huygenian eyepiece of 4-inch focal length that had a field lens 4" in diameter. This ocular, a modification of the usual design, was made by Arthur Leonard of Davis, California.

I feel that M35 is one of the greatest objects in the heavens. Observers with small telescopes will find it a superb object. The cluster appears as big as the Moon and fills the eyepiece with a glitter of bright stars from center to edge. In their very useful French-language observing manual, *Revue des constellations*, Robert Sagot and Jean Texereau comment that M35 is a splendid cluster, easy in even the smallest instruments. They note that a few of its stars can be seen in 6 × 20 binoculars, nearly 40 in a 2-inch refractor at 20×, and about 300 in a 12$\frac{1}{2}$-inch reflector at 80×.

In the summer of 1988, I happened to read what W. H. Smyth wrote about M35 in his 1844 *Bedford Catalogue*. While I had done this many times before, something new caught my eye. Smyth noted that the "center of the mass [was] less rich than the rest." This reminded me of the open cluster NGC 6811 in Cygnus, for it also has a dark center as seen with certain instruments. So that same summer I got up before dawn to look again at M35. With my 5-inch Apogee telescope at only 20×, M35 did appear like a starry smoke ring. This effect was far less apparent with the 4-inch Clark refractor at 60×, but was well seen with 15 × 65 binoculars. It was like a fat Life Saver candy, all white and glistening.

So M35 has no central condensation. There is as much dark sky between the stars at the center as at the edges. Large apertures can actually degrade the view by spreading the cluster out too much. In fact, M35 is one of the few clusters that loses its charm if you view it with too large a telescope. Anything bigger than a 12-inch spreads the image beyond the diameter of a single field, and the cluster seems to vanish. Several years ago I swept right over M35 with a 24-inch at 300× and did not realize it was there. But I recognized the field when I came upon NGC 2158.

To me M35 seems most lovely in a 6-inch at 40× — though I must admit that, through a 36-inch telescope and a wide-field eyepiece, this blaze of interwoven stars is an awe-inspiring sight. But I have probably viewed M35 the most with my homemade 10-inch reflector. This was my workhorse telescope years ago in Louisiana and Kansas. Wide-field eyepieces were rare during the 1940s and '50s, so, using a pair of achromats and fooling with the spacing between them, I made a wide-field eyepiece of passable quality. It was a copy of what 19th-century photographers called a landscape lens, and it wasn't far removed from the design now commonly called a Plössl.

With this eyepiece on the 10-inch I could get all of M35 into a single field. The view was too beautiful to describe with mere words. Bright stars were scattered with cosmic recklessness across the field, and it was difficult to establish where the cluster's edges dissolved into the stellar background. There were dozens of curving star chains. Everywhere I looked I could see between the stars into the black depths of infinity.

One thing always impresses me when I look at M35 — how young open clusters really are. Astronomers estimate that the solar system is about 4.6 billion years old. Yet we measure the age of these clusters only in millions of years. According to *Sky Catalogue 2000.0*, M35 is but 110 million years old. I can pick up rocks in my backyard more ancient than that. Only a handful of open clusters are believed to be older than Earth, and none is twice as old.

M35's "Comet" Companion

As we have seen, M35 in Gemini was Scotty's favorite open cluster. But he also liked to tell small-telescope users about the presence of another interesting object nearby. "About 0.4° southwest of M35," Scotty once explained, "lies a lasting monument to my early, somewhat careless, years of observing." He was referring to the faint open star cluster NGC 2158. Absent on some epoch 1950.0 and earlier star charts, the cluster is frequently mistaken for a comet. On the other hand, "many amateurs have scanned this area," he said, "without noting that little wisp seemingly floating between us and the greater M35. This cluster, which is not especially inconspicuous, illustrates the precept that a truly magnificent object tends to distract attention from a lesser one nearby."

The compiler of a sky atlas must call a halt at some magnitude limit, and this cutoff may control the experiences of several generations of observers. Such was the case with the star cluster M35 and its faint neighbor, NGC 2158. *Norton's Star Atlas* does not include the latter, and many amateur astronomers have grown up knowing nothing of it. The brief mention of NGC 2158 in T. W. Webb's *Celestial Objects for Common Telescopes* seems to have been generally overlooked.

The dim, arrowhead-shaped cluster (**Figure 3.3**) lies right on the outer edge of M35 and is a pitfall awaiting careless observers. In my youth NGC 2158 escaped my attention until one exceptional night. From the 1920s on I had looked at M35 many times, mostly with 4- to 6-inch telescopes but occasionally with the Milwaukee Astronomical Society's 13-inch reflector. Then, while observing with a 10-inch f/8.6 reflector in 1952 under the excellent skies of Manhattan, Kansas, I accidentally discovered a peculiar wedge-shaped object. For a few heartbeats I thought I had discovered a comet! Fortunately, before announcing my "comet" to the world, I checked the Skalnate Pleso *Atlas Catalogue* and found that it was the small star cluster NGC 2158.

A few days after I found the cluster, I received a letter from William Davey, in New York State, who saw it with an 8-inch. His description agrees with mine: "marvelous, glowing, and mysterious little cluster." Other observers I have asked have been unable to find it at all — not surprising if you consider that NGC 2158 is much smaller than M35, seemingly contains only one-fourth as many stars, and

Figure 3.3
The compact cluster NGC 2158 lies about 16,000 light-years away. Use at least a 4-inch telescope to spot it.

is some seven magnitudes fainter, according to the *Atlas Catalogue*. Why then, was NGC 2158 so conspicuous to my untrained eye in 1952, when I saw it easily in a 4-inch? Perhaps prolonged viewing (namely, 10 minutes) aided my perception.

The visibility of NGC 2158 is illustrated by Mark K. Stein's experience. This veteran observer writes, "When I lived under the polluted skies of Louisville, I considered some of your descriptions of deep-sky objects the result of a fine, experienced observer having a slightly overactive imagination. But now in the much darker skies of Bloomington, I can plainly see objects I wouldn't have attempted from Louisville. From there NGC 2158 could only be glimpsed, but from Bloomington it is quite easy in my 6-inch at 95×."

NGC 2158 is cataloged as magnitude 12.5, but different observers seldom agree on cluster magnitudes, and I think it is brighter. It is generally too difficult for apertures of less than 5 inches, and I have received only one other report of its being seen in a 4-inch. Different sources also cite the cluster's magnitude as anywhere between 11 and 12.5. Brian Skiff at Lowell Observatory estimates the cluster's magnitude as 8.5. I have racked my 4-inch Clark refractor out of focus until stars appear about 4' across (the same diameter as the cluster). This renders a star of 12th magnitude invisible, while one of magnitude 8.4 remains near the limit of averted vision, and appears similar to NGC 2158 seen with the telescope in focus. Thus I believe Skiff's estimate to be much closer to the truth than the fainter values. You can try the same experiment yourself.

In a 10-inch at 90×, NGC 2158 has a soft sheen, totally unlike the neighboring

Messier object, and resembles a diffuse nebula or a comet. It sparkles with stardust. Leonard B. Abbey, Jr., using a 16-inch at 300×, resolves this faint cluster into stars. Although the exact number of stars in NGC 2158 is not known, we are probably safe in assuming it is greater than the 200 in M35. Although most reference books list NGC 2158 as an open cluster, several call it a globular. Checking into the matter further I found that the group is very unusual and considered by some astronomers to be intermediate between the two types. At any rate, the stars of NGC 2158 are relatively old for an open cluster and, by one estimate, have been shining for 800 million years. It is similar in this regard to NGC 188 in Cepheus and NGC 7789 in Cassiopeia. NGC 2158 is perhaps 16,000 light-years from us, which would place it near the edge of our galaxy and about six times the distance of M35. If NGC 2158 were as near as M35, it would certainly be one of the most spectacular objects in the sky.

It's always been a mystery to me why I had not seen NGC 2158 earlier, even though I was using instruments large enough to show it. I believe the reason was simply that *Norton's Star Atlas* did not plot the object, and this caused my brain to reject the eye's message of "faint cluster here." We should not preprogram our brains before observing, and we should always be prepared for the unexpected.

Two more clusters occupy this corner of Gemini; both are challenging for small telescopes. About a degree west of NGC 2158 is IC 2157. It contains about 20 stars scattered across an area roughly 7' in diameter. The group, which lacks any central concentration, shines with the total light of a single 8.4- magnitude star. A 4-inch telescope will easily pick the cluster out of the rich galactic background. Since its brightest stars are 11th magnitude, the cluster appears best with magnifications of around 100×. Kenneth Glyn Jones, in his 1991 book *Messier's Nebulae and Star Clusters,* mentions that since IC 2157 "contains only three 10 mag. stars in a small triangle and a few others which are fainter, it is hardly recognizable as a cluster." What do you think?

While few amateurs seem to observe IC 2157, fewer still look at IC 2156, about 7' to its north and in the same field of view. It is visible in a 6-inch, and a 10-inch will show about half a dozen stars very loosely scattered. Here is a fine example of a group that is hard to tell from a random configuration of the stellar background.

The Domain of Castor

In early March, brilliant Castor and Pollux can be seen at the meridian after sunset, poised high like sentinels on the banks of the River Milky Way. Castor, the more northerly of the celestial twins, carries the proud label of Alpha (α) Geminorum, though it is 0.4 magnitude fainter than Beta (β) Geminorum (Pollux). This historical curiosity aside, Scotty often enjoined amateurs — especially small-telescope users — to turn their telescopes to Castor, which is a famous binary-star system. In 1969, the 1.9- and 2.9-magnitude components were at their minimum separation of 1.9", making the double a challenging object for small telescopes. In 1980 the components widened to 2.3". The pair is 4.0" apart in the year 2000 and continues to separate. A few decades after Scotty began writing his Deep-Sky Wonders column, this orbital motion caused confusion among amateurs who found the star difficult to resolve, especially during the 1960s. Scotty was quick to explain the beauty and wonder of the situation. He also learned a lesson during an exploration of Castor's environs that left him rather surprised.

What is the most impressive sight in the heavens? To me it is the whole sky seen with the dark-adapted naked eye on a moonless winter night, when the stars seem to sparkle just out of arm's reach against a velvet blackness. On such nights in March, the Milky Way stretches across the zenith with a soft gray, structured glow never seen on lesser nights. Just after twilight ends, the pale cone of zodiacal light may be noticed slanting upward in the west. These are nights for your telescopic old favorites: the first double star you ever identified or the grandest Messier objects. You might start in Gemini by looking at Castor, Alpha (α) Geminorum, one of the great show objects of the sky (**Figure 3.4**). It is also among the most famous of visual binary stars.

Figure 3.4 Castor, the famous binary star in Gemini, is both a visual treat and a challenge for small-telescope users. Success depends on the angular separation of its two brightest components (3.9" as seen here in 1999).

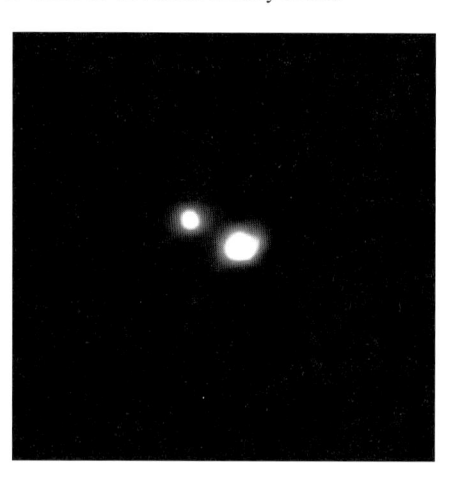

Observations of this star will be of special interest to your children and grandchildren if they develop a curiosity about the night sky, for Castor is a relatively fast-moving binary that will change visibly in coming years (**Figure 3.5**). John Herschel called Castor "the largest and finest of all the double stars in our hemisphere." In 1803 his father William had announced that the two stars were orbiting each other the way planets orbit the Sun. Observing handbooks from the 19th century and even up to the 1960s often say Castor's bright 1.9- and 2.9-magnitude components can be split with small telescopes. Indeed, I resolved them in the 1920s with a 1-inch homemade refractor. None of these books, however, mentioned that the stars would become a challenging pair around 1970, when the separation dropped to only 1.9".

By the mid-1960s I began receiving letters saying that Castor seemed to be a much more difficult pair than the books suggest. The observers were worried that their telescopes or eyes were failing them because they were unable to split Castor. But this is quite right. T. W. Webb's *Celestial Objects for Common Telescopes* mentions a measured separation of 4.4" in 1926, but by the mid-1960s the distance between the components was less than half that. It was then that I realized we were headed for a most unusual event. We would be the first generation of observers to see the famous double star at its minimum separation. Because Castor A and B have completed only a fraction of a revolution since they were first measured by the English astronomer James Bradley in 1718, their exact orbit is not well known but is believed to take about 445 years. One thing is certain, however: it will be more than four centuries before Castor's components appear closer than they do now.

No one saw Castor as a pair of stars at its previous minimum separation, for the invention of the telescope then lay more than a century in the future. In my

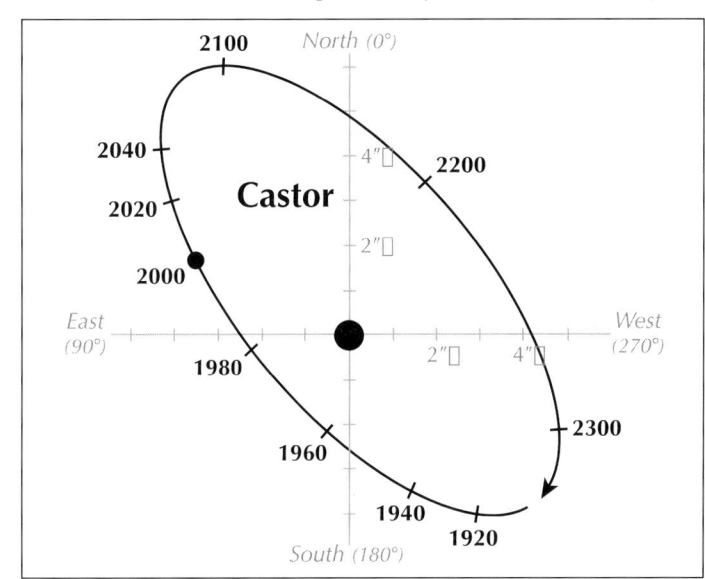

Figure 3.5
This chart shows the position angle and separation of Castor's companion from 1900 to 2300.

garden grows a white rose that came from a long line of cuttings brought to America by a sailor whose family had supported England's Duke of York during the 15th-century Wars of the Roses. The white rose originally was the Duke's symbol. When I look at the rose in my garden I think not only of the English civil wars, but also of Castor's minimum separation that passed unseen.

Even today, we witness the companion of Castor as no one will for another 400 years. According to Castor's apparent orbit, the companion travels in a highly elliptical path around the primary. The position angle is decreasing by about 5° per year, a change large enough to be detected with a simple homemade telescope attachment such as that described by Geoffrey Gleason on page 117 of the February 1967 *Sky & Telescope*. His micrometer consists essentially of a crossline attached to his telescope eyepiece, which can be turned to parallelism with the line joining the stars, and a protractor attached to the eyepiece holder. Presently the pair is widening to a maximum separation of about 7″ late in the 21st century. It will then shrink to a secondary minimum of about 4″ in the 23rd century. As far as I know, the secondary minimum point in a star's apparent orbit has never been dignified with a name.

Both of Castor's components are themselves close pairs of stars, resolved only by telltale Doppler shifts in their spectral lines. Another member of the Castor system, 9.5-magnitude Castor C, can easily be seen in a 4-inch tele-

Figure 3.6
Amateurs with 8-inch or larger telescopes can spot the four faint galaxies plotted here just south-southwest of Castor. Two more very faint ones are west of Castor.

scope. The reddish star is 73″ distant at position angle 165°. Castor C (also called YY Geminorum) is an Algol-type variable star — eclipsing red dwarfs orbiting each other every 9.8 hours. So there are actually six stars in the Castor system.

After writing about Castor and making a search of the surrounding sky, I was confident that no other deep-sky surprises lurked in its glow. Thus, I was amazed by a comment I ran across in Richard H. Allen's classic book *Star Names: Their Lore and Meaning*. At the end of the section on Castor it stated, "In 1888 Barnard found five new nebulae within 1° of Castor." Where had I erred?

In this case I made my search with atlases rather than on the sky itself. Nothing worthwhile showed on *Sky Atlas 2000.0*, and the only thing found on the *Uranometria 2000.0* charts (which have a fainter limiting magnitude for stars and deep-sky objects alike) was the galaxy NGC 2410 just 1° north of Castor. In 40 years of writing this column I'd never before mentioned this galaxy.

Reaching for a copy of *NGC 2000.0*, I began a search for other objects. This is easy to do since the objects are ordered by right ascension in the book. I simply turned to the section that covers Castor's right ascension and ran my finger down the list of declinations looking for ones close to that of the star. I promptly found six objects in addition to NGC 2410. All galaxies in the 14- to 15-magnitude range are certainly difficult for amateur telescopes. Nevertheless, a look through a 12-inch telescope at a star party showed the six quite clearly. They make a sweet group, which I suspect will be about as difficult as Stephan's Quintet in Pegasus. All six fit within a ½° field centered ½° south-southwest of Castor (**Figure 3.6**). The object nearest the center of the group is IC 2196. Has anyone seen them?

The "Leader of the Host of Heaven" and Its Neglected Entourage

In January 1960, Scotty thanked the many amateurs who continued to write to him about their observations. "The mail is too heavy to acknowledge each report individually," he said, "but they are all recorded and help shape the content and tone of this column. Without them, Deep-Sky Wonders would be handicapped." Scotty was comforted by the overwhelming response to his columns. Mail was a sign of appreciation. It assured him that his words were being read and that his audience was a willing one. The correspondence not only helped him generate new ideas about what to observe, it also helped him keep his finger on the pulse of visual astronomy, to follow what was being observed and what was being neglected.

In the early winter evening the atmosphere is turbulent with the day's accumulated solar energy radiating back into space. Stars twinkle vividly. It is at these times that I like to turn my telescope on brilliant Sirius — the blazing blue-white star that has been called the "leader of the host of heaven" — and watch it dance in a rainbow of colors (**Figure 3.7**).

Most amateurs know the story of how Alvan G. Clark, the Cambridge, Massachusetts, telescope maker, discovered Sirius's companion in January 1862 while he was testing an 18½-inch objective that his firm had just made (now at Dearborn Observatory in Illinois). The star, however, can be seen with much smaller telescopes. The companion, an 8th-magnitude white dwarf, takes 50 years to orbit the primary. I first split the pair in 1932 with the same 6-inch Clark refractor used earlier by the famous double star observer Sherburne W. Burnham. The pair reached a minimum separation of 2.7" in April 1994. But this orbital highlight was one best appreciated in the mind's eye. The previous periastron* occurred in March 1944.

Normally, such separations are well within the reach of amateur telescopes. However, this is no ordinary double star. With a magnitude of –1.46, Sirius is some 6,000 times brighter than the 8th-magnitude white dwarf. When the stars are closer than about 5", glare from Sirius makes the pair very difficult in even the largest telescopes. Indeed, even at their widest separation of more than 11", the pair can be extremely difficult to observe. It will be well into the 21st century before the pair again comes within reach of amateur telescopes.

One of the reasons Sirius appears so bright is that it is one of the Sun's closest neighbors. Almost every amateur knows that the nearest fixed star is Alpha (α) Centauri (actually a triple star system). However, with a declination of more than 60° south, few Europeans or North Americans ever get to view it. For us, the closest visible star is Sirius. At a distance of 8.6 light-years, it is almost exactly twice as far away as Alpha Centauri.

Figure 3.7
Canis Major is home to many naked-eye and telescopic splendors, including Sirius — the brightest star in the night sky — and 4° directly south of it, M41, a naked-eye cluster of 100 stars scattered across a Moon's diameter of sky.

Whether they see it as a double or not, many amateurs have turned their telescopes on brilliant Sirius on a crystal-clear cold winter evening, simply to watch the amazing procession of colors — reds, greens, violet-blues — as the mighty star twinkles in the heavy mantle of frigid air. Few, though, take the time to drop their finder 4° south of Sirius and wait for M41 to creep into the field of view. Yet this is a pleasant experience. In contrast to Sirius, the field below is dark and vacant, allowing the eye to regain some of its sensitivity. After a minute or two this mighty galactic cluster rides into view. Its stars shine with the total light of a sin-

---

* *The closest point of approach of two stars during their orbit around their center of mass.*

gle 4½-magnitude sun, which puts the cluster well within range of the naked eye. It would probably be better known as a naked-eye target were it not so low in the sky as seen from northern temperate latitudes. Its diameter is listed as 38' in *Sky Catalogue 2000.0*. While this is more than the Moon's width, it's very difficult for the eye to determine where the cluster ends amid the surrounding star field.

Although the discovery of M41 is generally credited to John Flamsteed in February 1702, it was referred to by Aristotle in his *Meteorologies* (fourth century B.C.) as a "star with a tail." In my early days of observing I never noticed M41 as a naked-eye object, even though it was described as such in T. W. Webb's *Celestial Objects for Common Telescopes*. W. H. Smyth, in his *Bedford Catalogue*, talked more about a double star within M41 than about the cluster itself. I suspect he would have been more enthusiastic had M41 been better placed above the horizon at his home in England.

A note in my travel diary made at Lake Atitlán high on the Guatemala plateau states that M41 outrivals the adjacent Milky Way. And my observing notes from years ago when I lived in Kansas say that M41 is easily visible to the naked eye, being brighter than M11 in Scutum. Try to estimate the visual magnitude of this cluster by comparison with surrounding stars, with the aid of binoculars racked far out of focus.

M41 is an excellent subject for small telescopes and low powers. Here more than a hundred stars are scattered across a Moon's diameter of sky. Because of its declination, observers in southern latitudes should see much more than an amateur in New York, for example. Indeed, they add up to a blaze that dazzled my eye when I viewed this open cluster with a 4-inch rich-field reflector from the jungles near Tikal in northern Guatemala. Many visual observers speak of seeing curved lines of stars in M41. Although they seem inconspicuous on photographs, the curves stand out strongly in my 10-inch, and the bright red star near the center of the cluster is prominent.

Glare and scattered light from M41's numerous bright stars mask the faintest cluster members that your telescope would otherwise show if the field were less crowded. This is one reason why a compact cluster with bright stars is a poor choice for determining the absolute limiting magnitude of a visual telescope. Telescopic fields are not the only ones affected by glare. Reports of people counting more than 17 or 18 stars in the Pleiades star cluster are rare, but many observers can reach magnitude 7.5 with the naked eye. Were it not for the bright Pleiades, these observers should be able to count upward of 30 stars in the group.

Tombaugh's Clusters and a Host of Galaxies

Most readers will know of Clyde Tombaugh, who discovered Pluto in 1930. Much more obscure, however, is the fact that Tombaugh discovered two open clusters in Canis Major about 4° east of M41. I first learned of them while reading an article by Brian Skiff in the Spring 1983 issue of *Deep Sky* magazine. Tombaugh 1 appears as a faint circular patch about 5' across in my 4-inch Clark refractor. I saw it as an unmistakable cluster in a 10-inch reflector, which also revealed a few individual stars, but it was with a 16-inch telescope that I got my first good look at the group.

Tombaugh 2 lies less than 1° southeast of Tombaugh 1, but I failed to see it with the Clark refractor. My 10-inch reflector at 150× showed a small, 3'-diameter blur that I estimated to be about magnitude 12.8. None of its stars were resolved. Skiff's article said that not much was known about this cluster, but he believed it was probably very old and composed mainly of yellow-giant stars.

There is also a small group of galaxies located just 6 degrees south of M41 (**Figure 3.8**). The three brightest are *NGC* objects — NGC 2292, 2293, and 2295 — and should all be within range of a 6-inch telescope under a good sky. They will be easier to distinguish from surrounding field stars if a magnification of 100× or more is used.

NGC 2293 is about 11th magnitude and has a listed diameter of 3' by 4'. Nearly the same brightness and slightly more oval is NGC 2292, just to the west of NGC 2293. Smaller, fainter, and 4' west of NGC 2293 lies NGC 2295. All three galaxies will fit within the field of a high-power eyepiece.

If you're really up for a challenge and have a large-aperture telescope, try sweeping the area around this trio. At least 16 galaxies are catalogued within a 1¼° radius centered on the group.

Figure 3.8
A small group of galaxies lies 6° south of M41. Its three brightest members are NGC 2292, 2293, and 2295. They should be within range of a 6-inch telescope.

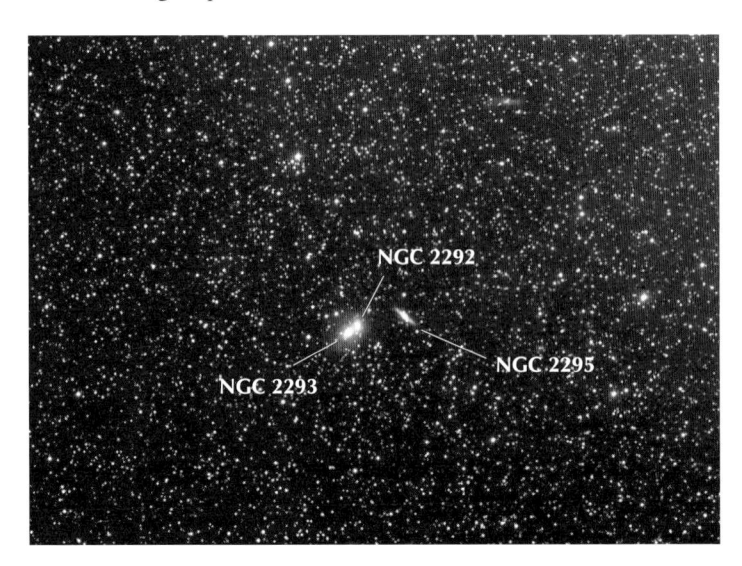

The Great Corridor of Open Clusters

Auriga is a wellspring of galactic treasures. Specifically, the constellation contains one of the greatest collections of open star clusters known. Of course, the favorites with beginners are M36, M37, and M38, three of the brightest galactic clusters listed by Messier. "All are at the threshold of naked-eye visibility," Scotty reminded us, "and the slightest optical aid will make them stand out as striking objects." But the galactic profusion only begins with these objects. In anticipation of the full drama, Scotty waxed poetic, making us look forward to nightfall: "The celestial actors are in place, a serene majesty washes over the stage, and I can almost hear the music of galactic trumpets in their opening bar. What better time to observe the splendor of the heavens?"

During March evenings eight 1st-magnitude stars sit in solemn conclave in the sky above my Connecticut home. Two are in Orion, while the others are arranged in orderly grandeur around the great Hunter. Three naked-eye star clusters — the Pleiades, Hyades, and Praesepe — are strung along the ecliptic carrying with them a wealth of ancient folklore. Near the meridian beams the

Figure 3.9
A multitude of open clusters adorns the Auriga Milky Way. The prettiest is M37, about midway between Theta (θ) Aurigae and Beta (β) Tauri.

great Orion Nebula, also visible to the naked eye. Galactic clusters are legion in the winter Milky Way, and overhead Capella shepherds a profusion of them in Auriga (**Figure 3.9**).

The prettiest of these is M37, dimly visible to the naked eye on extremely clear nights, a little southeast of the midpoint of a line from Beta (β) Tauri to Theta (θ) Aurigae. In binoculars or small finders M37 appears as a milky patch, while a 3-inch shows a fine swarm of stars. In his 5.9-inch refractor, W. H. Smyth saw "the whole field being strewed as it were with sparkling gold-dust." M37 is also the most striking of the three Messier clusters in the constellation, the brightest (visual magnitude 6.2), and has the most stars — 150, according to the *Atlas Catalogue*. The cluster is rather open, that is, the stars are not impressively concentrated toward the center. Whether your telescope is a small instrument or a large reflector, M37 is a treat. The view is further enhanced by the flickering background of the Milky Way.

Nearby is M36, a rich cluster of fainter stars, somewhat smaller than M37, but also impressive. The cluster measures 12′ in diameter and contains 60 stars. It is sparse compared to M37, and visually the stars tend to form a blunt cross. About 1° to the west is the faint reflection nebula NGC 1931. I have seen this 3′-diameter object in a 10-inch reflector, but have never looked for it with smaller instruments. How small an instrument is needed to see this faint nebula?

Moving "down" the Milky Way we run into such variegated star fields and clusters that it is almost impossible to know where to halt, but this might very well be at M38. Although this cluster is well within the star-strewn Milky Way, it is usually visible to the naked eye without much effort. It is certainly far easier than M33 (the Triangulum Galaxy), and probably easier than M11, the Scutum cluster. Evenly compressed into a glowing ball two-thirds the diameter of the full Moon are over 100 softly blazing stars. M38 is magnificent in any sized instrument. Photographs usually show a departure from circularity, a feature quite evident to visual observers. Older reports almost always mention a cross shape (like M36), which seems more pronounced with small instruments. A view with a 24-inch reflector on a fine Arizona night showed the cluster as irregular, and the host of stars made fruitless any effort to find a geometrical figure. Astronomer J. Cuffey places it at 2,804 light-years.

Nearly 3° to the west is small, delta-shaped NGC 1893, estimated to be only one million years old. When this object was born, the Allegheny Mountains were already vastly eroded and early humans were chipping away at their not-so-primitive stone tools. Attesting to its youth, this cluster is still involved with the dust and gas that gave it birth, the diffuse emission nebula IC 410. The cluster measures about 10′ across. *Sky Catalogue 2000.0* lists it as having a total magnitude of 7.5, but I would have estimated it to be slightly fainter. It contains a conspicuous Y pattern formed by four 8th-magnitude stars. British amateur Guy Hurst comments that NGC 1893 is "very rich with a haze of unresolved stars." But here is an interesting point. In small apertures, the cluster does show a haze of unresolved stars, but, as mentioned, NGC 1893 is involved with the nebula IC 410. Like many observers, I have looked at the cluster but not seen the nebula. Could the glow we attribute to stars just below our telescope's limit really be due to the nebula? Has anyone examined this group with a nebula filter? The results might be startling.

Other clusters lie nearby that will reward the earnest observer. They range from those with a dozen or so stars up to grand collections that must have narrowly escaped being included in Messier's famous catalog.

For starters, let's try hunting down NGC 1857 at the edge of the Milky Way within the pentagon of Auriga. William Herschel discovered it in 1785, and it is the 33rd object in his class VII (pretty much compressed clusters). He described it as "considerably rich," being a grouping of some 45 stars spread across about 9′ of sky. Its array attracted the notice of W. H. Smyth nearly 140 years ago because of a double star in its midst. Agreeing with John Herschel that the cluster's brightest member is orange-colored, Smyth also noted a faint triple star on the cluster's north side. A finder will show it as a hazy patch surrounding an orange 7th-magnitude star. At 20× my 5-inch Apogee telescope reveals a brighter haze but still offers no clue that NGC 1857 is a cluster. But the view through a 4-inch off-axis reflector immediately announces that this is a faint collection of stars. Writing in the Webb Society's *Deep-Sky Observer's Handbook,* Vol. 3, *Open and Globular Clusters,* Hurst reports seeing 17 stars mostly of 10th and 11th magnitude spread across a 7′ field. He observed the cluster with a 10-inch reflector at 120×.

Brian Skiff of Flagstaff, Arizona, remarks that it's easy to unwittingly sweep over NGC 1857 with a 6-inch telescope, but adds that he has seen about 40 stars here with a 10-inch. He also notes that glare from the 7th-magnitude star hampers efforts to view the cluster. Although 7th magnitude doesn't seem very bright, when you push a telescope to its limit even modestly bright objects in the field can present problems.

As a test, I made a small occulting mask (a paper punch was ideal for cutting a small circle out of black paper) and attached it to the field lens of an old Erfle eyepiece. Unlike many eyepiece designs, the front surface of this Erfle's field lens is almost exactly at the focal plane of the eyepiece and the perfect "support" for the mask. Again, I took on the challenge of NGC 1857. With the 7th-magnitude luminary hidden behind the mask, the sky appeared to darken and stars at least a half magnitude fainter were seen. I found this so surprising that I repeated the test three times. The result was always the same: a truly improved appearance of NGC 1857.

Several other Auriga clusters are seldom observed by amateurs. NGC 1664 tends to be difficult in my 4-inch Clark refractor when stars fainter than 11th magnitude cannot be seen. There seem to be 15 or 20 stars in a 15′ area, but I estimated several times that number with my 10-inch reflector in Kansas. The cluster appears wedge-shaped, with a foreground star of magnitude 7.5 that is at its eastern edge. NGC 1664 is in many ways similar to NGC 1857, albeit somewhat larger. It too would benefit from an eyepiece fitted with an occulting mask since a 7th-magnitude star lies just on the southeast edge of the group. In a finder, NGC 1664 manifests itself as a faint blur with no indication that there is a cluster hiding here. The familiar outlines of an open cluster emerge into view with a telescope somewhere between 4- and 6-inch aperture.

Observers working with 8-inch and larger telescopes comment enthusiastically on the richness of NGC 1664. Many mention numerous star chains. I once had an impressive view of these meandering strings of stars with a 12-inch telescope. It was almost as if small, roughly circular patches of dark material blocked the view in places. I even nicknamed the group the 4-H cluster because it had the appearance of a four-leaf clover that most farm children recognize as the symbol of the 4-H Club. This asterism, however, seems reserved for telescopes of 10-inch aperture and up; it does not show in my 4-inch Clark refractor.

NGC 1664 is estimated to have a rather middling age of 300 million years. There are more than 100 stars here brighter than 17th magnitude, so the group continues to show more detail as the viewing aperture increases.

For something more challenging, try NGC 2126, some 5° north of Beta (β) Aurigae. I had thought it might be troublesome for a small glass, but one fine night I spotted it with my 4-inch refractor as a coarse grouping of 20 or 30 stars. Curiously, my 5-inch Apogee telescope stubbornly refused to reveal the object at 20×. Perhaps I was less familiar with the appearance of the field at that low power and was not looking in the right spot.

For a change of fare, try the seldom-observed NGC 2281. This handful of stars, whose total light is equivalent to magnitude 6.7, is scattered over an area 0.3° across. To the northeast of the cluster, photographs reveal a lane about 0.6° long that contains very few stars. I have tried many times to discern this lane but with inconclusive results. My most encouraging observations were with 5-inch 20× binoculars; on at least a half dozen nights, I thought I saw this dark "tail" of NGC 2281.

Amateurs using larger telescopes can test their skill on the midget open cluster NGC 1883. Even William Herschel called it "very faint." Lying 1.7° east-northeast of brilliant Capella, NGC 1883 is only 3' in diameter and is listed at total magnitude 12.2. The *Atlas Catalogue* gives it 20 stars, but with a 16-inch in Texas I once estimated twice that number. When a spot of truly clean weather came through one fall, and my 4-inch was going easily to magnitude 13.5 on AAVSO variable-star fields, NGC 1883 was just held as an indistinct blur of starlight.

To complete our survey, look for NGC 1907. It's an easy object that will show in large binoculars and even finders. A 6-inch telescope will deal with it nicely. The group is somewhat elongated in the direction of the galactic plane. Most of us have heard that the stars in open clusters are being pulled apart by tidal forces in the galaxy. But it's seldom that the pundits point to specific clusters where this is evident from a brief look. In the case of NGC 1907, I've never been told that anything special is happening here, but it's fun to look and use your imagination.

Columba and the March Hare

Scotty was an opportunist. One morning, the mercury-vapor streetlight near his observing site went out. The sky was exceptionally clear, and he took the chance to observe some deep-sky wonders in Lepus, the Celestial Hare, as well as some sights in Columba (the Dove), very low in the south. This month the Celestial Dove transits the meridian around 8:30 p.m., so it is well-placed for viewing as long as the horizon remains free of atmospheric haze. As we shall see, Scotty didn't just lazily scan the horizon with his instruments — he experimented, too. He was a tinkerer and modified his 5-inch binoculars so that the right ocular provided 20× while the left ocular was boosted to 60× with a Barlow lens. He continues the story below.

Although binocular vision was impossible, I could readily compare the effect of changed magnification at the same aperture. Also, the 20× unit made an excellent finder for the other. The 10th-magnitude spiral galaxy NGC 1792 (**Figure 3.10**) on the Caelum/Columba border was easy. One way to locate NGC

Figure 3.10
The mighty twin galaxies NGC 1792 (lower right) and NGC 1808 in Columba hug the winter horizon in north temperate latitudes.

1792 is to sweep 3° south and slightly east of 4.6-magnitude Gamma (γ) Caeli. I saw the 3' by 1' object immediately at 20×. It has a neighbor, NGC 1808, two-thirds of a degree away, very similar in size and shape but a bit fainter.

The night was so transparent that I tried a really deep dip toward the southern horizon, looking for NGC 1851, a globular cluster in Columba. Actually, this 5' ball of stars is quite bright, about 7th magnitude, and must be a fine sight from the Southern Hemisphere. It was readily seen at 20×.

Other objects might require more aperture from our northerly locales. For instance, a bit northeast of the halfway point between the 3rd-magnitude stars Alpha (α) and Beta (β) Columbae lies NGC 2090. It was reported by John

Herschel to be a bright, partly resolved globular cluster, but it has been known for many years to be actually a spiral galaxy of the 12th magnitude. It is 2.5' long, and in a 7-inch refractor does indeed look rather like a globular. Try searching for it at about 100×. In 1974 I found it from Connecticut with the 4-inch Clark at 74×, but it was not visible in the brighter field of the Apogee telescope, which also means it might have been too difficult for my binocular experiment as well.

Figure 3.11
A distant 54,000 light-years away from our Sun, globular cluster M79 in Lepus can be seen under dark-sky conditions with a pair of binoculars.

Bordering Columba to the north is the famous celestial Hare, Lepus. For such a prominent constellation, it is one of the least visited by amateurs. Its principal highlight, of course, is the fine globular cluster NGC 1904, better known as M79 (**Figure 3.11**). It was actually first seen by Messier's colleague Pierre Méchain. This compact stellar swarm has a total brightness matching an 8th-magnitude star. The *New General Catalogue* called it a globular by oversight, copying a poor observation by John Herschel, while missing his later, correct description as "bright, irregular, gradually much brighter toward the middle."

Today the cluster is often overlooked. Some amateurs tell me they have trouble finding it, which is surprising because M79 is 3' in diameter and can be seen with binoculars under good conditions. A little more than a ½° to its southwest is an interesting double star, h3752, discovered by John Herschel. Its 5.5- and 6.7-magnitude components are separated by 3.1″, making it a fine object for 3-inch telescopes and larger.

Easier still to find is IC 418, a planetary in Lepus. Because of its 9th-magnitude central star, this object is readily located in 7 × 50 binoculars. The surrounding nebulous ring is about 14″ across and a bit brighter than 12th magnitude. I first saw it in 1945 with a 10-inch reflector from the bayous of Louisiana, and my most recent view was with a 4-inch rich-field reflector from the top of a Mayan pyramid in Guatemala, where IC 418 was high in the sky.

What is the southernmost faint deep-sky object you have ever observed from the United States? An observer in the Northern Hemisphere can, in principle, see into southern declinations as far as the corresponding co-latitude. That is, from 40° north latitude one should see to 50° south declination. Horizon haze, obstructions, and pure indifference usually keep the observer from reaching far south, but it is worth the effort.

MARCH OBJECTS

Name	Type	Const.	R. A. h　m	Dec. °　'	Millennium Star Atlas	Uranometria 2000.0	Sky Atlas 2000.0
Castor, Alpha (α) Geminorum	✶✶	Gem	07　34.6	+31　53	130	100, 101	5, 6
h3752	✶✶	Lep	05　21.8	−24　46	—	—	—
IC 410	BN+OC	Aur	05　22.6	+33　31	114, 136	97	5
IC 418	PN	Lep	05　27.5	−12　42	302	270	11
IC 2156	OC	Gem	06　04.8	+24　09	—	—	—
IC 2157	OC	Gem	06　05.0	+24　00	156	136, 137	5
IC 2196	Gx	Gem	07　34.1	+31　24	—	—	—
M35, NGC 2168	OC	Gem	06　08.9	+24　20	156	136, 137	5
M36, NGC 1960	OC	Aur	05　36.1	+34　08	113	97, 98	5
M37, NGC 2099	OC	Aur	05　52.4	+32　33	112, 134	98	5
M38, NGC 1912	OC	Aur	05　28.7	+35　50	113, 114	97	5
M41, NGC 2287	OC	CMa	06　46.0	−20　44	322, 346	318	19
M79, NGC 1904	GC	Lep	05　24.5	−24　33	350	315	19
NGC 1664	OC	Aur	04　51.1	+43　42	94	65	5
NGC 1792	Gx	Col	05　05.2	−37　59	396, 397	358, 392	19
NGC 1808	Gx	Col	05　07.7	−37　31	396, 397	358	19
NGC 1851	GC	Col	05　14.1	−40　03	417	392, 393	19
NGC 1857	OC	Aur	05　20.2	+39　21	93, 114	66, 97	5
NGC 1883	OC	Aur	05　25.9	+43　33	—	—	—
NGC 1893	OC	Aur	05　22.7	+33　24	114, 136	97	5
NGC 1907	OC	Aur	05　28.0	+35　19	113, 114	97	5
NGC 1931	OC+BN	Aur	05　31.4	+34　15	113	97	5
NGC 2090	Gx	Col	05　47.0	−34　14	394, 395	359	19
NGC 2126	OC	Aur	06　03.0	+49　54	71, 72	66, 67	—
NGC 2158	OC	Gem	06　07.5	+24　06	156	136, 137	5
NGC 2244	OC	Mon	06　32.4	+04　52	227	182, 227	11, 12
NGC 2281	OC	Aur	06　49.3	+41　04	89, 90	67, 68	5
NGC 2292	Gx	CMa	06　47.6	−26　45	346, 370	318	19
NGC 2293	Gx	CMa	06　47.7	−26　45	346, 370	318	19
NGC 2295	Gx	CMa	06　47.3	−26　44	346, 370	318	—
NGC 2410	Gx	Gem	07　35.0	+32　50	108, 130	100	—

Ast = Asterism; BN = Bright Nebula; CGx = Cluster of Galaxies; DN = Dark Nebula; GC = Globular Cluster; Gx = Galaxy; OC = Open Cluster; PN = Planetary Nebula; ✶ = Star; ✶✶ = Double/Multiple Star; Var = Variable Star

MARCH OBJECTS (CONTINUED)

Name	Type	Const.	R. A. h m	Dec. ° ′	Millennium Star Atlas	Uranometria 2000.0	Sky Atlas 2000.0
Rosette Nebula, NGC 2237–39	BN	Mon	06 30.3	+05 03	227	182, 227	11, 12
Sirius, Alpha (α) Canis Majoris	**	CMa	06 45.1	−16 43	322	272, 273, 317, 318	11, 12, 19
Tombaugh 1	OC	CMa	07 00.4	−20 30	—	—	—
Tombaugh 2	OC	CMa	07 03.4	−20 51	—	—	—

Ast = Asterism; BN = Bright Nebula; CGx = Cluster of Galaxies; DN = Dark Nebula; GC = Globular Cluster; Gx = Galaxy; OC = Open Cluster; PN = Planetary Nebula; ∗ = Star; ∗∗ = Double/Multiple Star; Var = Variable Star

APRIL
CHAPTER 4

The Intergalactic Wanderer and Some Extragalactic Wonders

Where's the best deep-sky observing in the heavens? Scotty always pointed observers to the area high overhead, where the sky is usually darkest. "Moving our telescopes away from the zenith," he explained, "we begin to look through more atmosphere, dust, and smog that not only dims the view but also reduces contrast and curtails the amount of detail visible in faint, diffuse objects." For those of us who spend most of our time viewing the northern skies, early April evenings might seem to be a lost cause, because that's when Lynx, a rather dull naked-eye celestial gathering, rides high overhead. But Scotty found it to be a repository for galaxies and a host of other deep-sky objects. They are scattered generously across the constellation, and all are within the range of 6- or 8-inch telescopes. Here Scotty investigates the history of the celestial feline and introduces some of its intriguing globular clusters and galaxies.

Centuries of folklore mark April as the month when the world shakes free of winter's frozen bondage. For some amateurs it is also a time to dust off the telescope and pay more attention to weather forecasts that predict clear skies. For those of us at northern latitudes, it's refreshing to step outside into evening temperatures that seem like a heat wave and find winter's steel-gray cirrus clouds scrubbed from the sky.

High in the evening sky in early April lies one of the great vacant spaces of the celestial vault. Between the cluster-studded sparkling of Auriga and the hordes of galaxies in Ursa Major is a void that was ignored by celestial cartographers until the late 17th century. Then Polish astronomer and instrument maker Johannes Hevelius, unable to resist the temptation of a blank space, filled the area with his constellation Lynx.

Lynx endured the test of time to become one of the 88 constellations officially recognized by the International Astronomical Union since 1930. Hevelius created several others, and for those who have an interest in such things, I recommend George Lovi's history of celestial cartography that appears as the introduction to Volume I of *Uranometria 2000.0*. Lovi includes a reproduction of an early 19th-century American star chart by Elijah H. Burritt. Just below Lynx you'll find

Telescopium Herschelii, a short-lived attempt to honor the great British astronomer William Herschel.

According to William H. Smyth, Hevelius defended his creation of Lynx by noting that the space it occupied was one that celestial globe makers usually filled with "Title and Dedication." If so, then we owe thanks to Hevelius. His original depiction of Lynx had only 19 stars. Over the years the sophistication of cartographers and some tampering with constellation boundaries increased this number. Dalmiro F. Brocchi included 63 stars in Lynx in his *American Association of Variable Star Observers Star Atlas*, published in 1936.

Brocchi's charts had a rigorous magnitude limit of 6.05. Since they were intended for star-hopping to faint variable stars, he adjusted the size of his star disks for every tenth of a magnitude. I have seen and used a specially-made drop-bow pen with its adjusting knob calibrated for star magnitudes. It resulted in a chart that looked like the sky. Devoid of bright stars and lacking a decent trace of a geometrical pattern on which to pin some celestial creature, Lynx pays us back with interesting deep-sky objects.

Figure 4.1
Called the Intergalactic Wanderer, NGC 2419 in Lynx is a very remote globular cluster that may be independent of our Milky Way system.

Perhaps the most famous is the globular cluster NGC 2419, located some 7° north of brilliant Castor in Gemini (**Figure 4.1**). When William Herschel discovered it in December 1788, he called it a bright nebula (his class I of objects) and continued with his survey of the sky. Little could he have imagined that it would bedevil generations of future astronomers.

Lord Rosse, observing with his huge reflectors in Ireland during the mid-1800s, suggested that NGC 2419 was a globular cluster, but not until 1922 was its globular nature established beyond dispute. Although this cluster's sky location nearly opposite the center of our Milky Way is unusual, what is really of interest to astronomers is the cluster's distance. According to *Sky Catalogue 2000.0*, NGC 2419 is more than 300,000 light-years from the Sun. Lying nearly twice as far away as the Large Magellanic Cloud, the cluster drifts through intergalactic space. Thus, it has been referred to as an "intergalactic wanderer."

Scanning the distances for globulars listed in *Sky Catalogue 2000.0* reveals several other very distant objects, but none is even close to the brightness of NGC 2419. Of the approximately 100 known globular clusters, almost all lie within a 65,000-light-year radius of the galactic center.

Despite its great distance, NGC 2419 shines at about 10th magnitude and appears a little less than 2′ across. Under good observing conditions the cluster should be visible with a 3-inch telescope. I once saw it from Kansas with a 4-inch refractor stopped to 2 inches and 100×. The cluster should always be within reach of a 6-inch glass, and a 12-inch may start to show some hint of individual stars around its edge. It is a beautiful object for a 17-inch. More distant globular clusters have been discovered on photographs made with the 48-inch Schmidt telescope on Palomar Mountain. However, it is unlikely that any would be within the visual reach of amateur instruments.

Host of Galaxies

Burnham's *Celestial Handbook* lists 13 galaxies in Lynx, but most are faint. One exception is NGC 2683, a nearly edge-on spiral with a visual magnitude of about 9.3. Under good sky conditions, this galaxy can be seen with moderate-sized binoculars firmly held on a tripod. Its distinctive cigar shape can even be picked up while sweeping. In general, a loss of 1.5 or 2 magnitudes occurs when a rigid stand is not used. I was amazed at how much better my 20-power Apogee telescope performed after a solid support was made for it. Experienced observers know that bright objects can be seen during a sweep, while those near the telescope's magnitude limit require the field of view to be steady. It helps to know exactly where to look. In this way I was able to locate NGC 2683 with a 3-inch aperture at 60×. The galaxy is also a pleasant sight in a 10-inch at 120×. NGC 2683 was aptly termed by Leland S. Copeland "the forerunner of the galactic host of the spring and early summer." Curiously, it is not mentioned by W. H. Smyth or Thomas Webb in their guides. Yet it is an easy object for a 4-inch telescope on just about any night.

Three galaxies form a nice little triangle, with 3rd-magnitude Alpha (α) Lyncis at its center (**Figure 4.2**), and are plotted in Wil Tirion's *Sky Atlas 2000.0*. They are fine for beginning observers, because the bright star serves as a good point to start sweeping, and the galaxies are never more than one or two low-power diameters away. Tiny NGC 2832 lies 3/4° southwest of Alpha. I first viewed this 13th-magnitude galaxy in 1932 with the 6-inch Clark refractor at Washburn Observatory in Wisconsin. This was the same telescope that astronomer Sherburne W. Burnham had used around the turn of the century to discover many double stars. Try searching for it with 100×, as this round patch, only 0.6′ in diameter, may be missed at lower magnifications. Robert Schmidt in Nebraska found NGC 2832 with a 6-inch reflector at 80×, but he was unable to glimpse its companion galaxy, NGC 2831, located just 48″ southwest and a magnitude fainter. How small an aperture will show this companion galaxy?

A much easier object is the 10.7-magnitude spiral NGC 2859, the largest and brightest of the three galaxies plotted around Alpha. It, too, is only ¾° from

Figure 4.2
The three galaxies NGC 2859, NGC 2793, and NGC 2832 frame the bright star Alpha (α) Lyncis.

Alpha Lyncis, but to the east. It is easy to locate. Center your telescope on Alpha Lyncis and wait about 3½ minutes with the drive not operating. The galaxy should then be just north of center in the eyepiece's field of view. You should be able to see NGC 2859 with a 4-inch aperture. Try keeping Alpha out of the field, as its glare may hinder your ability to find the galaxy's slightly oval disk measuring roughly 4′ across. Some observers mention seeing a bright core surrounded by a faint outer ring, which is consistent with the galaxy's barred spiral form on photographs. Arizona observer Ron Morales found the object with an 8-inch f/5 reflector at 60×, and he reports using averted vision to see the outer ring. When W. H. Smyth published his famous *Cycle of Celestial Objects* in 1844, this was the only nebula he listed for the constellation Lynx. However, with the official reorganization of constellation boundaries in 1930, NGC 2859 became part of Leo Minor.

The third spiral galaxy is NGC 2793, which is slightly less than 1° west of Alpha. It is about 1′ in diameter. Since it is small, try sweeping with powers of 70× to 100×. For observers with very good skies, a 10-inch or larger telescope, or unbounded confidence, should show NGC 2793. In my 10-inch reflector it appeared about magnitude 12.6, with a featureless disk a little less than 1′ in diameter.

In northern Lynx there are five more interesting galaxies which are all about 12th magnitude. Perhaps the easiest is NGC 2500. Its oval outline is about 2′ in diameter. A trifle more difficult is circular NGC 2537, which looks like a faint planetary nebula 1′ in diameter. When searching for this tiny object, use at least 70× to avoid missing it. The faintest galaxy in the collection is 12.5-magnitude NGC 2541. It is roughly 5′ long and about half as wide. NGC 2549 is listed as photographic magnitude 12.2. Visually I make it 11.8, and have glimpsed it with my 4-inch Clark refractor. In a 10-inch reflector its 1.8'-long spindle shape is unmistakable. Our final galaxy of the five is NGC 2552. Its roughly circular outline is about 2½' in diameter, and I make its visual magnitude to be about 12.3.

Another galaxy in the area is NGC 2782. I found it easy with the 4-inch, and place it at magnitude 11.9. It is of great interest because it is a Seyfert galaxy, a spiral with a compact, energetic core. I estimate it to be 1½' in diameter with a magnitude of 11.9. In photographs it shows a small but especially bright nucleus. The diffuse outer spiral arms visible in photographs are reminiscent of M51, the famous Whirlpool Galaxy in Canes Venatici.

After you look at NGC 2782, let the field drift for 7¾ minutes with the drive off. You should then have NGC 2844 nearly centered in the eyepiece. This galaxy has two 7th-magnitude stars less than 10' to the northeast and northwest which help to locate this 12½-magnitude wisp, only 1.0' × 0.5' in size. Since the position of this galaxy is readily pinpointed, averted vision is very easy to use.

The Dynamic Duo

Circumpolar objects are special because they can be observed all night. But that statement is somewhat misleading. That they do not set from a given location certainly sounds auspicious, but how useful is it to look at a galaxy, say, 5° above the horizon? Unless we're after a temporal event, such as a comet or supernova, a deep-sky object lower than 10° above the horizon is usually an unattractive option for a night's observing. The object is dimmed because we are looking through the densest part of Earth's atmosphere, which often includes a lot of pollution. Usually, it's better to set our sights higher in the sky. "Near the horizon," Scotty says, "an altitude change of only a degree or two can drastically affect the appearance of a celestial sight. But when objects are much higher up, an altitude change of 10° or so is necessary before image quality is appreciably changed." He then points out that of the 110 Messier objects, only about four may be considered circumpolar at latitude 40° north, when an allowance of 10° is made for horizon haze. "Two of these," he says, "M52 and M103, barely fall within 30° of the north pole, but M81 and M82, in Ursa Major, are easily visible every night of the year. This pair of show objects should be familiar to every amateur."

Our telescopes clanked as the car tires bounced over each expansion joint on the narrow approach road to the Golden Gate Bridge. The windshield wipers assaulted the accumulation of San Francisco fog, but we could hardly see the pavement. Instead we followed tail-lights on the cars ahead of us. It didn't seem like a very good night for a star party.

Soon we were on a long gentle climb up twisting roads when suddenly we broke through the fog layer. Overhead stars were sprinkled across heaven's darkened vault. Good, but not great, I thought as I looked back expecting to see a flood of light from the city across the bay. But there was no San Francisco — not even the faintest trace. It was hidden by the mist blanketing the valley below us. The fog also trapped the ground's heat. No obnoxious thermals rose to wobble our view. The air above was locked in elaborate calm, as we soon learned when we pointed our telescopes skyward.

Figure 4.3
The famous dynamic galaxies M81 (bottom) and M82 (top) in Ursa Major.

My first target for the evening was the stunning pair of galaxies M81 and M82 in Ursa Major (**Figure 4.3**). Both are visible in the same low-power field, and both can be seen in a 2-inch finder. A favorite of mine, M81 is the brightest and most famous of the circumpolar Messier galaxies as seen from 40° north latitude. A beautiful spiral, it lies nearly 70° north of the equator. At such a location altitude changes very slowly with the passage of time. The galaxy is just a little brighter than 7th magnitude, and there are people who can see such stars with the naked eye. Has anyone glimpsed M81 without optical aid? It is certainly easy in even the smallest pair of binoculars.

Sky Catalogue 2000.0, Vol. 2, lists M81's oval disk as being about 26′ long. This implies an object nearly a Moon's width across, but the catalog measurements include faint extensions of the spiral arms that are not visible in amateur telescopes. Thus, observers will see an object that appears only about half as large. M81's arms are wound tightly around its large core, and it would appear that these are beyond the range of most amateur telescopes. Indeed, observer Ronald Buta was only able to trace out one arm visually with a 30-inch telescope at Texas's McDonald Observatory. Yet, on those infrequent nights when transparency is particularly good, M81's sharply sculptured grace stands luminous in a sable sky. Its edges are crisp, and observers with a 6-inch telescope will see traces of several arms, while 16-inch and larger telescopes can likely pick out one or more of the full spiral arms.

While M81 is a textbook example of a spiral galaxy, its companion, M82, is anything but. It is, in fact, one of the most unusual galaxies within the range of small telescopes. At magnitude 8.4, it is also within the grasp of binoculars. Like its neighbor, M82 was discovered on the last day of 1774 by astronomer Johann Bode at Berlin Observatory. Some older books even refer to the two galaxies as "Bode's nebulae."

Visually M82 appears as a spindle of light about 10′ long and relatively bright. That night under the Golden Gate skies I could not resist pretending that I was looking at the hulk of an ancient space cruiser torn to shreds in some long-forgotten galactic war. Alas, admitting to such fantasies can lead to trouble. In a lecture I once likened M82 to an Egyptian mummy wrapped in ragged linen strips. I then got letters insisting that this proved the Egyptians were immigrants from outer space!

Larger amateur telescopes will show traces of the central mottling clearly recorded in long-exposure photographs, giving M82 the appearance of some brimstone hell where dark filaments jostle with bright fragments of the galaxy. For many years astronomers attributed M82's turmoiled appearance to an explosion of unparalleled magnitude occurring within the galaxy. During the 1980s, however, a somewhat gentler process of intense star formation peppered with supernovae became recognized as the cause — somewhat like the Orion Nebula on a galactic scale. Regardless of the events behind M82's appearance, its close proximity to M81 makes a wonderful example of how dissimilar galaxies can look in amateur telescopes.

Several interesting but lesser-known galaxies lie nearby. NGC 2976 is a bit less than 1½° southwest of M81. It is an easy 10th magnitude and 4′ across. I have seen it in a Skyscope reflector as well as a 2.4-inch Unitron refractor. To my eye this spiral has a slight diamond shape. NGC 3077 is an elliptical galaxy similar in size and brightness to NGC 2976. It lies about ³⁄₄° east-southeast of M81. Not far away lie two 11th-magnitude galaxies. NGC 2787 is a barred spiral 2′ in diameter. A spiral of type Sb is NGC 2985. While the *Shapley-Ames Catalog** of galaxies gives the diameter as 3′, I see this object as 1′ in a 4-inch.

By the way, I should mention that M81 took on special significance for me in 1986. Its NGC number, 3031, is shared by an asteroid that was named in my

---

* A survey of bright galaxies completed by Harlow Shapley and Adelaide Ames in 1932, this catalog was incorporated in the Revised Shapley-Ames Catalog of Bright Galaxies by Allan Sandage and Gustav A. Tammann and published by the Carnegie Institution of Washington in 1981.

honor. It's a delightful coincidence, prompting me to "adopt" M81; it is a showpiece and much easier to see than my asteroid.

Seeing Double, and a Mysterious Planetary in Lynx

When Scotty first began writing his Deep-Sky Wonders column, amateur telescope making was in its prime. To test the optical quality of their handmade marvels, telescope makers often turned them to known double stars to see if they could resolve their components at the Dawes limit — the theroetical minimum separation observable with a telescope of a given aperture. But the Dawes limit is not an ironclad measure of performance. Seeing conditions may well affect double-star detections. One way to minimize any atmospheric effects is to look at doubles in deep twilight when they're high overhead, such as those found in Lynx on cool April evenings.

Lynx, the celestial feline, is perhaps most noteworthy for its fine double stars. Double stars are frequently a test of observers' seeing conditions rather than the optical quality of their telescopes. For this reason, I must first mention the difficult double Beta (β) Delphini, which never gets more than 0.7″ apart. A flood of mail has arrived from amateurs who split the pair. Typical is the report of Charles Cyrus of Baltimore, Maryland, whose 12 1/2-inch f/7.2 reflector has no clock drive. At 572× he saw the components clearly separated. After placing a 10-inch aperture mask on the telescope he could still make out the pair but this time the images were in contact. An 8-inch mask showed the companion as only a lump in the side of the primary.

In applause of this sighting, I will first describe three very different doubles in Lynx. Even the smallest telescopes will be suitable for 5th-magnitude 19 Lyncis. The primary (almost always the brightest star in a multiple system) is magnitude 5.5, and some 15″ to the northwest is the 6.5-magnitude secondary. Because of the much wider separation, many double-star catalogs do not list an 11th-magnitude star more than 1′ to the west-northwest, but a 3-inch telescope should show the three stars nicely.

Our next double star, 10 Ursae Majoris, is now well within the border of Lynx because of the new constellation boundaries set by the International Astronomical Union in 1930. The two components, of magnitudes 4 and 6, are very close, and I suspect they will require a 10-inch telescope with very good optics and a magnification of 300×. The fainter star completes an orbit of the brighter one every 22 years; in 1981 the companion was 0.7″ due north of it.

The double star Burnham 576 presents a different problem for the observer. As the components are separated by 1.5″, it would normally be an easy object for a 4-inch telescope. However, the companion is 4 1/2 magnitudes fainter than the 7th-magnitude primary and may easily be lost in the brighter star's glare. Joe Ashbrook's discussion of Harold Peterson's diagram plotting telescope performance on double stars, on page 380 of the November 1980 issue of *Sky &*

Telescope will be valuable to anyone interested in observing close doubles with such large magnitude differences.

One good technique on such doubles takes advantage of the diffraction spikes caused by a telescope's spider (the secondary mirror support). Place the companion between the spikes from the brighter star. Doing this will often help improve the visibility of the fainter companion star. But what about refractors and the popular Schmidt-Cassegrain telescopes, which have no spiders? Here it is customary to make a hexagonal aperture mask, like the one described on page 407 of the June 1975 issue of *Sky & Telescope*. I find it even simpler to place a cross of black electrical tape over the front of the objective (or corrector plate) cell. Be careful not to let the sticky tape actually touch the glass surface. On a 4-inch aperture I have used tape $^1/_4$- and $^1/_2$-inch wide with good results. One-inch-wide tape carefully placed on the spider vanes of my old 10-inch reflector also enhanced the diffraction spikes.

In the early 1930s I placed a cross of $^1/_2$-inch-wide tape on the 6-inch Clark refractor that had been used by the great double-star observer S. W. Burnham. At 500×, I was able to split Sirius in the winter and Antares in the summer.

A Planetary Mystery Solved

Lynx has another interesting object in the form of the planetary nebula PK 164+31.1, which is often mistaken for the nearby galaxies NGC 2474 and 2475. The mystery surrounding the confusion was neatly told by Nancy and Ronald Buta on page 368 of the April 1981 issue of *Sky & Telescope*. In a nutshell it goes like this: Both William Herschel and his son John separately viewed the tiny galaxy NGC 2474. Later a close companion was found by Rosse, and both galaxies were listed in the *NGC*. So far, so good.

In 1939 two researchers at Harvard Observatory were scanning photographs when they discovered a faint planetary nebula only $^1/_2°$ north of the galaxies (**Figure 4.4**). As fate would have it, the planetary had two bright knots on opposite

Figure 4.4
The eerie-looking planetary nebula PK 164+31.1 lies in the constellation Lynx.

sides of its ringlike structure. The researchers assumed these were the objects seen by the Herschels and Rosse, apparently overlooking the slight difference in position. It didn't take long before catalogs and star charts were a confused mess regarding the positions and proper identifications of these objects.

Today, however, all is well, due in part to the work of Nancy and Ron Buta. Charts in both *Sky Atlas 2000.0* and *Uranometria 2000.0* show the objects correctly. *Sky Catalogue 2000.0* goes as far as saying that the planetary "is not NGC 2474-5."

Although the planetary has been mentioned several times in this column, I have never received a visual observation of it. From my mail I know that amateur interest in observing planetaries is alive and well. Has anyone looked at PK 164+31.1 with a nebula or O III filter?

About 3° north of the planetary lies a 6th-magnitude star, and just a few arcminutes north of that is a string of faint galaxies running east to west. These may prove a challenge for 16-inch telescopes. They are plotted on the *Uranometria 2000.0* charts, but only one, NGC 2469, is listed in *Sky Catalogue 2000.0*. NGC 2469 is 13th magnitude and about 1' in diameter. It should be within range of a 10-inch and is a good starting point to look for the others, which are one or two magnitudes fainter.

The Beehive Challenge

Is there a relationship between the clarity of the naked-eye sky and telescopic limiting magnitude? It would seem so, but for one variable — the stability of the atmosphere. For most of us living at or near sea level, transparent nights don't necessarily translate into stable nights in the atmosphere. The leading edge of a cold front, for instance, can push out haze and pollution, making the night sky appear crisp and transparent, but the turbulence associated with the moving front can also cause the stars and planets to "boil." To the naked eye, the night sky looks clear, but through a telescope, the views are a nightmare: stars swell to the apparent size of golf balls and planets appear as if seen through a stream of running water. Scotty introduced one test to tell us what to expect telescopically on clear spring nights. "You'll need only your naked eye to see it," Scotty said. "But if you wear glasses it's best to clean them before heading out under the celestial fires passing quietly overhead."

There are quite a few deep-sky objects near the limit of normal naked-eye vision that serve as guides to atmospheric conditions overhead. One of the oldest and best known involves the Beehive, M44, in Cancer (**Figure 4.5**). With some experience it is possible to estimate the limiting magnitude of a telescope from a naked-eye view of M44.

For those who have read the Old English heroic poem *Beowulf,* one of the most stirring passages is the short exaltation to the coming of spring. Translations don't quite make it. Early northerners had a tough time just surviving the winter, especially during the brutal cold spells that drove the Danes out of Greenland and thereby gave Columbus his later chance at becoming a great historical figure. Such harsh conditions left humans with an emotional reaction to spring that is all out of proportion to today's general climate. Nevertheless, when the warming does come and the streams and rivers chorus once more, we celebrate in many ways. Amateur astronomers clean their telescopes and start making impossible observing plans.

I've always considered the beautiful open star cluster M44, also known as Praesepe, to be symbolic of spring. The ancients knew it as a dimly glowing cloud. If the sky were veiled with the slightest trace of cirrus clouds at the leading edge of a storm system, M44 could not be seen with the naked eye. As early as several centuries B.C., the invisibility of M44 was considered an omen of coming rain. Today amateurs can use the same observation to judge the night sky's transparency.

The name "Beehive" is apparently of fairly recent origin. To Hipparchus it was the Little Cloud; Aratus called it the Little Mist; and Johann Bayer termed it Nubilum

Figure 4.5 Resolving the brightest stars in the Beehive cluster (M44) with the naked eye is a challenge, even for skilled observers.

(cloudy sky). Astronomers of the 16th and 17th centuries called it the Nebula, and R. H. Allen (in *Star Names: Their Lore and Meaning*) says that it was the only universally recognized object that could not be resolved by ordinary vision. Its true nature was first discerned by Galileo, who described it in his 1610 astronomy pamphlet *Sidereus Nuncius*, as "The nebula called Praesepe, which is not one star only, but a mass of more than 40 small stars. I have noticed 30 stars, besides the Aselli."*

Indeed, only the slightest optical aid is needed to resolve the cluster. In fact, there are 11 stars brighter than magnitude 7.0, which puts them within range of many naked-eye skywatchers. (Remember that the naked-eye limit of 6th mag-

---

*The Aselli are two stars — Asellus Borealis (Gamma (γ) Cancri) and Asellus Australis (Delta (δ) Cancri) — which bracket the Beehive to the north and south, respectively.

nitude usually cited in books is an average value. There are well-documented cases of observers seeing fainter than 8.0 without optical aid.)

Although many skilled observers have tried to resolve M44 with the naked eye and failed, there are some notable exceptions. In the early 1980s, Swedish amateur Paul Schlyter wrote me saying that he had seen the cluster as "a patch of light with little bright dots in it" while looking from the cabin window of an aircraft cruising at 37,000 feet. He used a coat over his head to block the aircraft's interior lights during the observation. I suspect his window was a lot better than most of the ones I've looked through on commercial aircraft.

A bit more down to Earth, albeit not quite at sea level, were the observations of *Sky & Telescope*'s Stephen J. O'Meara. He was at the 9,000-foot level of Hawaii's Mauna Kea when he counted and mapped at least 11 of M44's stars with the naked eye. He saw more from the mountain's summit nearly a mile higher, but then he was "aided" by breathing pure oxygen from a tank.

Lest anyone doubt these sightings, consider what the legendary Canadian explorer Thomas James penned in his journal on the night of January 31, 1632, while searching for the Northwest Passage to the Pacific: "There appeared, in the beginning of the night, more Starres in the firmament than ever I had before seene by two thirds. I could see the Cloud in Cancer full of small Starres, and all the [Milky Way] nothing but small Starres; and amongst the Plyades, a great many small Starres." We can only imagine what the conditions must have been like on that cold winter's night more than 350 years ago!

I was also interested in James's comments on the Milky Way. While living in Kansas in the 1950s, I had watched how our galaxy's naked-eye appearance changes with varying atmospheric transparency. Under average conditions it is a smear of uniform luminosity within which stars to 5th magnitude are sharply defined. With increasing transparency the luminous background becomes very pronounced and tends to drown out 4th-magnitude and fainter stars. (These conditions are probably the best most observers see.) On very rare occasions, however, when the air is extremely clear, the luminous surface vanishes, to be replaced with myriads of 6th- to 8th-magnitude stars. I have seen this spectacle from Kansas and, in the 1930s, from Arizona, but never from my home in Connecticut.

Thus, it seems pretty clear that M44 can be resolved with the unaided eye, but it must take an exceptional night. You might also try viewing M44 through a cardboard tube, blackened on the inside, so interference from stray light will be eliminated and the chance of success improved. If any individual stars are seen, their locations relative to more conspicuous objects should be sketched. It would be a worthwhile project for someone to record the times when M44 is and is not visible to the naked eye, and at the same time the faintest star that is seen nearby. A series of such observations would allow a threshold magnitude to be determined for the cluster's visibility.

In low-power fields, finders, and binoculars, M44 is a brilliant show object. It has no sharp boundary. No one can say for sure where the cluster's faint glow merges into the placid sky background. And the center is hardly brighter than the edge. The cluster appears as a ghostly sheen of cobwebs at least a degree in diameter, sometimes maybe two. Through a large telescope, the view is not particularly impressive, because the stars are widely scattered. But the cluster is an exciting object for binoculars and rich-field telescopes. The best instrument for viewing M44 is one that has a field at least $1\frac{1}{2}°$ across with the largest aperture that will still give an exit pupil no more than 7 millimeters in diameter. I had an excellent view of the object with my 4-inch Clark refractor and a special eyepiece of 4-inch focal length designed by Arthur Leonard.

There's no trace of nebulosity here, though current astronomical theory tells us that there must have been material mixed with the stars when they were born some 650 million years ago. M44 is one of the closest open clusters, at a distance of only about 500 light-years. Some 200 stars of between 6th and 14th magnitude are believed to be true members of the group, and over 350 stars to as faint as 17th magnitude have been listed in its vicinity. The average distance between M44's stars is rather large, and thus the group would normally be quite susceptible to the galactic forces that break up all open clusters. However, this process is probably slowed by the fact that M44 is located well outside the galactic plane.

Contrasting with the modest age of M44 (fish were already swimming in Earth's oceans when it formed) is that of M67. Studies indicate that this open cluster (**Figure 4.6**) is not much younger than the solar system, having been born perhaps 3.2 billion years ago. As far as open clusters go, only NGC 188 in Cepheus is known to be older. When conditions are right, M67 can be seen with the naked eye in the barren sky of Cancer about 8° south of M44. Its apparent

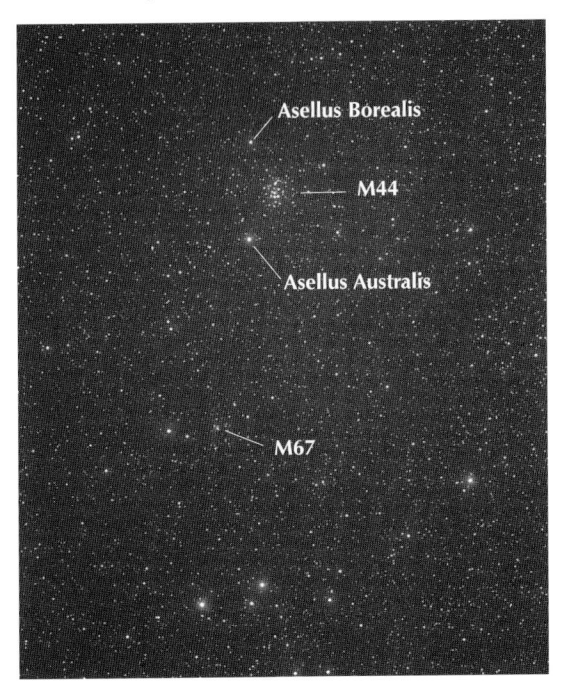

Figure 4.6
Cancer, home to naked-eye open clusters: bright M44 (above center) and the dimmer M67 about 8° below it.

diameter is roughly the same as the Moon's, and I estimate its total light to equal that of a 5.9-magnitude star. The brighter stars in this aged cluster have evolved to the red-giant stage. W. H. Smyth suggested that M67 showed the form of a Phrygian cap (somewhat like a "liberty cap"). The French astronomer Camille Flammarion likened the cluster to a sheaf of corn. What is your opinion?

Galaxies Near M44

Owners of large telescopes, which cannot fit all of M44 into a single field, should not bypass the cluster. The *New General Catalogue* and its two supplemental *Index Catalogues* list at least eight galaxies here that should be within the grasp of a 12-inch instrument. Brian Skiff called to my attention the two brightest members of this group, NGC 2672 and 2673. NGC 2672 is the brighter (visual magnitude 11.6) and larger of the pair and lies about 2° east-southeast of M44.

Figure 4.7
The close pair NGC 2672 and NGC 2673 are elliptical galaxies about 2° east-southeast of M44.

NGC 2673, just ½' east of its companion, is slightly more difficult to see at magnitude 12.9 (**Figure 4.7**). It should be visible easily in a 10-inch telescope at about 150× and a fair test in most skies for anything under 8 inches. One exceptional night I was able to fish it up with the 4-inch Clark at 125×, obtained with a 25-mm eyepiece and a Barlow.

A few degrees eastward are two more galaxies that are plotted in *Sky Atlas 2000.0* but have never been mentioned in this column. At magnitude 13.2, NGC 2749 is the more difficult of the pair, but under good skies a 4-inch should reveal its tiny oval disk less than 1' across; an 8-inch will almost always uncover it. NGC 2764 is a bit larger and brighter. A 4-inch should easily show its disk, which I estimate to be 11th magnitude.

The remaining galaxies are too faint for the Clark and may even challenge those who observe with 17-inchers. NGC 2667A was discovered by d'Arrest with an 11-inch refractor at Copenhagen. The *Revised New General Catalogue of*

Nonstellar Astronomical Objects (RNGC) lists it as 15th magnitude, but I have always held the faint magnitudes in the *RNGC* somewhat suspect. It also rates the companion, NGC 2667B, as magnitude 15.5. Rounding out the group is another 15th-magnitude smudge of light, NGC 2677. Have any amateurs seen these?

If you're really into challenges and have access to a 16-inch telescope, try searching the area about 5° due west of M44. The region is peppered with faint galaxies. I thumbed through Roger W. Sinnott's *NGC 2000.0* catalog and tallied up more than 60 within a 3° window centered on the coordinates $8^h\ 20^m$, +20°. It's almost a celestial gridlock, with dozens of galaxies scattered like confetti. Just identifying what you find will be a monumental task, since only a handful of the brighter objects are plotted on the *Uranometria* charts. Good hunting!

Hydra Hysteria

Crossing the meridian on early April evenings, 10° due south of the famous Beehive Cluster, is the familiar naked-eye asterism of six stars forming the head of Hydra. It is the small but conspicuous beginning of the sky's largest constellation. To the naked eye the sinuous constellation does not appear very appealing, just a stretch of dim stars and imagination. But it is rich in deep-sky objects. To Scotty, Hydra was a paradise for the determined galaxy hunter. He needed only to point to the table of Hydra's deep-sky objects in Burnham's *Celestial Handbook* to explain why. "More than 60 galaxies are listed," he said, "but most are fainter than magnitude 12.5 and difficult for 8- or 10-inch telescopes. In a larger instrument these faint galaxies not only are easier to see but often take on individual personalities." So when Scotty went seeking a region of sky that would test the mettle of trained deep-sky observers, something "a little different" from the Messier marathon, Hydra was the perfect place to go mad with desire. Thus was born Scotty's famous celestial voyage, which he coined "Hydra Hysteria."

The idea of a Messier marathon — an all-night session to view as many of the Messier objects as possible — sprung up independently in several locations. According to Harvard Pennington, president of California's Pomona Valley Amateur Astronomers (PVAA), the first marathon dates to the late 1960s and a group of observers in Spain. On this side of the Atlantic, it was the mid-1970s before amateurs in Florida and Pennsylvania took up the challenge. Unaware of the earlier efforts, California comet hunter Don Machholz suggested a Messier marathon in an article published in the San Jose Amateur Astronomers' newsletter in 1978. Pennington claims that the cat got out of the bag when I wrote about the Florida and Pennsylvania projects in my March 1979 column. After that, marathons became increasingly popular.

But I began thinking about other voyages across the heavens. I wanted something a little different from the original marathon, something that would sharpen the skills needed to star-hop to small and faint galaxies as well as bright and easy objects. Hydra turned out to be the perfect place for such a trek. Indeed, this celestial sea monster stretches across nearly 100° of sky and, with an area of 1,303 square degrees, is the largest of the 88 constellations. (Virgo, with 1,294, is a close second.)

Hydra contains a host of 8th- to 12th-magnitude galaxies. While it is not naked-eye heaven, there is an excellent supply of 5th- to 8th-magnitude stars that are perfect for star-hopping with a finder. I planned my "Hydra Hysteria" using *Sky Atlas 2000.0*. But another good atlas, especially for the fainter objects, is *Uranometria 2000.0*.

We can begin our hunt at the extreme western edge of Hydra with one of the constellation's two Messier objects. The open star cluster M48 was long believed to be a "missing" object until Harvard astronomer Owen Gingerich linked it with NGC 2548, which Caroline Herschel discovered in 1783. If Gingerich is correct, the original published position for M48 was about 5° in error. Seemingly, Messier made a mistake of 5° in declination, but his right ascension is correct. But the identification seems pretty certain since there is no other nearby candidate matching Messier's visual description of M48.

This sparse sprinkling of stars has roughly the angular size of the Moon. Because it is a very loose group, it is best viewed with low magnification or the finder. W. H. Smyth with his 6-inch refractor saw this cluster as "a splendid group, in a rich splashy region of stragglers, which fills the field of view, and has several small pairs, chiefly of the 9th magnitude." In my big 5-inch binoculars, its shape appears distinctly triangular. This cluster contains about 60 members brighter than 13th magnitude. The total magnitude is about 5.8, and the English author and observer Kenneth Glyn Jones notes that many people can see the cluster's glow with the naked eye. Being so bright and large, it would seem that M48 would be easy to find. Experience suggests otherwise. The cluster is sparse and the background rich. I've never been convinced that it is visible to the naked eye, but it does show nicely in small telescopes.

A prominent group of three stars, including 1 and 2 Hydrae, lies about 3° northeast of M48. Sweeping another $3^1/_2$° east of the trio brings us to the spiral galaxy NGC 2642. Although I have never mentioned it in this column before, I swept it up with my 4-inch Clark refractor at 120× and estimated its 2′ disk to be a little brighter than 12th magnitude. The spiral galaxy NGC 2713 is nearly the same magnitude. It lies 8° north-northeast of NGC 2642, near the stars forming Hydra's head.

NGC 2962 is a 12th-magnitude spiral almost on the Hydra/Sextans border less than 1° northeast of 2 Sextantis. This galaxy is incorrectly labeled NGC 2967 on the first (1981) printing of *Sky Atlas 2000.0*, but that was corrected along with nearby NGC 2967 (originally labeled NGC 3067) in later printings. I have seen NGC 2962 easily with a 4-inch rich-field reflector.

We can take a break from observing galaxies by swinging our telescopes southwestward to where the constellations Hydra, Puppis, and Pyxis intersect. There we will find the planetary nebula NGC 2610. It appears about as large as the planet Jupiter, but is only 13th magnitude. In the early 1970s I mentioned that I had never seen NGC 2610 in a telescope smaller than 16 inches. This brought a lot of mail, and I can now tell you that under dark, transparent skies I have viewed it with apertures as small as 6 inches. A 10-inch telescope is probably needed for a comfortable view of this tiny but distinct object. My old observing notes made while using a 10-inch reflector do mention NGC 2610, but I may have never searched for it. My records describe collecting it with a 16-inch reflector, and it appeared very bright with the 20-inch Clark refractor at Wesleyan University in Connecticut. Ron Morales, observing from Arizona's Empire Mountains, had no trouble seeing NGC 2610 with a 10-inch f/5.6 reflector and 16-mm eyepiece. He also fished up the central star, which some catalogs list as 16th magnitude.

Next on our travels across Hydra is a clump of galaxies that lie within a field approximately 10° in diameter with 2nd-magnitude Alpha (α) Hydrae (Alphard) at its northernmost edge. NGC 2763 is the westernmost of the group. It's a 12th-magnitude spiral about 2' across that should be easy in a 6-inch telescope at 100×.

NGC 2781 is a little brighter and appears like a thin oval some 3' long. Brighter still is NGC 2811, another spiral galaxy that appears oval in the eyepiece.

NGC 2848 is about 2' across and 12th magnitude. If you can fish it out, try for its 14th-magnitude companion NGC 2851, located 7' to the northeast. It was missed by the Herschels when they made their great visual sky surveys and owes its discovery to the keen eye of Lewis Swift. He picked it up while observing with the 16-inch refractor at Warner Observatory in Rochester, New York, during the 1880s. It's the hardest object on this month's list, but I felt that those of you who need a dump truck to get your telescopes to a dark site shouldn't have it totally easy.

NGC 2855 is easily found about ½° east-northeast of 26 Hydrae. Its 11th-magnitude disk is about 2½' wide and should prove to be a nice target for a 4-inch telescope. Just 1½° to its east is NGC 2889, 12th magnitude and distinctly rounder.

The galaxy pair NGC 2992-93 is almost as easy to locate since it lies about halfway between 38 and 39 Hydrae. You should have no problem identifying which is which, because NGC 2992 is distinctly cigar-shaped while NGC 2993 is more oval. William Herschel noted them as about the same brightness, but I do not. How do they look to you?

Next comes one of my favorite galaxies, though few other authors ever make much mention of it. NGC 3109 is a long, spindle-shaped, irregular galaxy about half a Moon diameter long. Its ends appear squared off, and while the surface brightness is rather uniform I see a strong hint of curdling with the 4-inch Clark. I'm sure NGC 3109 would be a grand sight in a large-aperture telescope.

NGC 3145 lies just a few arcminutes southwest of 4th-magnitude Lambda (λ) Hydrae. The trick here will be to pick it out from the glare of Lambda. High magnification will certainly help, and the bright star certainly keeps you from getting lost.

NGC 3200 was discovered by William Holden with the 15½-inch Clark refractor at Washburn Observatory in Madison, Wisconsin. I estimate its oval, 4'-long disk to be magnitude 11.1.

If you have a particularly good night, with the sky very clear down to the horizon, you might try for the southern galaxy NGC 2997 in Antlia; it's a little more than 10° southwest of NGC 3200. It's a small deviation but a great challenge. Located about 31° south of the celestial equator, this 11th-magnitude object is theoretically within reach of every observer in the continental United States. My old observing notes made with a 10-inch reflector in Louisiana during the war years state that NGC 2997 was visible at 40×. I now wish I had tried a higher power, because the record does not comment on any detail being seen.

Now, as we work our way to the eastern end of Hydra, we come to the beautiful globular cluster M68. It was discovered by Pierre Méchain in 1780 and described by Messier as a "nebula without stars." It took the larger telescopes of William Herschel to resolve M68 into a cluster. Today, the individual stars should be easily visible in a 6-inch telescope. M68, about 4' in diameter, is located below the "sail" of Corvus. You may have to wait until late evening before this object is well above the southern-horizon haze. Only 45' southwest is a 5½-magnitude star, making this cluster easy to identify. Its total magnitude is 8.2, putting it within reach of 7 × 50 binoculars under favorable conditions. In my 4-inch Clark refractor M68 appears distinctly oval, with tattered streamers winding out from a central disk. The late John Mallas, who did so much to popularize observing the Messier objects, called this cluster a beauty, with a bright central region fading outward to a ragged edge. Older writers have also commented on the peculiar shape of M68, W. H. Smyth likening it to a bishop's miter, Flammarion to a sheaf of wheat. The Mallas description fits what most amateurs see. About 5' northwest of the globular cluster but not associated with it is the Mira-type variable FI Hydrae, which has a period of 324 days. This red star can become as bright as visual magnitude 9.

NGC 5085 is an 11th-magnitude spiral galaxy about 3' in diameter. It lies 1¼° south-southeast of 3rd-magnitude Gamma (γ) Hydrae. NGC 5101 is a big but faintish spiral about 5' in diameter. I estimate it to shine with the total light of a 10.9-magnitude star, but the surface brightness is low because of its large size.

NGC 5150 is an elliptical galaxy only about 1' across and 13th magnitude. The *Uranometria 2000.0* charts show it to be one member of a faint trio, but I have not included the other two because, at 14th magnitude, they are too faint for the Hysteria.

This brings us to the last galaxy on the list, M83. This fine galaxy went undiscovered until the French astronomer Nicolas-Louis de Lacaille sailed south to study the heavens from the Cape of Good Hope in 1751–52. Messier heard of the galaxy's discovery 30 years later, and he set out to find it in the polluted skies of Paris. Despite the poor atmospheric conditions, he glimpsed the galaxy enough times to add it to his celebrated catalog. The object barely climbed more than 10° above his Paris horizon. It was such a difficult object for him that even the faint illumination from his telescope's micrometer wires overpowered the galaxy's image.

Several years ago while in Central America, I saw M83 as a beautiful object in a 4-inch rich-field reflector with a magnification of 20×. This is clearly a delightful spiral for small telescopes. Its 8th-magnitude disk is 10′ in diameter and seen nearly face on. Amateur photographs reveal its spiral structure, but visually only a trace of this can be seen in a 5-inch Apogee telescope at 20×. A 10-inch reflector in Kansas showed portions of the arms. I've never looked with binoculars for M83, but it would be worth a try on a good night. In the *Messier Album*, John Mallas's drawing, made using a 4-inch refractor, suggests that a surprising amount of detail can be seen in M83's inner core by a skilled observer working with a small telescope.

That Messier saw M83 at all should encourage users of large binoculars and small telescopes, especially those who live in more southerly latitudes. On a clear dark night, averted vision and patience will enable nearly all of M83 to be seen in a 10-inch or larger telescope. However, at low power one can sweep past this galaxy, since its bright core is easily mistaken for a star. Messier's catalog lists seven objects at more southerly declinations than M83. Four were discovered by other observers, but Messier found the remaining three from Paris — quite a feat for his small telescopes. His record was the globular cluster M70 at declination –32° 30′ (1780 coordinates). Messier, of course, knew that the southern heavens had their share of deep-sky splendors. We can only speculate that his inner soul yearned to get there somehow and reap more immortality. Yet had he spent a couple of years at the Cape, imagine how many comets his rivals at home would have been able to "steal" from him.

Several degrees south of M83 is another interesting galaxy, NGC 5253 in Centaurus. There is some question as to whether this is an irregular or elliptical galaxy, for it would be considered an unusual object in either class. A 6-inch telescope will show it as an evenly illuminated oval some 4′ by 2′. Apart from the famous 1885 supernova in the Andromeda Galaxy, which may have reached magnitude 5.4, NGC 5253 has produced the brightest extragalactic stellar outburst on record. A photograph taken on July 8, 1895, recorded a 7.2-magnitude supernova very near the center of NGC 5253. Another such event occurred in the same galaxy in 1972, and, although first seen at magnitude 8.5 after maximum light, this star may have also been as bright as magnitude 7.2.

Our final object is the globular cluster NGC 5694. With a diameter of more than 3′ and shining at 10th magnitude, it should be relatively easy to find. This is surprising since it is one of the more distant globulars listed in *Sky Catalogue 2000.0*, which places it some 100,000 light-years from the Sun. Good luck with your Hydra Hysteria.

The Ghost of Jupiter

In the summer of 1980 Scotty had cataract surgery on his right eye, which removed the ultraviolet-absorbing lens and replaced it with a plastic insert that transmits ultraviolet light. Some feared Scotty's observing was nearing an end. But such fear was unfounded, for Scotty returned to the eyepiece with literally a new outlook on the heavens. His surgery made him eager to test the effectiveness of his new eye lens, for he knew that as we age, the eye's lens begins to yellow and acts as a filter, blocking light from the blue end of the spectrum. "Generally, younger observers see planetaries as blue," he said, "while the older ones find them more green." So, in a sense, Scotty, after surgery, was reborn. His views of planetary nebulae and their hot blue central stars proved that his right eye with its artificial lens was receiving more ultraviolet radiation than his natural left eye.

There is a fine planetary nebula in Hydra, NGC 3242. It was discovered in the western half of this sprawling constellation by William Herschel in 1785. It is located about 2° south and slightly west of 4th-magnitude Mu (μ) Hydrae. W. H. Smyth mentioned that NGC 3242 is similar in size and color to the planet Jupiter. Accordingly, it still is sometimes called the Ghost of Jupiter (**Figure 4.8**). "Whatever be its nature," wrote Smyth, it "must be of awfully enormous magnitude." Today we know that it is indeed large, by some estimates, about half a light-year across.

Figure 4.8
The planetary nebula NGC 3242, nicknamed the Ghost of Jupiter, is one of the sky's most famous non-Messier objects.

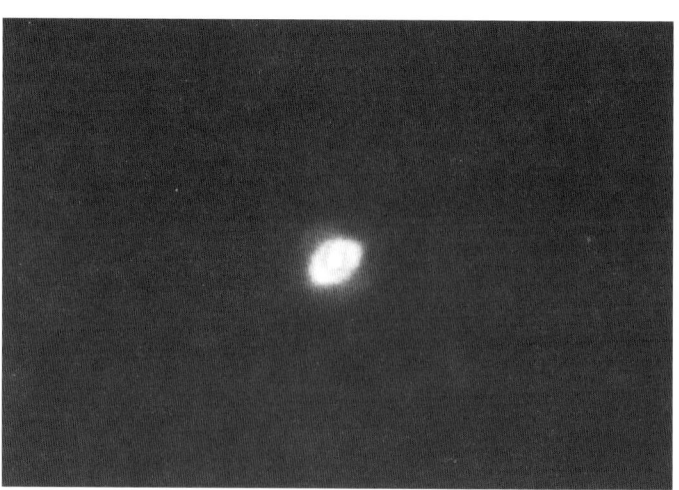

Oddly enough, Smyth, who was a great one for writing about the apparent color of celestial objects, especially double stars, calls this planetary "pale greyish-white." Many modern observers see it as a striking blue. Indeed, at the Cape of Good Hope John Herschel inspected this planetary several times between 1834 and 1837 with his 18¼-inch reflector. His notes include: "Colour a decided blue; at all events a good sky-blue. Elliptical; position angle of axis 140° . . . 30" long, 25" wide; uniform and very bright; but not quite sharp at the edges."

The total light of NGC 3242 roughly equals that of an 8th-magnitude star. With a disk only 0.5' in diameter, the surface brightness of this planetary is quite high, averaging about 10 times greater than the Ring Nebula in Lyra. NGC 3242 is slightly oval. While the central star is said to be of photographic magnitude 11.4, it is much fainter visually. Using a 14½-inch reflector, California amateur Tokuo Nakamoto estimated the star as magnitude 12.5. (To increase the contrast between the nebula and central star, use as high a magnification as seeing conditions permit.) Nakamoto also noted the suggestion of two rings. And indeed, these were very apparent with the 30-inch reflector at the amateur-operated Stony Ridge Observatory near Mount Wilson.

What does the planetary look like in smaller telescopes? In their *Revue des constellations*, R. Sagot and J. Texereau cite observations by four French amateurs using a variety of instruments: "Easily visible, starlike in a 27-mm 13× finder. Readily recognized as a planetary of appreciable size and with shaded edges in a 55-mm refractor at 50×. Central part uniform and very bright in a 95-mm refractor at 95×. Elliptical in a 200-mm reflector at 200×. Central part grayish, and squared into a bright lozenge with a dimming outer edge, in a 215-mm reflector at 375×. Color bluish or yellow."

Ron Morales found NGC 3242 easily with a 6-inch f/5 telescope at 50×. Recently I looked at it with my 5-inch Apogee telescope and a 20× eyepiece. It appeared slightly oval but without the pointed ends so prominent in photographs of the object. The central star was easily seen with my eye that had its lens removed during cataract surgery. The star appeared almost as bright as the entire planetary in this eye, while it was hardly visible at all in my normal eye. This was surely due to a greater amount of ultraviolet (UV) light reaching the retina of the eye without its natural lens. Central stars in planetaries are generally strong emitters of UV.

I also examined NGC 3242 with a Lumicon UHC nebula filter. The results were impressive. I did not, however, attach the filter to the eyepiece, but rather slipped it in and out of the light beam between my eye and the eyepiece. I learned this method from Mike Mattei of the Amateur Telescope Makers of Boston. Because the eye responds more strongly to a changing image than to a static one, more detail is apparent when the filter is flipped in and out of view than if the scene is held steadily either with or without the filter. Others who have tried this "flicker" method with a filter all report superior results compared with when the filter remains fixed.

One final note about the color of planetaries. The aging of the eye might not be the only factor that determines what color an observer sees. On several occasions I have mentioned that different people see the same planetary as having different colors. So other factors must also affect the color of planetaries. Bruce Chapin of New York City, for example, wrote saying that NGC 7662 in Andromeda at first appeared green in his Celestron 8, but after continued gazing it turned blue. Such is the mystery with these ghostly glows.

APRIL OBJECTS

Name	Type	Const.	R. A. h m	Dec. ° '	Millennium Star Atlas	Uranometria 2000.0	Sky Atlas 2000.0
10 UMa	**	Lyn	09 00.6	+41 47	620	70	6
19 Lyncis	**	Lyn	07 22.9	+55 06	52	42	1, 5
Beehive Cluster M48, NGC 2548	OC	Hya	08 13.8	−05 48	810	230, 231, 275, 276	12
Beta (β) Delphini	**	Del	20 37.5	+14 36	1217, 1241	209	9, 16
Burnham 576	**	Lyn	08 21.6	+33 56	643, 644	101, 102	6
FI Hydrae	Var	Hya	12 39.9	−26 40	—	—	—
M44, NGC 2632	OC	Cnc	08 40.1	+19 59	712, 713	141	6, 12
M67, NGC 2682	OC	Cnc	08 50.4	+11 49	736	186, 187	12
M68, NGC 4590	GC	Hya	12 39.5	−26 45	868, 869, 892	329	21
M70, NGC 6681	GC	Sgr	18 43.2	−32 18	1413	378	22
M81, NGC 3031	Gx	UMa	09 55.6	+69 04	538, 549, 550	23	1, 2
M82, NGC 3034	Gx	UMa	09 55.8	+69 41	538	23	1, 2
M83, NGC 5236	Gx	Hya	13 37.0	−29 52	889	370, 371	21
NGC 2419	GC	Lyn	07 38.1	+38 53	87, 107, 108	69, 100	5, 6
NGC 2469	Gx	Lyn	07 58.1	+56 41	37, 51, 566, 582	43	—
NGC 2474	Gx	Lyn	07 57.9	+52 51	51, 582	43	—
NGC 2475	Gx	Lyn	07 58.0	+52 51	51, 582	43	—
NGC 2500	Gx	Lyn	08 01.9	+50 44	51, 67, 582, 602	43, 69	1, 2, 5, 6
NGC 2537	Gx	Lyn	08 13.2	+46 00	602	69	5, 6
NGC 2541	Gx	Lyn	08 14.7	+49 04	602	69	5, 6
NGC 2549	Gx	Lyn	08 19.0	+57 48	566	43	1, 2
NGC 2552	Gx	Lyn	08 19.3	+50 01	602	43, 69	5, 6
NGC 2610	PN	Hya	08 33.4	−16 09	857	276, 321	12
NGC 2642	Gx	Hya	08 40.7	−04 07	808, 809	231	—
NGC 2667A	Gx	Cnc	08 48.3	+19 01	—	141, 142	—
NGC 2667B	Gx	Cnc	08 48.4	+19 02	—	141, 142	—
NGC 2672	Gx	Cnc	08 49.3	+19 04	712	141, 142	6, 12
NGC 2673	Gx	Cnc	08 49.4	+19 04	712	141, 142	—

Ast = Asterism; BN = Bright Nebula; CGx = Cluster of Galaxies; DN = Dark Nebula; GC = Globular Cluster; Gx = Galaxy; OC = Open Cluster; PN = Planetary Nebula; * = Star; ** = Double/Multiple Star; Var = Variable Star

APRIL OBJECTS (CONTINUED)

Name	Type	Const.	R. A. h m	Dec. ° ′	Millennium Star Atlas	Uranometria 2000.0	Sky Atlas 2000.0
NGC 2677	Gx	Cnc	08 50.0	+19 00	712	141, 142	—
NGC 2683	Gx	Lyn	08 52.7	+33 25	642, 664	102	6
NGC 2713	Gx	Hya	08 57.3	+02 55	760, 784	232	12
NGC 2749	Gx	Cnc	09 05.4	+18 19	711	142	6, 12
NGC 2763	Gx	Hya	09 06.8	−15 30	831, 855	277	12
NGC 2764	Gx	Cnc	09 08.3	+21 27	687, 711	142	—
NGC 2781	Gx	Hya	09 11.5	−14 49	831, 855	277	12
NGC 2782	Gx	Lyn	09 14.1	+40 07	619	70, 71	6
NGC 2787	Gx	UMa	09 19.3	+69 12	539, 550, 551	23	1, 2
NGC 2793	Gx	Lyn	09 16.8	+34 26	641	103	—
NGC 2811	Gx	Hya	09 16.2	−16 19	855	277, 278, 322, 323	12, 20
NGC 2831	Gx	Lyn	09 19.7	+33 44	641	103	—
NGC 2832	Gx	Lyn	09 19.8	+33 44	641	103	6
NGC 2844	Gx	Lyn	09 21.8	+40 09	619	71	—
NGC 2848	Gx	Hya	09 20.2	−16 32	854, 855	277, 278, 322, 323	12, 20
NGC 2851	Gx	Hya	09 20.6	−16 28	854, 855	277, 278, 322, 323	—
NGC 2855	Gx	Hya	09 21.5	−11 55	830	277, 278	12
NGC 2859	Gx	LMi	09 24.3	+34 31	641	103	6
NGC 2889	Gx	Hya	09 27.2	−11 38	830	278	12
NGC 2962	Gx	Hya	09 40.9	+05 10	757	188, 233	12, 13
NGC 2967	Gx	Sex	09 42.1	+00 20	781	233	12, 13
NGC 2976	Gx	UMa	09 47.3	+67 55	550	23	1, 2
NGC 2985	Gx	UMa	09 50.4	+72 17	538	8, 23	1, 2
NGC 2992	Gx	Hya	09 45.7	−14 20	829, 853	278	—
NGC 2993	Gx	Hya	09 45.8	−14 22	829, 853	278	12
NGC 2997	Gx	Ant	09 45.6	−31 11	900	365	20
NGC 3077	Gx	UMa	10 03.3	+68 44	538, 549, 550	23	1, 2
NGC 3109	Gx	Hya	10 03.1	−26 09	876	324	20

Ast = Asterism; BN = Bright Nebula; CGx = Cluster of Galaxies; DN = Dark Nebula; GC = Globular Cluster; Gx = Galaxy; OC = Open Cluster; PN = Planetary Nebula; ✶ = Star; ✶✶ = Double/Multiple Star; Var = Variable Star

APRIL OBJECTS (CONTINUED)

Name	Type	Const.	R. A. h m	Dec. ° ′	Millennium Star Atlas	Uranometria 2000.0	Sky Atlas 2000.0
NGC 3145	Gx	Hya	10 10.2	−12 26	828	279	13
NGC 3200	Gx	Hya	10 18.6	−17 59	852	279, 324	13, 20
NGC 3242	PN	Hya	10 24.8	−18 38	851	324, 325	13, 20
NGC 5085	Gx	Hya	13 20.3	−24 26	866, 867	330	—
NGC 5101	Gx	Hya	13 21.8	−27 26	866, 867, 890	330, 370	21
NGC 5150	Gx	Hya	13 27.6	−29 34	889, 890	370	—
NGC 5253	Gx	Cen	13 39.9	−31 39	889	370, 371	21
NGC 5694	GC	Hya	14 39.6	−26 32	862, 863, 886	332, 333	21
NGC 7662	PN	And	23 25.9	+42 33	1120	88	4, 9
PK 164+31.1	PN	Lyn	07 57.8	+53 24	51, 582	43	1, 2, 5, 6

Ast = Asterism; BN = Bright Nebula; CGx = Cluster of Galaxies; DN = Dark Nebula; GC = Globular Cluster; Gx = Galaxy; OC = Open Cluster; PN = Planetary Nebula; ✶ = Star; ✶✶ = Double/Multiple Star; Var = Variable Star

MAY

CHAPTER 5

The Grandeur of Omega Centauri

"Amateur astronomers face a variety of perplexing limitations in their humble observing stations upon the Earth's surface," Scotty once lamented. He was referring not so much to the turbulence of our atmosphere (not to mention clouds), but to the fact that Northern Hemisphere observers cannot see deep into the southern sky. Scotty often remedied this situation by traveling southward. But he also liked to "graze the grass" visually from his homeland. In doing so, he discovered — to the surprise of many — that the great southern globular cluster Omega Centauri is not out of reach for observers at mid-northern latitudes. By discussing this seemingly impossible target in his column, Scotty introduced what would arguably become his most famous visual challenge. Of course, nothing beats seeing this globular from more southerly locales. Shortly before his death in 1993, Scotty traveled to the Winter Star Party (WSP) on Florida's Big Pine Key, which offers southern views to a declination of −65°. And it was at WSP that Scotty saw Omega Centauri for the first time through one of amateur astronomer Bob Summerfield's impressive Dobsonian reflectors. The eyepiece was so low to the ground that all Scotty had to do to enjoy the view was to stroll to the waterfront in the predawn hours and sit in a chair. There, within earshot of rustling palms and waves lapping gently on the shore, Scotty sat in silent contemplation, staring at this blazing wonder as if it were the key to all understanding. He looked for countless minutes, until a tear moistened his eye and misted the starlight. Scotty truly loved the heavens.

When the jet stream bulges southward, it allows Canadian air to pour across the United States and cover all but the far West with a stable mass of cold dry air. Amateur astronomers benefit with dark nights of crystalline transparency and better astronomical seeing. Under these conditions it is no problem viewing 5th-magnitude stars only 1° above the horizon. Globular cluster fans should wait for that special evening to try for Omega (ω) Centauri (**Figure 5.1**), the finest of all globulars. The search must be done when the cluster is at its highest point in the sky. On May evenings the cluster lies near the meridian. It culminates at the same time as Spica; just look for the cluster 36° below the brilliant star.

Figure 5.1
Although the great globular cluster Omega Centauri is an easy naked-eye object for viewers in more southern latitudes, it is a challenge to observers at north temperate locales. The higher-power view at bottom shows millions of stars.

To the naked eye, Omega shines with the light of a 3.7-magnitude star (**Figure 5.2**); however, its image is rather soft, not crisp like that of a real star. The cluster covers about a Moon's diameter of sky but does not appear that large to the eye. Ptolemy cataloged it as a star over 18 centuries ago, and Johann Bayer did the same in the early 1600s (hence the Greek-letter designation). In 1677, Edmond Halley was the first to record the object as a cluster. With moderate optical aid, the cluster is enthralling: an area more than half as broad as the Moon is thickly strewn with glittering stardust. And John Herschel gave a vivid description of Omega's appearance in his 18¼-inch reflector in South Africa on March 3, 1837: "All clearly resolved into stars ... this most glorious object fills the whole field with its most condensed part, and its stragglers extend ¾ of a field beyond it either way." If we were located inside this cluster, we would see the night sky ablaze with many thousands of 1st-magnitude stars, and it would be twilight all night long.

In theory, an observer in the Northern Hemisphere can see into southern declinations as far as the corresponding co-latitude (down to 90 minus the latitude of your location). From geometry alone we can calculate that Omega Centauri should be visible from as far north as 42½° north latitude. In practice that value is too small, because atmospheric refraction at the horizon lifts starlight by ½°, so Omega might be viewed from 43°. The challenge is to see it through terribly dense and contaminated air.

Figure 5.2 Despite its distance of 18,000 light-years, Omega Centauri can be seen with the naked eye as a 3.7-magnitude fuzzy "star."

Ordinarily horizon mists, smoke, and dust take a good 10° or 15° off this figure. But on those special nights when extinction is small and stars can be seen down to the horizon, Omega Centauri should theoretically be visible to amateurs living near New York City, Pittsburgh, Cincinnati, and San Francisco. If you live near the northern limit of visibility, try 65-mm binoculars from a high hilltop. The cluster will appear as a large patch of light, dimmed by atmospheric extinction.

Record telescopic sightings from the Northern Hemisphere give us a clue to the task's difficulty. For instance, Harry Koken of Bird City, Kansas, bought a second-hand 6-inch objective and fabricated a telescope around it in the 1950s. At his latitude (39° 45') Omega Centauri was indeed horizon-scratching. Yet Koken included it in his list of the 10 greatest sights through that 6-inch. We can also trust E. D. Flynn's sighting from near Pittsburgh, Pennsylvania (40.4° north). With a 10-inch reflector at 50× he found it an "unmistakable white haze, pretty large, not

MAY 101

resolved but with a couple of stars seen on its borders." At latitude 41.8° in the very northern reaches of California near Yreka, Russell Milton viewed Omega between the trees on a distant hill. His homemade 8-inch f/4.7 reflector showed the cluster as "very bright, resembling a comet, with one or two of the brightest cluster members near the edge visible." It has also been seen from near the California/Oregon border.

Here in East Haddam, Connecticut, I have climbed a broad hill in a cow pasture and with binoculars have seen the giant cluster passing between tree branches. I have yet to bring a telescope to this site, but it is unlikely the cluster could be resolved at such a low altitude. Years ago in Manhattan, Kansas, I had a visitor from New York who said he had never seen it, and, trusting to theory and my growing faith in Kansas skies, I offered to show him this cluster. And there it was in the binoculars, its lower edge just tangent to the top of the Flint Hills. In the 6 × 30s its immense size was obvious, and while it had lost about three magnitudes by extinction, the brighter stars came through, clearly making it look more like a bright galactic cluster than the globular it really is. The next night at the same time a 6-inch rich-field telescope showed clearly the glow of its thousands of unresolved stars. The apparent size of Omega Centauri depends not only on the size of your telescope but also on your reaction to seeing scattered light. It's rather like the better-understood "cocktail-party effect" — the more sensitive you are to hearing, the more conversations you will overhear.

When I was in Campeche, Yucatán, I was far enough south to make a fair comparison between four great globular clusters. Omega Centauri has an appreciable diameter to the naked eye — not the 30′ listed in some books, but certainly 15′. Seen in a 4½-inch f/4.5 Tuthill Star Trap reflector, this cluster looked much the way M13 does with my 4-inch Clark refractor back in Connecticut. Although M13 was also visible to the naked eye, in binoculars it was a pale imitation of Omega Centauri. Also with binoculars, M13 seemed a shade brighter than M3 in Canes Venatici. However, the 18× Star Trap made the difference in brightness appear less obvious to me. Omega also surpassed M5 in Serpens, though both were riding high in a black-velvet sky. Certainly, if it were well placed in the northern sky, Omega Centauri would be as popular as the Orion Nebula.

Galaxy Visibility

Although Scotty enjoyed "rummaging" through Leo Minor's celestial closet of galaxies, he also found it a decent proving grounds for optics. For example, in his May 1951 Deep-Sky Wonders, he pondered the problem of galaxy visibility through various apertures. He suggested his readers train their telescopes on a gathering of three visually disparate galaxies in Leo Minor — NGC 3414, NGC 3504, and NGC 3486 — and report on the difficulty of seeing each one in various sized telescopes (including their finders). "Does the difference in brightness seem comparable to their listed difference in magnitude?" he asked. "How does magnification affect their visibility? Does the total light seem equivalent to that of a star of the same magnitude? Try expanding a star out of focus until it is the same diameter as the galaxy (that is, compare the object in focus with a star out of focus)." Scotty was weaving his magic, prompting us to go out and give thought to the quandaries of the visible universe. It is through such exercises that he helped us grow as observers.

In May 1951, this column listed three galaxies in Leo Minor, NGC 3414 at magnitude 12.2; NGC 3504 measured at 11.7; and NGC 3486 (**Figure 5.3**) showing magnitude 11.4. Observers were asked to report on the visibility of these objects under various apertures and powers. The several amateurs who collaborated merely add evidence that the problem of nebular visibility is not a simple one. We are especially indebted to William Galbraith of Lemon Grove, California; James Corn of Phoenix, Arizona; and to Fred Grabenhorst of O'Fallon, Missouri.

The observers all agree that it requires more than three inches to see 11th-magnitude objects. With a 12½-inch instrument, Corn tried reduction masks until the visibility was nil. But on the effect of powers, they differ. With a 5-inch telescope Galbraith met "increasing difficulty" as he raised his magnification, whereas Corn found 3414 and 3504 as easy at 400× as at 80×. But Corn says that 3486 seemed to fade off with increasing powers; and all three were invisible at 1,000×. Grabenhorst was unable to see any of these galaxies with a 3-inch telescope.

The Arizona observer suggests that 3486 has "luminosity per unit area evidently much lower than the other two." Galbraith, a variable-star observer, estimated the magnitudes differently from the catalogs: 3414, 12.0; 3504, 12.0; 3486, 12.3.

From these and other reports it seems likely that brightness holds fairly steady as magnification increases until a certain point, after which it falls off very fast.

On an exceptionally clear night in January 1976, I re-observed these galaxies with a 4-inch Clark refractor, 5-inch Moonwatch Apogee telescope, and a pair of 5-inch 20× binoculars (having the same optics as the Apogee telescope). In the refractor, the galaxies were easily visible. Even when it was stopped down to three inches, all of them were seen at 400×, but at 50× only NGC 3504 (the easternmost of the trio) could be detected. The 20× Apogee telescope failed to show any, but the binoculars steadily held NGC 3504. Lately, I have become increasingly aware that more can be seen with two eyes than with only one. (Microscopists have known this for centuries.) Therefore, my own experience has been about halfway between those of Corn and Galbraith. The optical system may have a considerable effect. It might be interesting to compare Newtonian and Cassegrainian arrangements.

The visibility of galaxies raises some important practical observing problems. Suppose an observer whose eye is properly dark-adapted has a transparent, moonless sky. How large a telescope is needed to see some particular galaxy? The answer depends on the total magnitude of the object, its angular size, and its surface brightness. A closely related question is: If an amateur wishes to view a galaxy as distinctly as possible in a certain telescope, what magnification and what diameter of field are best? What observing techniques can aid detection?

Figure 5.3
By observing these three visually disparate galaxies in Leo Minor — NGC 3414, NGC 3504, and NGC 3486 — amateurs can learn about the practical problems of observing galaxies. North is to the left.

Wondering what contemporary amateurs would find using their more varied and generally larger telescopes, I repeated the Leo Minor galaxy cluster challenge in 1980. More than a dozen persons responded by calling the group "easy." So I now offer you another, perhaps more difficult group of galaxies in Leo Minor, located about halfway between 21 Leonis Minoris and 34 Mu (µ) Ursae Majoris.

NGC 3158, at magnitude 12.8, is the brightest member of this group, and should be visible in a 6-inch telescope. There are seven other galaxies nearby which carry *NGC* numbers. (As a good rule, experienced observers can usually find any *NGC* object with a 12-inch aperture.) In addition to NGC 3158, William Herschel discovered NGC 3163 here. The keen eye of Guillaume Bigourdan, coupled with a 12-inch refractor at the Paris Observatory, added four more galaxies to the group in the late 1800s. The last two members of the cluster that carry *NGC* numbers were discovered with the large telescopes built by Lord Rosse in Ireland. Good luck!

The Dwarfs that Dwell in Leo

During a trip to California in 1980, Scotty learned of an interesting observation from Gerry Rattley, president of the San Jose Astronomical Association. He told Scotty that another West Coast observer, Lee McDonald, had been examining one of the Palomar *Sky Survey* plates (made with the 48-inch Schmidt telescope) when he found a very faint nebulosity in Leo just north of brilliant Regulus. "However," Scotty notes, "he expected the object to be beyond the reach visually of amateur-size telescopes." But one night Rattley and McDonald pointed a 17-inch altazimuth-mounted reflector (belonging to the San Francisco Sidewalk Astronomers) toward it. "Using powers of 50× and 100×, and by keeping Regulus just out of the field," Scotty continues, "both observers were able to see a faint glow without averted vision, filling more than half the field of view." These amateurs were looking at the dwarf spheroidal galaxy Leo I — an inconspicuous member of the Local Group of galaxies. This observation led Scotty, in April 1990, to ponder the incredible changes he had witnessed in amateur astronomy in his long career as an observer and columnist for *Sky & Telescope*.

As we begin the last decade of the 20th century, I'm flooded with the realization of how much astronomy has changed in my own lifetime and how rapidly it continues to change. In the 1930s I remember when the first photoelectric measurements of starlight were made using an electronic amplifier. Back then we only dreamed of space rockets. But today those rockets loft telescopes into space with detectors thousands of times more sophisticated than that crude photometer of the 1930s.

I've watched as the study of meteorite craters on Earth evolved from a heated debate among scientists to a respectable science. (Some people claimed that all craters on Earth were volcanic.) Recently the topic of celestial bombardment has reappeared in a new dress, and there's growing belief that impacts have greatly affected the course of life on Earth.

Nearly as impressive as the spacecraft missions to the planets has been the growth of amateur astronomy. During the 1920s articles in *Scientific American* got the telescope-making fad launched in North America, which greatly increased the number of amateur astronomers. This, in turn, made it economically feasible to publish a magazine devoted to popularizing astronomy; it wasn't long before the forerunners of *Sky & Telescope* were rolling off the presses. These magazines provided a successful place for commercial telescope-makers to advertise. It was an upward spiral that continues to this day.

Amateurs work very differently now than they did only a few decades ago. For example, in the early years of deep-sky observing I would set up a small refractor near my home in Milwaukee's Bay View. With a copy of *Norton's Star Atlas* in hand (the only deep-sky reference commonly available at that time), I would sweep the sky. Today's beginners are likely to have an 8-inch or larger telescope and access to detailed charts showing hordes of galaxies. Tirion's *Sky Atlas 2000.0* gave us 2,500 deep-sky objects. *Uranometria 2000.0* plotted four times as

Figure 5.4
The 10th-magnitude Leo I dwarf galaxy would be an easy target for most amateur telescopes, if brilliant Regulus weren't so close; the star lies a mere ⅓° due south of the galaxy.

106   DEEP-SKY WONDERS

many. And it's easy to find reference material for others. Even the most skilled observers don't have to look too far to find information on difficult objects.

Not all the sky's challenges have to be found with sophisticated charts. Leo and its beacon Regulus, at magnitude 1.35, are well placed for viewing during evening hours in April. Only 20 stars shine brighter than Regulus. If you center that dazzling celestial jewel in your telescope and nudge the instrument just ⅓° due north, you come to the dwarf galaxy Leo I. It's a member of the Local Group — a neighboring galaxy like the Magellanic Clouds, M31 in Andromeda, and one of my favorites, NGC 6822 in Sagittarius.

Leo I (**Figure 5.4**) shines with the total light of a 10th-magnitude star. But because this glow is spread across a disk some 10' in diameter, the actual surface brightness is rather low. Nevertheless, were it not for glare from Regulus, Leo I would be an easy target for a 10-inch telescope. California amateurs Gerry Rattley and Lee McDonald managed to snare it with a 17-inch reflector. They used magnifications of 50× and 100× and kept Regulus outside the field of view.

The 13th-magnitude galaxy IC 591 lies only ¼° west of Leo I. It's about 1' in diameter. Although smaller and fainter than Leo I, IC 591 must be easier to see since it was found visually at the close of the 19th century and Leo I wasn't discovered until 1950, when it was photographed with the 48-inch Schmidt telescope at Palomar Mountain.

Even more challenging would be dwarf galaxy Leo II (**Figure 5.5**). It too is a member of the Local Group. Leo II is slightly smaller than Leo I, and about 1½

Figure 5.5
After spotting Leo I, try your hand at finding Leo II (center). This dwarf galaxy is smaller and fainter than Leo I.

magnitudes fainter. I have not heard of any amateurs observing this system, but judging from the experiences of Gerry and Lee, I suspect it could be seen with a 16-inch aperture.

Galaxy Hunting in Leo

Tirelessly observant, Scotty found that a quick glance at Wil Tirion's *Sky Atlas 2000.0* revealed some interesting facts about Leo. There are no planetary or diffuse nebulae plotted within the confines of the constellation, nor are there any globular or open star clusters. Observers will, however, find the celestial Lion a rich hunting ground for galaxies, with some 60 of them plotted in *Sky Atlas 2000.0*. Of course, the *Millennium Star Atlas* reveals many more. So when Northern-Hemisphere amateurs eagerly venture out under the springtime sky, they can enjoy many hours of galaxy viewing within the boundaries of this attractive constellation. "Leo's prominent Sickle of bright stars is almost as easy to locate in the starry sky as the Big Dipper," Scotty said. "The beacon Regulus at the tip of the Sickle's handle helps guide the way, since there are only 20 brighter stars on all of heaven's vault. None of the Milky Way's silvery star dust is strewn across the constellation, and the naked-eye stars are rather few, so you do not easily get lost in Leo."

Spring is a time when the sky emerges from winter clouds. The Big Dipper is upside down above the pole, and if we follow the Pointer Stars to the south (rather than northward to Polaris), we can find Leo towering on the meridian.

To many of us, Leo symbolizes spring, just as Scorpius, Pegasus, and Orion do summer, fall, and winter. From Leo we can branch out to find all the other, lesser-known star patterns in the northern spring sky.

While the deep-sky objects of Leo might seem a little drab compared with the brilliant star clusters scattered across the winter Milky Way, there are some remarkable sights here for 8-inch and larger telescopes. Burnham's *Celestial Handbook* lists over 70 deep-sky objects in Leo. All are galaxies from 9th to 13th magnitude. I wouldn't even try to guess the number a 17-inch telescope could find. Within the boundaries of the constellation there is not one open or globular star cluster or planetary nebula suitable for amateur telescopes. This is interesting because Leo is the 12th largest constellation, covering just under 947 square degrees of sky.

About 10° east of Regulus lie M95, M96, and M105 — a trio of galaxies (**Figure 5.6**) all listed in modern versions of Messier's catalog, but actually discovered by his fellow countryman Pierre Méchain. The latter was the more mathematically inclined and occasionally calculated the orbits of Messier's comets. The two freely shared their deep-sky discoveries. As a case in point, two galaxies in the trio, M95 and M96, were first seen by Méchain on March 20, 1781. He passed the information on to Messier, who located the pair four nights later. M96, at magnitude 9.2, is about half a magnitude brighter than M95. Both spirals (M95 is a barred spiral) are about 7' in diameter. They are not breathtaking (to say the least) and for users of small telescopes, a good description is that they are bright enough to be seen. I have seen M96 as a pale white blur in instruments as small as 8 × 30 binoculars and a 1-inch, 40× refractor. Larger telescopes do not add much detail.

M95 is fainter but still within easy reach of a 4-inch. In fact, many observers consider M95 to be the more eye-catching galaxy, and some find M95 the more obvious of the pair. What do you think?

Figure 5.6
Leo's famous trio of Messier galaxies — M95, M96, and M105 — is packed into an area of sky slightly larger than one square degree. M95 and M96 are spirals while M105 is an elliptical system. NGC 3384 and NGC 3389 lie to the east of M105 and form a little triangle with it.

Both have bright central cores. Indeed, the nucleus of M95 can even be seen in large binoculars. Close scrutiny will reveal it to be a gently squared-off circle. Long-exposure photographs made with large telescopes record two straight arms extending from the nucleus and a faint ring surrounding the nucleus. Some observers have also suggested that the central bar of M95 is visible in amateur telescopes. Can you confirm or deny the visibility of this feature?

Just to the northeast of the pair lies NGC 3379, which is number 105 in modern extended lists of Messier objects. It is an elliptical about 4' across and shines with the total light of a magnitude 9.3 star. It is located about 1° north-northeast of M96. It too has a bright nucleus and a faint halo that fades into the background. It is curious that Méchain failed to notice M105 on the night he discovered M95 and M96. I suspect he was comet hunting when he found the pair of galaxies, and was probably making east-west sweeps parallel to the horizon with his refractor. Thus he could have easily missed the more northerly object.

M105 is visible in large binoculars. Two fainter galaxies lie just east of it. NGC 3384 (magnitude 10.0) is northwest of NGC 3389 (11.8), and together with M105 they form a little triangle that is about 8' on a side; they were discovered by William Herschel. Interestingly, even though M105 is the prominent object, William H. Smyth mentions the three under a heading for NGC 3384 in his 1844 *Bedford Catalogue*. Smyth further notes that he could not definitely see NGC 3389, though he suspected something at its approximate location. He was using a 5.9-inch refractor at his private observatory in England. There are at least half a dozen other 11th- and 12th-magnitude galaxies suitable for amateur telescopes spread across a few degrees of sky to the north of the M105 group.

Figure 5.7
For some unknown reason, Charles Messier missed discovering NGC 2903. Look for this large, 9th-magnitude spiral near the tip of the Sickle, about 4° southwest of Epsilon (ε) Leonis.

At least five of Leo's galaxies are large and bright enough to reveal internal detail in an 8-inch telescope. With a 17-inch you can spend half an evening on some of them. One of the most noteworthy is NGC 2903, a big 9th-magnitude spiral (**Figure 5.7**) about 4° southwest of Epsilon (ε) Leonis, the star at the tip of the Sickle, and hangs like a misty jewel 1½° south of Lambda (λ) Leonis. NGC 2903 seems distinctly oval, but not as much as the catalog dimensions 11' × 4.6' suggest. At magnitude 9 it should be visible even in a good 2-inch finder. Long-exposure photographs show it as a single galaxy, but this object's ownership of two NGC numbers, 2903 and 2905, is a reminder that early visual observers thought otherwise. In his *Cycle of Celestial Objects*, Smyth tells us that William Herschel described it in 1784 as "a double nebula, each having a seeming nucleus, with their apparent nebulosities running into each other." William's son, John, and Smyth himself later observed duplicity. Smyth had some difficulty in making this out with his 6-inch refractor: "The upper or south part is better defined than the lower; it requires, however, the closest attention and most patient watching, to make it a bicentral object." It would be an interesting experiment to view NGC 2903 with a long-focus instrument at medium to high magnification, gradually reducing the aperture until the object becomes nearly invisible in the hope of inducing the "double" appearance. I myself have never been able to see such an appearance, despite experiments with various apertures. Has any reader of this column observed NGC 2903 as double?

I've always wondered why Charles Messier missed this galaxy, especially since he cataloged fainter ones in nearby Virgo. Ron Morales of Tucson, Arizona, found it "easy and impressive" as seen with an 8-inch reflector. California amateur Tokuo Nakamoto reports seeing a faint oval halo surrounding the "much brighter" center of NGC 2903 with his 14-inch reflector. Years ago in Kansas I viewed the galaxy with a 10-inch reflector at 120×. The arms were rather ill-defined, but several knots of material dotted the extensions around the core. These knots are bright clouds of ionized hydrogen (called H II regions) similar to the Orion Nebula within our own galaxy. Photographers might try recording NGC 2903 with a series of exposures. Long ones should show the galaxy's arms in all their splendor, but shorter exposures may better reveal the knots and the nucleus. Astronomers have identified over 70 such glowing clouds within NGC 2903. In the 1960s I saw several of them with the 20-inch Clark refractor at Connecticut's Wesleyan University. They give the galaxy a mottled appearance.

Tucked under the "triangle" of eastern Leo, yet passed over with slight attention by most handbooks, lie three galaxies so close together that some telescopes will show them in the same low-power field (**Figure 5.8**). The two brighter ones, M65 and M66, were discovered by Méchain. In his original catalog, Messier writes that the comet he discovered in 1773 must have passed through the field of these galaxies on the night of November 2, but that he missed them because of the comet's light. It wasn't until 1780 that Méchain found the two galaxies and relayed the discovery to Messier.

M65, the westernmost of the two brighter galaxies, is a lenticular object measuring 8′ by 2′, of visual magnitude 8.9 according to Johann Holetschek, while Owen Gingerich assigns 10. It is well worth inspecting, especially with the technique of averted vision, whereby its quite respectable dimensions will become apparent. The Webb Society handbook on galaxies mentions a dark lane just west of M65's nucleus, as seen with an 8½-inch telescope. The lane is about 2½′ long. Visually, with my 10-inch reflector and a fine sky, this arm was only suspected, and probably not seen at all. Have others noted it?

Also an Sb galaxy but with a more regular and curdled appearance is M66 (NGC 3627). Its published dimensions are 8′ by 2.5′, and the visual magnitude is

Figure 5.8
Search out this clump of three spiral galaxies in eastern Leo. M65 and M66 are each like miniature versions of the Andromeda Galaxy, while NGC 3628 is seen nearly edge on.

8.6 (Holetschek) or 9 (Gingerich). M66 is more conspicuous than M65 and is usually visible in a 2-inch finder. It is interesting to compare these galaxies for brightness, to see if the difference of magnitude given above varies with the instrument, and whether it is the same for visual observations as for photographic. Generally, the visual magnitudes should be about a magnitude brighter than the photographic ones.

The skilled observer John Mallas called M65 "beautiful" in a 4-inch refractor, while M66 had a "mottled or clumpy appearance, reminiscent of the Orion Nebula" in the same telescope. This is due to numerous stellar condensations in the galaxy's spiral arms.

Most photographs of M66 do not show this patchy appearance because the bright central region of the galaxy is overexposed. Compared with M65 in my 10-inch, M66 is hardly recognizable as a spiral, resembling a stray diffuse nebula, but photographs prove otherwise. It's been my experience that many observers tend to overlook details in bright objects, perhaps because they think these easy-to-find things are for beginners. These same people will write me long letters about glimpsing a subtle brightening at the edge of some 13th-magnitude galaxy.

W. H. Smyth speaks of a third nebula close by, but this is a mistake copied from John Herschel. However, there is another spiral 1° west of the Messier pair, NGC 3593, which many people overlook because of the brighter galaxies just east of it. The galaxy is "bright" in a 16-inch. Look for an object about 3' long.

All three of these galaxies are easily seen with a 3-inch telescope.

There is another galaxy just north of M65 and M66, and apparently a little too difficult for Messier's telescopes: NGC 3628. It is reported to have been seen in a 3-inch f/6 glass at 18×. Although it has about the same total light as M65, the galaxy is larger and thus has a lower surface brightness. Nearly 15′ long, NGC 3628 spans half the Moon's diameter and can stretch clear across the field of a medium-power eyepiece. It is an Sb galaxy, 12′ × 1.5′ in extent. Holetschek called its visual magnitude 10.2, while the photographic value is 11.3. It is only a few tenths of a magnitude fainter than M65 and M66, but because the light is spread over a larger area the surface brightness is lower. A dark lane runs the length of the galaxy and can be seen in 16-inch telescopes. The feature is usually difficult for an 8-inch, but on a particularly good night I caught a glimpse of it with my 4-inch Clark refractor and a 9-mm Nagler eyepiece. It might serve as an indicator of seeing conditions. Indeed, these three galaxies, so conveniently located, can be used for tests of visual magnitude estimates.

Leo Challenges

For observers with access to large telescopes, there is a cluster of galaxies in Leo known as Abell 1367. It is one of the richest nearby galaxy clusters cataloged by George Abell. Near its center is NGC 3842, about 1′ across and 13th magnitude. While this is the brightest of the group, nearly 30 other galaxies in a one-square-degree area have been seen with 16-inch telescopes. At 120× I can hold NGC 3842 steadily with the Clark. A good chart of this cluster is in Vol. 5 of the Webb Society *Deep-Sky Observer's Handbook*.

Those who enjoy searching for difficult objects will probably find NGC 3588 to their liking. This faint (perhaps 15th-magnitude) galaxy is located only 8′ south of 2.6-magnitude Delta (δ) Leonis, the northern star of the triangle marking the Lion's hindquarters. The galaxy was discovered by Lewis Swift with a 15-inch refractor and can probably be seen in smaller instruments. The bright star in the same field of view, however, will make this small object difficult to see.

In the Sickle of Leo is an interesting sprinkling of faint galaxies between Gamma (γ) and Zeta (ζ) Leonis. NGC 3162 looks 2′ in diameter and of magnitude 11.4 to my eye. Farther south and east is tiny 12th-magnitude NGC 3177. Visually it appears as an almost circular disk less than 1′ across, which seems to bear high powers well. A bit farther east and nearly on the line between Gamma and Zeta is a fancy threesome of small galaxies arrayed almost end to end. The middle and brightest is 11th-magnitude NGC 3190, about 3′ × 1′. To its southwest is NGC 3185, magnitude about 12.3 and 1.5′ × 1′. At the other end of the chain, NGC 3193 is brighter (11.4) but only 1′ in diameter. These three objects are an attractive sight on a good night in the low-power field of a large amateur telescope.

My old notes contain a reminder to look for the very faint galaxy NGC 3130 that should be in the same field as 31 Leonis, a 4th-magnitude star 2° south of Regulus. (By the way, 31-A Leonis is mislabeled as 41 in the first printing of Tirion's *Sky Atlas 2000.0*.) A word of warning, however: the naked-eye star in the same field will make this observation very difficult at best. NGC 3130 is a small, faint galaxy of about 14th magnitude. It has eluded me for years, and I have never met an amateur who has seen it. I suspect that the naked-eye star hinders the view. Certainly many amateurs routinely observe fainter galaxies, so perhaps someone using an occulting bar in an eyepiece will be able to see NGC 3130.

Finally, for those of you who really want an observing challenge, try searching for any of the numerous galaxies that surround Iota (ι) Leonis. A chart from an out-of-print star atlas published by the Smithsonian Astrophysical Observatory shows almost every square (each is about 1° on a side) containing a half dozen or more galaxies. With just a few exceptions, these objects are listed in the second *Index Catalogue* to the *NGC*. Most were discovered photographically by Max Wolf using the 16-inch Bruce refractor at the Heidelberg Observatory in Germany around the turn of the century. While many of the objects are small and faint, I suspect that many are within the visual grasp of a 17-inch telescope. Happy hunting!

Lure of the Little Lion

Leo Minor is tiny, covering only 232 square degrees of sky. It is one of the "most barren parts of the sky to the naked eye," Scotty said, yet he seemed particularly attracted to this dim constellation, which contains a rich assortment of galaxies within the range of small and modest-sized telescopes. Interestingly, before such cartographic wonders as Wil Tirion's *Sky Atlas 2000.0* and the *Millennium Star Atlas,* observers had to work with less-detailed star charts, which were effective and helpful for their time but became problematical as the hobby grew. The Skalnate Pleso *Atlas of the Heavens* (a product of the 1940s and '50s), for example, plotted but did not number galaxies fainter than around magnitude 11. Furthermore, unless a backyard observer had access to an astronomical library, data on many of these objects were simply unobtainable. When large-aperture telescopes became increasingly available to photon-thirsty amateurs, Scotty was among the first to recognize our need for more adequate deep-sky star charts and catalogs. It is in part thanks to his constant dipping into the inner sanctum of the night that amateurs enjoy better reference materials today.

When William Tyler Olcott wrote his classic *A Field Book of the Stars* around the turn of the century, he began the text with Ursa Major. Not only was this "the best known of the constellations," but the stars of the Great Bear could act as guideposts to other parts of the sky. As an example, consider the undistinguished constellation Leo Minor. It was created by Johannes Hevelius during the wild proliferation of constellations that occurred during the 17th century, when no part of the sky was immune to celestial cartographers. Some, such as Scutum Sobieski (now called Scutum), filled a visual void and still survive today. But others — Musca Borealis, Noctua, and Bufo, for example —

have long since been forgotten. Leo Minor is unusual in that it contains no bright star designated Alpha (α), though it does have a Beta (β). Of the 88 constellations currently recognized, only three others — Norma, Puppis, and Vela — are without an Alpha. But it can easily be tracked to its shady lair by following the stars in the Bear's hind legs. If we begin with Gamma (γ) Ursae Majoris, in the Big Dipper's Bowl, we can sweep down a line of stars curving southwestward and ending at the pair of 3rd-magnitude stars Mu (μ) and Lambda (λ) Ursae Majoris.

Just 1° due west of Mu is the 10th-magnitude spiral galaxy NGC 3184. Although still in Ursa Major, when the galaxy is centered in a low-power eyepiece, the stars at the southwest edge of the field are in Leo Minor. An 8-inch reflector shows NGC 3184 as a pale disk about 5' across with a brighter center. At 30× it looks much the way the well-known spiral M33 does in my 2-inch finderscope. Ronald Morales comments that NGC 3184 reveals a rather uniformly illuminated disk as seen with a 6-inch reflector at 35×. Tokuo Nakamoto of Temple City, California, glimpsed a hint of spiral structure and saw one star superimposed on the galaxy with a 13-inch telescope.

Figure 5.9
The largest and brightest galaxy in Leo Minor, 10th-magnitude NGC 3344, displays a tight core with loosely wound spiral arms.

The largest and brightest galaxy in Leo Minor is NGC 3344. This spiral **(Figure 5.9)** stretches across more than 6' of sky and has a total visual magnitude of 9.9. My observing notes from 1956 mention a wide, 9th-magnitude double star about 1' east of the galaxy which interfered with the view. NGC 3344 lies midway between the naked-eye stars 40 and 41 Leonis Minoris, about 6° east of Zeta (ζ) Leonis in the Sickle, or about 3° west of the 4th-magnitude star 54

Leonis. One cold morning I found the galaxy easily in my 4-inch Clark refractor. It is a nearly circular blob about 7' in diameter, being photographically a spiral galaxy seen nearly face on, but the arms cannot be seen in small amateur instruments. With his 6-inch telescope at 50×, Morales found it to have a hint of a starlike nucleus.

Another large and bright galaxy in Leo Minor is NGC 2859. It is also one of the easiest to locate since it lies less than 1° east of the brightest star in neighboring Lynx (Alpha (α) Lyncis). If you point your telescope at Alpha and wait 3¼ minutes with the drive turned off, the galaxy will be just north of the center of the eyepiece field. The 11th-magnitude barred spiral galaxy appears as a somewhat irregular grayish patch 4' × 3' or smaller. In 1978 I estimated its visual magnitude to be 10.7. A 4-inch instrument shows NGC 2859 to have a bright center, and there may be just a hint of the bar visible. My old 10-inch reflector revealed the bar and traces of an outer ring. Florida amateur Brenda Branchett reports that her 6-inch reflector at 70× also shows hints of the galaxy's outer ring. In a 36-inch reflector at Tucson many years ago this object was bright and fascinating. It is said to appear as a faint patch in a 3½-inch telescope at 45×.

NGC 3245 is an elliptical that's about as bright as NGC 2859 and slightly smaller. It is reported to be faint in a 3-inch at 18×. There are two moderately bright telescopic stars nearby, 10' southwest and 7' southeast. Branchett also notes a string of stars in the field that enhances the beauty of the view. Visual observers have commented on the oblong shape (2' × 1') of this 10.5-magnitude elliptical ever since its discovery by William Herschel in 1785. The galaxy is not difficult for my 4-inch Clark refractor at 100×. I once viewed it with the 20-inch Clark refractor at Wesleyan University, and, while it appeared larger and brighter than in the smaller telescope, I did not notice much additional detail. Nearby are some very faint galaxies.

The *Atlas of the Heavens* does not give the NGC numbers of galaxies that are fainter than magnitude 11, roughly, and of the 19 in Leo Minor 14 are thus unlabeled. The sizes of the ellipses representing these galaxies, though a good guide to angular extent, give little idea of whether they are easy or hard to spot.

Take, for example, the unlabeled pair NGC 3003 and NGC 3021. The former is plotted as a larger ellipse than the latter, yet a 10-inch will show NGC 3021 much more easily. I have known amateurs to see only one galaxy here and assume it is NGC 3003. The explanation is that NGC 3021 is both brighter (12th-magnitude) and more compact (1' × 0.5'), so its great surface brightness makes it a fairly easy object if one knows where to look. I have seen it in Connecticut with a 4-inch. NGC 3003, on the other hand, is a full magnitude fainter, and its light is spread over an area of 5' × 1', making its surface brightness much lower. Here in the mediocre skies of Connecticut, NGC 3021 is easy to acquire in the 4-inch Clark, but NGC 3003 requires an excellent night and careful searching. Once when NGC 3021 was seen at 60×, it took 150× to establish with certainty the presence of NGC 3003.

Even today there is still some confusion about the relative brightnesses of these galaxies. My estimates of 12th magnitude for NGC 3021 and 13th for NGC 3003 agree well with the values listed in Beçvár's old *Atlas Catalogue*. However, the *Revised New General Catalogue* and Burnham's *Celestial Handbook* reverse the order. Perhaps this is merely a typographical error that has been copied. Many instances of such errors are known in the popular observing reference books. As more amateurs adopt larger-aperture telescopes for deep-sky observing, and fainter objects are examined, more of these discrepancies will be found.

About 2° east-northeast of Beta (β) Leonis Minoris is the 12th-magnitude galaxy NGC 3294, which is easy to find since it is slightly west of the midpoint of a line joining the 6th-magnitude stars 35 and 38 Leonis Minoris. There is a star just southeast of the galaxy, which is some 3' × 1' in extent. Branchett comments that NGC 3294 required averted vision with her 6-inch at 43×. Joseph Schmidt in Ely, Minnesota, however, called it "bright" through his 6-inch f/11 reflector. The 12-inch Porter turret telescope at Springfield, Vermont, can show the central core, but no individual arms, at 200×. A comprehensive variable-star catalog, the *General Catalogue of Variable Stars*, prepared at Sternberg Astronomical Institute by Boris V. Kukarkin and fellow Russian astronomers, lists an unconfirmed 11th-magnitude supernova visible in NGC 3294 in 1955.

NGC 3395 is another 12th-magnitude spiral, about 2' in diameter. Amateur telescopes will show it almost in contact with NGC 3396 at its northeast edge. These are interacting galaxies, but the bridge of material between them does not show in small telescopes. Has anyone viewed them with a 30-inch aperture?

Navigating Sextans

The southern sky is littered with constellations that celebrate the scientific renaissance of the 17th and 18th centuries "like debris from an old-time science lab hit by a tornado," Scotty once quipped. Among these time-honored instruments you will find a set of navigational tools — namely an octant, a sextant, and a mariner's compass. "I suspect if a drift meter had existed when the constellations were named," Scotty continued, "we would even have included it in current lists of star patterns." As a former celestial navigator with the U. S. Army Air Force, Scotty was naturally partial to the constellation of the sextant, claiming that "at one time I'm sure I took better care of mine than I did of my wife and children!" Ironically, Sextans is a challenge for beginners to navigate. Spanning an area of approximately 314 square degrees, Sextans ranks 47th in size of the 88 constellations. Its brightest star, Alpha (α) Sextantis, shines meekly at magnitude 4.5 and is easily lost in modest moonlight or light pollution, leaving that region of sky appearing totally blank.
So Scotty figuratively pulls up a chair, sits beside you at the telescope, takes a puff of his pipe, then explains how to star-hop to the sole prize of Sextans, the bright lenticular galaxy NGC 3115.

The first constellations were modeled after familiar objects that reflected the life and times of people millennia ago. There were farm carts, boats, bears, fishes, and sea monsters. In the beginning each village may have had its own set of constellations. But as the villages developed into cities and the cities into nations, individual lists became homogenized into one. The sky patterns we know today are an amalgamation of those from dozens of long-ago empires.

After the chaotic early history of the constellations, they remained essentially unchanged in the Western world from about 275 B.C. to A.D. 1500 thanks to the wide acceptance of lists compiled by Aratus and Ptolemy. But then began another period when new constellations appeared on star charts and celestial globes. No recognized constellation was eliminated, but many had their edges nibbled away to make room for the new and usually tiny groups. For each of these patterns we know the creator and in many cases even the exact words he used to defend his actions.

Furthermore, during this time explorers opened routes to the Southern Hemisphere where a whole new sky ripe for European constellation makers awaited. Scientists couldn't resist filling the sky with star patterns commemorating their scientific instruments. They were acting very much like their agrarian ancestors — putting the tools of their trade into the sky.

One of my favorite constellations from this scientific renaissance is Sextans, the device navigators use to measure star altitudes. Hevelius formed the group in 1687 to commemorate the huge instrument he used to measure star positions from his rooftop observatory in Danzig (now Gdansk). There is, by the way, another such instrument at the south celestial pole. Octans, named for the octant of John Hadley, is a slight variation of the older sextant. Both instruments do essentially the same job, and some navigators have used the terms interchangeably.

On modern star charts, Sextans is merely a neatly bounded sky area, almost a perfect square. Inside this square the Skalnate Pleso *Atlas of the Heavens* shows nine galaxies, most of which are fairly difficult objects. By far the most prominent one is NGC 3115. This is an elliptical galaxy about $4' \times 1'$ across. Its visual magnitude is given as 9, but this seems a little bright to me. NGC 3115 is bright enough to be seen in a 2-inch finder and even large binoculars. The problem is trying to locate it in such a star-poor region. This is a perfect place for the experienced beginner to practice star-hopping, that essential technique of correlating a star chart with the view in a finder.

Amateurs can never become too proficient at finding their way around the sky. Observing can lose some of its appeal when it takes 20 minutes to find an object that you may look at for only a minute or two. Some years ago I kept a record of how long I and others spent searching for variable stars compared with the time we spent making the magnitude estimates. I was amazed to discover that roughly 90 percent of our time was devoted to the search. It made little difference whether the observer used an altazimuth or equatorial mount since even an equatorial with setting circles usually just puts you close to the object of interest. You still must star-hop to center the field in the eyepiece.

Burnham suggests trying to locate NGC 3115 by sweeping 20° south from brilliant Regulus in Leo. Even with setting circles, I wouldn't bet money on making a 20° leap to land on a 10th-magnitude galaxy. Philip Harrington, in his book *Touring the Universe Through Binoculars*, suggests finding the galaxy by sweeping 8° east from Alphard (Alpha (α) Hydrae) to pick up the stars 17 and 18 Sextantis. NGC 3115 is about 1½° northwest of the pair. The problem here, however, is that you can confuse 17 and 18 Sextantis with another pair nearby and end up in the middle of nowhere. Each of these methods is handicapped by trying to make too-long jumps using too few stars.

It is much better to plan a series of small moves, each covering no more than the field of your finder. This way you can always check your star chart to be certain where you are. One word of warning: star-hopping is nearly impossible if you have a finder that produces a mirror image of the sky — and such is the case with virtually all right-angle finders. Very few people can mentally transpose a mirror image to make it match a chart.

So where do we begin our hop to NGC 3115? Regulus is too far away, as is Alphard. Alpha (α) Sextantis would be a good choice except that there are several vacant spots along the way to the galaxy, making this a dubious ploy. Instead, let's begin at Lambda (λ) Hydrae (**Figure 5.10**). Not only is this an easy naked-eye star, but a distinct asterism formed by several surrounding stars will offer positive identification even in small finders.

Figure 5.10
To locate the elliptical galaxy NGC 3115 in dim Sextans, try star-hopping from either Alphard (α Hydrae) or Lambda (λ) Hydrae.

From Lambda, move about 1½° north-northwest to a 7th-magnitude star easily identified by two slightly fainter companions to its southwest. About 2° northwest of this group is another 7th-magnitude star, this one at the northern end of a sprawling W-shaped asterism about 1° long and oriented roughly north-south. From here it's an easy 2° jump northeastward to the distinct east-west pair formed by 17 and 18 Sextantis. Now a 1° hop to the northwest brings you to the other pair of stars mentioned earlier, and this puts you within striking distance of the galaxy, ½° due west.

Another method for locating NGC 3115 is especially good if you are ready for a short coffee break during your observing schedule. Select an eyepiece which shows at least a Moon's diameter of sky, and place the 5th-magnitude star Gamma (γ) Sextantis near the southern edge of the field. If you leave the telescope stationary for 12¾ minutes (turn the drive off if the telescope has one), the galaxy will be centered near the northern half of the eyepiece field.

Figure 5.11
The sole extragalactic prize of Sextans is the bright lenticular galaxy NGC 3115, also known as the Spindle Galaxy. It appears edge on; no dust lanes or spiral arms are visible.

And what a splendid sight NGC 3115 is (**Figure 5.11**), especially when you realize that few amateurs ever observe this island universe. Aptly named the Spindle Galaxy, it is an evenly illuminated elliptical about 4' long and 1' wide. Powers as low as 20× will show the spindle shape. Though I have observed NGC 3115 with a 4-inch telescope from Connecticut, averted vision was occasionally needed to find it. In a 5-inch Moonwatch Apogee telescope the galaxy appeared small but sharp. Smyth writes in the *Cycle of Celestial Objects* that it was very clearly distinguished in his 6-inch refractor. Many years ago I viewed it with my 10-inch reflector under the diamond-clear skies of Kansas. Burnham notes in his *Celestial Handbook* that the galaxy appears much the same to the eye in an amateur telescope as it does in photographs made with large telescopes. It has been classed both as a highly flattened elliptical galaxy and as an edge-on spiral. It lacks any hint of a dust lane, however, which shows frequently in edge-on systems. I think it fascinating to find an object that shows well in 5-inch binoculars, looks similar in a 12-inch instrument, and still perplexes professional astronomers working with the largest telescopes in the world.

MAY OBJECTS

Name	Type	Const.	R. A. h m	Dec. ° '	Millennium Star Atlas	Uranometria 2000.0	Sky Atlas 2000.0
Abell 1367	Gx	Leo	11 44.5	+19 50	—	—	—
IC 591	Gx	Leo	10 07.5	+12 16	732	189	—
Leo I	Gx	Leo	10 08.5	+12 18	732	189	13
Leo II	Gx	Leo	11 13.5	+22 09	681	146	—
M65, NGC 3623	Gx	Leo	11 18.9	+13 05	729	191	13
M66, NGC 3627	Gx	Leo	11 20.2	+12 59	728, 729	191	13
M95, NGC 3351	Gx	Leo	10 44.0	+11 42	730	190	13
M96, NGC 3368	Gx	Leo	10 46.8	+11 49	730	190	13
M105, NGC 3379	Gx	Leo	10 47.8	+12 35	730	190	13
NGC 2859	Gx	LMi	09 24.3	+34 31	641	103	6
NGC 2903	Gx	Leo	09 32.2	+21 30	686, 710	143	6
NGC 2905	Gx	Leo	09 32.2	+21 31	—	—	—
NGC 3003	Gx	LMi	09 48.6	+33 25	639, 640, 661, 662	104	6
NGC 3021	Gx	LMi	09 51.0	+33 33	639, 640, 661, 662	104	6
NGC 3130	Gx	Leo	10 08.2	+09 59	732	189	—
NGC 3158	Gx	LMi	10 13.8	+38 46	638	72, 104	6
NGC 3162	Gx	Leo	10 13.5	+22 44	684	144	6
NGC 3163	Gx	LMi	10 14.1	+38 39	638	72, 104	—
NGC 3177	Gx	Leo	10 16.6	+21 07	684, 708	144	6
NGC 3184	Gx	UMa	10 18.3	+41 25	617	72	6
NGC 3185	Gx	Leo	10 17.6	+21 41	684	144	6
NGC 3190	Gx	Leo	10 18.1	+21 50	684	144	6
NGC 3193	Gx	Leo	10 18.4	+21 54	683, 684	144	6
NGC 3245	Gx	LMi	10 27.3	+28 30	660	105, 145	6
NGC 3294	Gx	LMi	10 36.3	+37 20	637	105	6
NGC 3344	Gx	LMi	10 43.5	+24 55	682	145	6
NGC 3384	Gx	Leo	10 48.3	+12 38	730	190	13
NGC 3389	Gx	Leo	10 48.5	+12 32	730	190	13
NGC 3395	Gx	LMi	10 49.8	+32 59	637, 659	105	6
NGC 3396	Gx	LMi	10 49.9	+32 59	637, 659	105	6

Ast = Asterism; BN = Bright Nebula; CGx = Cluster of Galaxies; DN = Dark Nebula; GC = Globular Cluster; Gx = Galaxy; OC = Open Cluster; PN = Planetary Nebula; ✶ = Star; ✶✶ = Double/Multiple Star; Var = Variable Star

MAY OBJECTS (CONTINUED)

Name	Type	Const.	R. A. h m	Dec. ° '	Millennium Star Atlas	Uranometria 2000.0	Sky Atlas 2000.0
NGC 3414	Gx	LMi	10 51.3	+27 59	659	105, 145	6
NGC 3486	Gx	LMi	11 00.4	+28 58	658	105, 106	6
NGC 3504	Gx	LMi	11 03.2	+27 58	658	106, 146	6
NGC 3588	Gx	Leo	11 14.0	+20 24	681, 705	146	æ
NGC 3593	Gx	Leo	11 14.6	+12 49	729	191	13
NGC 3628	Gx	Leo	11 20.3	+13 36	728, 729	191	13
NGC 3842	Gx	Leo	11 44.0	+19 57	703	147	6, 13
NGC 6822	Gx	Sgr	19 44.9	–14 48	1339, 1363	297	16, 22
Omega (ω) Centauri, NGC 5139	GC	Cen	13 26.8	–47 29	953	403	21, 25
Spindle Galaxy, NGC 3115	Gx	Sex	10 05.2	–07 43	804	279	13

Ast = Asterism; BN = Bright Nebula; CGx = Cluster of Galaxies; DN = Dark Nebula; GC = Globular Cluster; Gx = Galaxy; OC = Open Cluster; PN = Planetary Nebula; ✶ = Star; ✶✶ = Double/Multiple Star; Var = Variable Star

JUNE
CHAPTER 6

The Bowl of Night

Never setting from mid-northern latitudes, the Big Dipper endlessly circles the North Star and is arguably the most sought after star group in the heavens. For centuries it has been one of humanity's most important celestial guideposts and remains so today. Tyros of all ages turn first to it to get oriented to the sky and to determine the scale of the patterns they see on their star charts. Telescope users soon learn that the Big Dipper and its parent constellation, Ursa Major, harbor a fleet of deep-sky objects, several of which belong to the Messier catalog. Scotty has already introduced us to two of these splendors — the Dynamic Duo of M81 and M82. Now he leads us into and around the Dipper's Bowl, which abounds in deep-sky delights for beginning and advanced amateurs.

One of the nicest pieces of celestial real estate for hunting down cosmic treasures is the area around the Bowl of the Big Dipper (**Figure 6.1**). Aside perhaps from Orion, the Big Dipper is the sky's best-known star pattern, with bright, easily found starting points for star-hopping to deep-sky objects. And what a wonderful selection of objects there is, for it is here in the polar region that the great stream

Figure 6.1
The Big Dipper is among the most familiar star patterns in the sky. Its bowl is filled with deep-sky wonders.

JUNE   123

of galaxies reaching northward from Virgo and Coma comes to a brilliant conclusion. Several bright galaxies from the Messier catalog bedeck the Dipper, amid scores of others that are easy targets for 6-inch telescopes.

Unlike objects far to the south that can only be seen to advantage for an hour or two when they cross the meridian, those in the polar sky can be profitably studied for many hours each night and for many months of the year. In late spring, the flower basket of the Big Dipper's Bowl rides nearly overhead during evening hours.

Figure 6.2 M97, the famous Owl (lower left), is a planetary nebula with two dark "eyes." Less than a degree northwest lies the remarkable spiral galaxy M108, which is inclined only 8° from edge on.

Close to the southern edge of the Big Dipper's Bowl is the remarkable planetary nebula M97, popularly known as the Owl (**Figure 6.2**). At magnitude 11.2, M97 is second only to the planetary nebula M76 in Perseus (magnitude 11.5) as the faintest object in the French comet hunter's catalog. Nevertheless, M97 is no problem for 4-inch telescopes, and I have easily seen it in 15 × 65 binoculars. The two dark spots that form the Owl's "eyes" are more challenging, but under good skies they might be within range of a 4-inch telescope. The nebula measures only 3', and if it is not seen at once, let your eye wander aimlessly over the field of view until the disk springs into view. Barns, in his *1001 Celestial Wonders,* calls this object dull-toned, but memorable. William Herschel long ago thought it was a cluster of stars, unresolvable because of its immense distance. Even though today we know of planetaries three times farther, this object's distance of about 10,000 light-years will make many share Herschel's feeling of its remoteness.

Strangely enough, William H. Smyth could see nothing but a pale, uniform disk "about the size of Jupiter," yet a 6-inch today will show something of the two dark holes in M97. But Smyth wrote before Lord Rosse found them, indicating that it is much more difficult to discover something than to confirm it. My 10-inch, under clear Kansas skies, shows the full 3' diameter of the Owl.

It's only a short hop of 0.8° northwest from M97 to the spiral galaxy M108, which can be seen in the same low-power field. Another way to find it is to start by aiming the finder at the Dipper's Pointer Stars, Alpha (α) and Beta (β) Ursae Majoris. Using their 5.4° spacing as a gauge, shift 1½° southeast of Beta and try to spot a faint elongated streak. Although Messier did not include this object in his original, 18th-century catalog, handwritten notes in his personal copy of the work make it clear he was aware of it. During the 20th century, several astronomers who studied Messier's work proposed that this galaxy and six others known to Messier be formally added to the original list, bringing the total to 110. The galaxy is a bright strip of a spiral, 8' × 2' in extent, of magnitude 10, with a faint foreground star superimposed. Herschel listed it as Number 46 in his class V (very large nebulae), and therefore *Norton's Star Atlas* marks it as H46⁵. It is easily seen in 50-mm binoculars.

Photographs reveal it as an edge-on galaxy. The dim central condensation has been reported as just visible in a 3½-inch refractor. Regardless of the telescope's aperture, try a high magnification on this spiral. At 30×, the 12-inch Porter turret telescope showed lots of delicate internal structure. In some ways, M108 is similar to M82. I wonder if the excessive hydrogen-alpha light known to come from M82 would cause the galaxy to appear different from M108 when both are observed through various nebula filters.

Now move to Gamma (γ) Ursae Majoris, the southeastern star of the Bowl. Just ⅔° southeast of this star is a 9½-magnitude galaxy, M109. Oval in outline and several arcminutes long, this one is bright and easy in my 5-inch Apogee telescope. Older catalogs call it only NGC 3992. Its Messier designation dates from 1953, after Owen Gingerich of Harvard Observatory discovered a previously overlooked note concerning it in Charles Messier's personal copy of a 1784 French almanac (discussed on page 127 in "The Mystery of M102"). In my 5-inch Apogee telescope at 20×, this galaxy seems brighter than its cataloged magnitude of 10.8. I can even catch its 6' × 10' shape in 15 × 65 binoculars. I know of no amateur sightings of the galaxy's arms or central bar, which show well on photographs. Visually NGC 3992 looks like a featureless oval glow surrounding a brighter core. There is a 13th-magnitude star near its northern edge. In 1956 a supernova in this galaxy reached magnitude 11.2, which was well within the grasp of amateur telescopes. This galaxy is thus a good candidate for inclusion in a visual supernova-search program.

Returning to Gamma, slew the telescope south 6° and center on 3.7-magnitude Chi (χ) Ursae Majoris; then go east another 6°. In the telescope's low-power field, there should be seen an oblong diffuse patch, the 8th-magnitude galaxy M106, just across the boundary of Ursa Major in Canes Venatici. If the night is good, you may also glimpse NGC 4217, a smaller galaxy ½° west of M106.

The Bowl's Inner Sanctum

The Dipper's Bowl also contains a fair number of 11th-magnitude and fainter galaxies which generally go unmentioned in amateur observing guides. *Norton's Star Atlas* plots seven galaxies inside the bowl, and the Skalnate Pleso *Atlas of the Heavens* adds another five. During a visit with members of the San Jose

Astronomical Association in California, I found that Gerry Rattley and comet discoverer Don Machholz are quite familiar with these objects. As none of these galaxies have ever been mentioned in this column before, let's take them up now.

One of the brightest objects is the spiral galaxy NGC 3982. It is surprisingly easy to spot in my 4-inch Clark refractor, and appears as an 11.3-magnitude blur, about 1' in diameter and slightly oval.

Another spiral, almost as bright at magnitude 11.4 and nearly twice as large, is NGC 3898. NGC 3610 is similar in brightness to NGC 3898 and is a tiny elliptical galaxy only about 1' in diameter. My only observation of it was from Kansas in May 1956, with a 10-inch f/8.5 Newtonian reflector. NGC 3613 is a distinctly oval galaxy about 1½' long. My magnitude estimates of it average 11. And at photographic magnitude 13, NGC 3683 is a difficult object. NGC 3894 is very faint, while NGC 3690 is a little easier as its visual magnitude is about 12.0.

NGC 3642 is an interesting galaxy. Although its total light, equivalent to an 11.4-magnitude star, is spread over a patch some 5' across, this galaxy is not too difficult to find. Indeed, in the same 1956 observing journal in which I recorded NGC 3610, I noted NGC 3642 as a beautiful sight in the 10-inch, even at 200×. I use a Barlow lens rather than a short-focal-length eyepiece for highly magnified views of deep-sky objects. The Barlow seems to add less scattered light to the field, thereby improving the contrast between the sky and galaxy or nebula. NGC 3619 is about magnitude 11.7 and 1' in diameter. In my 4-inch Clark it is best seen at 100× or more. With lower powers it could be mistaken for a faint star. Slightly fainter but easier to find is NGC 3738. Still dimmer is 12th-magnitude NGC 3756. This spiral galaxy is about 3' long and half as wide.

I have also selected five galaxies located north and east of the Bowl. All are moderately faint, but surprisingly easy in average skies. NGC 4036 is a large diffuse glow, 2.4' × 0.9' according to Sidney van den Bergh's revision of the *Shapley-Ames Catalog*. Some searching was needed to find it in a 5-inch richfield Apogee telescope, but once located it was readily visible as 11th-magnitude, fading off at the edges. In the same field is NGC 4041. Of about the same size and brightness, these galaxies look like twins, but actually NGC 4036 is elliptical, NGC 4041 spiral.

Brightest and easiest to find in our quintet is NGC 4605. This 10th-magnitude spiral lies nearly on the extension of a line joining Gamma (γ) and Delta (δ) Ursae Majoris. It is obvious in 65-mm binoculars, and a large telescope makes it a fine sight, extending across a 5' × 1.2' area of sky. Equally nice is NGC 3945, an 11th-magnitude barred spiral. Covering an area measuring 5.2' × 2.2', it is fairly easy to find and most interesting in large telescopes.

The final galaxy, NGC 4814, is a barred spiral a little more than 2' in diameter. But do not be surprised if you should come across other objects in and around the Bowl. There are more galaxies here within reach of today's larger amateur telescopes. Observers with 8-inch and larger telescopes will find an almost endless supply of galaxies in the area. Indeed, the *Uranometria 2000.0* atlas plots 50 just within the Dipper's Bowl. Hunting some of them down is a pleasant way to spend an evening under the stars.

The Mystery of M102

One of the oldest and most unrelenting mysteries to haunt the pages of astronomical literature is Charles Messier's enigmatic 102nd catalog entry. For many years M102 had no proper identity — no deep-sky object exists in the terribly rough (and obviously wrong) position provided to Messier by its discoverer, Pierre Méchain. Over the years, however, a handful of celestial detectives have challenged themselves to solve this centuries-old mystery. Some sleuthed the sky for clues. Others perused the dusty pages of old journals searching for an answer. Indeed, as Scotty explains here, a major discovery in Canada finally resolved the matter. But did it? Alas, the answer is yes, but also no. Despite indisputable evidence, many amateurs to this day do not accept the Canadian discovery, opting instead to espouse and promote their own theories for the identity of M102. In the end, in the inimitable words of Scotty, "one wrong assumption coupled with immaculate logic produced a can of worms."

Amateur astronomers are a persistent lot. Give them a problem, and they can spend years chipping away at it. The late Leslie Peltier checked the star T Coronae Borealis on thousands of nights beginning in 1920. In 1866 it had briefly become a naked-eye nova, and for decades thereafter it remained a rather unremarkable 10th-magnitude object. Peltier's vigil was an act of faith, since the star had no known history of repeated outbursts. But on February 9, 1946, the star again rose to easy naked-eye visibility at 3rd magnitude. Although Peltier wasn't the first to note the brightening, the outburst vindicated his years of effort.

In the 1930s Joseph Meek noticed a slight, rapid cycle in the light curve of the dwarf nova SS Cygni and began wondering if the star was a close binary. It would be 20 years before spectroscopic observations at Mount Wilson Observatory proved him right.

Texas amateur Oscar Monnig had taken an early interest in photographing the spectra of meteors. Before professional astronomers turned their attention (and vastly greater resources) to this difficult task, Monnig took about half of all the world's meteor spectra.

Recent history abounds with similar stories. Most have never been written down and are preserved only in the fragile vessel of oral tradition, which has a lifespan usually limited to that of the people who were there.

But the enthusiasm of amateurs sometimes leads them astray. Take, for example, the case of M102. In 1781, Charles Messier prepared the last supplement to his list of nebulae and clusters and published it in the French almanac *Connaissance des temps*. As in earlier installments, Messier included several objects discovered by Méchain. These were listed as numbers 101, 102, and 103.

Unlike Méchain's previous objects, however, Messier did not observe these three himself. Furthermore, he lacked precise positions for M102 and M103; Méchain only gave positions relative to surrounding stars.

Soon after the publication of Messier's final supplement, Méchain wrote a letter that appeared in Bode's *Astronomisches Jahrbuch* for 1786. In it he stated that object 102 was actually a duplicate observation of 101, but due to an error on his star chart he originally thought it was a separate nebula. At the time it was considered no big deal — mistakes sometimes happen.

What turned this mistake into a big deal was that most astronomers remained unaware of Méchain's letter. Later editions of Messier's catalog failed to correct the error. Thus, M102 became one of the "missing Messier objects." As with M47, M48, and M91, its absence fascinated observers ever afterwards. Amateurs, especially, have delighted in trying to track them down, offering all sorts of explanations for why the positions in the original listing were wrong.

In his 1844 *Bedford Catalogue,* Smyth made a case for M102 being the galaxy NGC 5866. He argued that Méchain's original position for M102 — between Iota (ι) Draconis and Omicron (ο) Boötis — was too vague even for a skilled observer to find it. Instead, Smyth assumed that Méchain had misread his own handwriting and the star he meant was not Omicron but Theta (θ) Boötis (as you can see, omicrons and thetas can look somewhat similar). Smyth's "revised" position defined a much more precise location and pointed the way to NGC 5866, which is the brightest of five galaxies in the area.

Thus one wrong assumption coupled with immaculate logic produced a can of

Figure 6.3
The once mysterious 102nd entry in Messier's famous catalog is now known to be a duplicate observation of M101, a fantastic open-faced spiral galaxy.

worms. Smyth was one of the most respected amateurs of all time, so you can see why his proposal that M102 was really NGC 5866 gained stature. So much so that when the Canadian astronomer Helen Sawyer Hogg uncovered Méchain's letter in the 1940s, her "proof" that M102 does not exist met with amateur resistance. For whatever reason, in the hearts of many observers NGC 5866 remains as M102 even to this day. But there is little doubt that M102, as published by Messier, is really just M101 repeated with the wrong position (**Figure 6.3**).

NGC 5866 could well have been included in Messier's catalog. Méchain probably would have seen NGC 5866 had he happened upon it with his telescope. With a magnitude of about 10.5 it is brighter than some of the catalog's other entries, and it is easily recognized in a 3-inch telescope. On excellent nights I have seen its rather narrow, 3'-long shape through the 2-inch finder of my Clark refractor.

The growing popularity of Messier clubs, however, causes the following question to be asked more often: If an amateur aims at personally observing each of Messier's 100-odd nebulae and clusters, what should be done about M102?

The Deep-Sky Triangle

The night sky is replete with deep-sky wonders. "It's difficult to imagine anyone viewing every object within reach of a 6-inch telescope," Scotty said, "not to mention a 12-inch." One problem he noticed with beginners who wrote to him was that they often wondered where to start. Should they work like bird watchers who refer to life lists and methodically check off each item? One obvious reply would be for them to start out by viewing the 109 objects on the Messier list, then graduate to the catalog compiled by William and John Herschel. Others would suggest seeking out all the objects of a particular class. Do you like globular clusters? Well, you can obtain lists of all the globulars visible to a certain magnitude limit and start checking them off. The same goes for planetaries, galaxies, and double stars. Scotty, however, was not your mainstream correspondent. If you wrote asking for advice, his response would not be trite. Scotty was keen on devising new ways to explore the heavens, and he welcomed the challenge of blazing new trails for beginners. Thus was born, for lack of any "proper Scottyism," his Deep-Sky Triangle.

The memory of winter begins to ebb in June as mild but crisp nights complement the celestial riches now in the sky. Arcturus shines overhead, and Corona Borealis, the Northern Crown, is at its dainty best. Draco coils in pinpoint stars about the ecliptic pole, and the great globular star cluster M13 is climbing up the eastern sky. It doesn't matter if you use binoculars or a 20-inch telescope, there is so much to see that you wish for an impossible succession of crystal-clear nights — but where to begin?

One way to start your assault on the night sky is by selecting an area defined by several bright stars. With the aid of a good star atlas, we can then seek out all the objects in this area. For example, this month let's take the triangle of stars

Figure 6.4 Seeking out all the astronomical treasures within the Deep-Sky Triangle — formed by Eta (η) Ursae Majoris, Alpha (α) Canum Venaticorum, and Gamma (γ) Boötis — is a good way to start your "assault" on June's night sky. North is to the upper left.

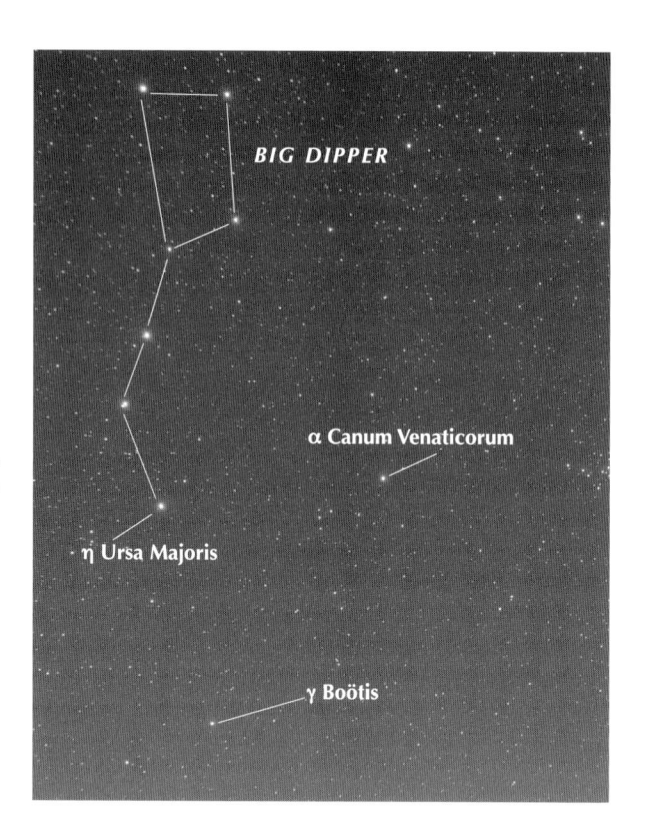

Figure 6.5 M51, the Whirlpool Galaxy, is popular among photographers and visual observers because its spiral structure is so easily resolved. The small companion galaxy is NGC 5195.

(**Figure 6.4**) formed by Eta (η) Ursae Majoris, Alpha (α) Canum Venaticorum, and Gamma (γ) Boötis, found hanging on the end of the Big Dipper's handle. A glance at Tirion's *Sky Atlas 2000.0* shows that at least a dozen galaxies lie within the triangle. Just inside and outside its western edge, several noteworthy examples are M51, M63, and M94, which can all be seen in binoculars under good conditions.

This part of the sky is far from the plane of the Milky Way so, as we might expect, there are no prominent star clusters nearby. M51, the Whirlpool Galaxy (**Figure 6.5**), lies just outside the triangle about a quarter of the way from Eta Ursae Majoris to Alpha Canum Venaticorum. Messier found the galaxy in October 1773, while observing a comet he had discovered. Eight years later Méchain noted that the galaxy was double. He saw the two components as clearly separated centers with halos touching one another. Today, nearly every amateur telescope will show M51 (with its companion NGC 5195). Even through smog and thin clouds a 6-inch will reveal the two components that such 19th-century observers as Smyth and Thomas W. Webb wrote so enthusiastically about.

The Whirlpool offers challenges for any telescope. For example, what is the smallest aperture required to reveal the spiral structure? Lord Rosse first detected spiral structure when he turned his giant 72-inch reflector on the galaxy in the spring of 1845. Today, with our vision sharpened by knowledge, the spiral features of M51 are visible in instruments as small as 10 inches, and some observers have glimpsed them with 6-inch telescopes in very dark skies. An 8-inch is sufficient for me, but John Mallas needed a 12½-inch in a dark desert sky. He correctly noted that experience and exceptional transparency are important for success. In 1936 I had a very good view of the spiral structure using the University of Arizona's 36-inch reflector in Tucson.

In clear, dark skies, an 8-inch at 40× will show the brightest portion of M51 to be about 11′ long and 8′ wide. Increasing the magnification to 200×, especially when the seeing is good, may reveal considerable structure in the disk, which at first glance appears rather uniform. Observers can also try to detect the "bridge" of light that connects the two galaxies. Again, I have seen it clearly with an 8-inch, but Mallas drew it as it appeared through only a 4-inch. Yet he recalled that perhaps he knew the field too well from photographs, and his observation may have been biased. Some amateurs have followed the entire bridge using a 17.5-inch in the clear desert skies of the American southwest. I have even seen traces of it here in Connecticut. The real key to viewing it is sky transparency, since the slightest moonlight or light pollution renders it invisible.

A third challenge was proposed by Ronald Morales of Tucson, who observed from the Santa Rita Mountains southeast of the city. Using a 10-inch f/5.5 reflector with a 16-mm König eyepiece, he steadily held with direct vision a faint star superimposed on the southwest quadrant of M51's disk. I wonder how small a telescope will reveal this star. I failed to see it with a 4-inch in good skies. However, I suspected the bridge, which apparently was not seen by Morales on the night he noted the star. West of M51's center, and superimposed on the galaxy's disk, is a foreground star that is occasionally mistaken for a supernova.

About ½° south of M51 is a much more difficult object, the small elliptical galaxy NGC 5198. It is round, 2′ in diameter, and I estimate it as magnitude 12.2. High magnification is useful when searching for this small object, as it can look rather starlike at low powers.

Moving southward, the bright galaxy M63 is about two-thirds of the way from Eta Ursae Majoris to Alpha Canum Venaticorum, and as far inside the triangle as M51 is outside. Méchain discovered it in 1779 (**Figure 6.6**). At magnitude 8.6 it is apparent in any telescope, and I have seen it well in a 2-inch refractor. On another occasion, while using a 29-inch, I caught sight of the faint spiral arms surrounding M63's bright core.

Figure 6.6
M63 is a rare type of galaxy whose many spiral arms bear a rash of star clouds. Its dappled appearance gave rise to its nickname, the Sunflower Galaxy.

Alpha Canum Venaticorum is popularly named Cor Caroli, the heart of Charles. Legend attributes the name to Edmond Halley, who was honoring his benefactor, England's King Charles II. There is less than solid proof of this, however, and some question remains as to whether the honor was intended for Charles I. Sweeping just over 1° southeast of Cor Caroli brings us to a pair of faint galaxies. NGC 4914, at 12th magnitude, is brighter and larger than NGC 4868. Both should be within range of an 8-inch, given a dark enough sky.

NGC 5005 is located about 1½° farther to the east-southeast. At magnitude 9.8 and 5′ long, this slender spindle of light is better known than either NGC 4914 or NGC 4868. Look for another galaxy, NGC 5033, roughly 1° southeast of NGC 5005. It would be especially nice to observe this object with an aperture of 20 or more inches. I suspect that some internal detail is visible, though none is apparent in an 8-inch telescope.

Near the midpoint of the triangle's southern edge is a loose swarm of galaxies. We can begin with a group of three that lie in a north-south line within the same low-power field of view. Even casual sweeping of the area should turn them up. The largest (nearly 3' in diameter) is the 11.4-magnitude spiral NGC 5350. It shows up clearly at 20× in my 5-inch Apogee telescope, and I once saw it at the annual Stellafane convention in Vermont with a 10-year-old girl's home-made 4-inch off-axis reflector.

Just 5' south of NGC 5350 — roughly half the distance between the famous double star Mizar and Alcor in the handle of the Big Dipper — lies a pair of galaxies almost in contact with each other. The southern one is the 11.1-magnitude elliptical NGC 5353, which is cataloged as a bit over 2' in diameter. Just 1' to its north is NGC 5354, about the same size and perhaps half a magnitude fainter. Like NGC 5350, it is a spiral.

About ½° to the east of this triplet is the spiral NGC 5371. At magnitude 10.8 and some 4' in diameter, it should be an easy object in just about any telescope. An 8-inch telescope will show hints of internal detail, while my notes on the galaxy tell of one observer, using a 36-inch, who could see faint arms and a prominent central bar. My 8-inch reflector shows what is either a bright stellar nucleus or a foreground star superimposed on the galaxy's center.

NGC 5371 is the subject of an interesting and instructive tale. The galaxy was discovered by William Herschel, who called it "pretty bright, large, round, brighter middle with a faint nucleus." It was also observed by his son John in May of 1831. John recorded no description of the object on this date, but during one of his sky sweeps two months earlier he "discovered" NGC 5390 located 20' east of NGC 5371. (For the sake of clarity I'm using NGC numbers, but these were actually added later. The Herschels used other designations.) John recorded the new object as "faint, large, very gradually brighter middle, with a 9th-magnitude star to the northeast." He does not mention seeing NGC 5371 on the night he observed NGC 5390.

Today we know that there is no object at the position John gave for NGC 5390. I've even checked photographs of the area made with the Palomar 48-inch Schmidt telescope. A quick review of modern references, including Vol. 2 of *Sky Catalogue 2000.0*, reveals that NGC 5390 is a duplicate observation of NGC 5371, but I don't know who first pointed this out.

There is little doubt that it was a duplicate observation. First, John didn't see NGC 5371 on the night he "discovered" NGC 5390. Furthermore, the real NGC 5371 has a star just to its northeast that matches John's description of NGC 5390. A slight error in recording the object's position would make it seem like a discovery. But why did John call the galaxy "faint" and his father say it was "pretty bright," when both were using the same telescope with an 18.7-inch speculum-metal mirror? Perhaps the mirror's surface was tarnished and in need of polishing on the night John observed the galaxy.

All this goes to show that the descriptions in the *New General Catalogue of Nebulae and Clusters of Stars* (*NGC*) are a rough approximation of what we can expect to see in a telescope. They were never intended to be more than that, in any case.

The Wonder of M106

Aside from the "missing" Messier objects (M47, M48, and M91), the most intriguing aspect of Messier's catalog — one that arouses amateur interest, as Scotty believed — is the origin of its final seven entries. Messier's original catalog, published in installments from 1774 to 1787, listed only 103 deep-sky objects, including one duplicate observation (M102), a double star (M40) and an asterism (M73). Indeed, some observers resist the so-called "final seven" and argue that the true Messier catalog contains only 103 deep-sky objects. Earlier this century it was proposed and widely accepted that seven more objects be added to the Messier catalog for various and sundry reasons. "Of these additions," Scotty proffered, "the galaxy M106 in Canes Venatici is particularly interesting."

Charles Messier discovered a comet drifting through Ophiuchus on September 27, 1793. As he had done in the past, the French astronomer called upon his mathematician friend Jean-Baptiste-Gaspard Bochart de Saron to calculate its orbit. Readers with a bent for astronomical history may recall the unusual circumstances surrounding this event. Bochart de Saron had been imprisoned during the French Revolution's Reign of Terror and was awaiting execution when he received word of Messier's new comet. Nevertheless, he worked out an orbit, which Messier used to recover the comet after its conjunction with the Sun. Messier was able to slip his friend word of the success before the guillotine's blade fell on April 20, 1794.

I wonder if other aspects of the Reign of Terror have trickled down through history to affect today's amateur astronomers. Certainly Messier's work must have been affected during the Revolution. Had it not been, he might have caught and corrected some of the errors that appeared in his original catalog — errors that continue to fuel star-party debates as amateurs speculate on the true identity of several "missing" Messier objects.

Another aspect of Messier's catalog that arouses amateur interest is the final seven entries. An original paper by H. S. Hogg, in which she describes five of these seven objects, and which deserves to be read by every amateur as a lesson in industry, appears in the Journal of the Royal Astronomical Society of Canada, September 1947. Of these additions, the galaxy M106 in Canes Venatici is particularly interesting (**Figure 6.7**). Descriptions of its visual appearance vary considerably. Photographs are of little help in resolving the discrepancies since the difference between them can vary almost as much. (Even prints made from the same negative will look different depending on how they are exposed and developed.)

Hans Vehrenberg's *Atlas of Deep-Sky Splendors* includes a shot of M106 made with the 200-inch Hale telescope at the Palomar Observatory, as well as a

Figure 6.7
The peculiar galaxy M106 in Canes Venatici has had a violent past. Its arms are scarred with material blown out of the galaxy's nucleus during vast interstellar storms.

wide-field view made with a 12-inch Schmidt camera. The 200-inch photograph gives the impression of a barred spiral seen almost edgewise, but in the image made with the smaller telescope M106 looks more like a regular spiral. The truth probably lies somewhere in between, since M106 is classified as a peculiar Sb spiral in *Sky Catalogue 2000.0*.

Ronald Morales viewed M106 with his 10-inch Newtonian reflector. Using 87× he described it as "extremely large; very bright with a bright, compact center; extended in a north-south direction with a large, fuzzy outer envelope." Years ago in Kansas I viewed the galaxy with a 10-inch reflector at about the same magnification and saw a "very bright parallelogram shape with fragile spiral arms at the ends of the major axis." The nucleus appeared uniform with little variation in brightness. Other observers using 8-inch telescopes have reported M106's appearance as long and needle-like, and one saw a dark area near the nucleus. So much for consistency!

Forgotten Corridors

Say the words "Virgo" or "Coma Berenices" to a seasoned observer and a knowing look will come over his or her face. If there is one area of the heavens beloved by galaxy hunters and hated by comet hunters (because it's so hard to spot a comet against all the other "faint fuzzies"), it is the great Coma-Virgo cluster of galaxies. "Amateurs today tend to acquire larger and larger telescopes," Scotty explained, "and the constellations of Coma Berenices and Virgo offer challenges that take us beyond merely re-observing well-known objects. Since the North Galactic Pole is nearby, there is very little interstellar dust in this direction, giving observers a clear window to the universe." But tackling this region for the first time can be a daunting task. Star-hopping is an acquired skill, one best practiced in a less galaxy-polluted region of sky. Naturally, Scotty was not at a loss for a better way. Here he introduces a fun and pragmatic technique practiced in the late 1700s, but largely forgotten or avoided today (because of the necessary time commitment). There are also several underdog objects among the opulence of galaxies. For instance, Scotty reveals a neglected corridor of globular clusters in the region. Another object, the largest and brightest in this vast area of sky, is obviously itself one of the sky's most uncelebrated hidden treasures.

What open cluster is visible to the naked eye, has no Messier or NGC number, and was called "gossamers spangled with dewdrops" by the 19th-century astronomy popularizer Garret P. Serviss? Almost every skygazer has seen this group at one time or another, and its official designations are Melotte 111 and Collinder 256. The answer will surprise many, for the cluster is the shimmering haze of 5th- and 6th-magnitude stars we call Coma Berenices. It is a real cluster and not just a chance alignment of stars. There are about 80 members scattered across 5° of sky. A camera with high-speed film can record the cluster with an exposure of just a few seconds. Coma Berenices is only 260 light-years away and is one of the nearest open clusters. Therefore its stars appear well separated. If they are a bit too faint for your naked eye, a simple 2× or 3× opera glass gives a wonderful view.

Of course to observers of intergalactic depths, this region of sky belongs to the great clouds of spirals that dominate Virgo, Coma Berenices, and Canes Venatici. Less realized, perhaps, is that in these same regions we get the first onrushers of the globular clusters which finally swarm in their greatest profusion around the galactic center in Scorpius/Ophiuchus/Sagittarius. Three globulars are within the narrow limits of Coma itself, including two that lie near Alpha. NGC 5024 (M53) is 14.4′ in diameter with a magnitude of 8.7. Unimpressive in a 3-inch refractor, its star sprinklings are magnificent in a 12½-inch reflector, where faint streams of curving stars run out from the central blaze in all directions. Smyth called it an "interesting ball of innumerable worlds."

Almost in the same field is NGC 5053, diameter 8.9′, magnitude 10.9. In large instruments it is a little gem of woven fairy fire. Hogg's *Bibliography of Individual*

Globular Clusters shows that it was first observed by William Herschel on March 14, 1784. It is remarkable for its position in space, perhaps 50,000 light-years above the galactic plane.

Considerably to the west of these two lies NGC 4147, magnitude 11. Its faintness misled Herschel, on the same date as that he first observed NGC 5053, into listing it as something external, and only with the great reflectors of the 20th century could its individual stars be seen well and studied. We now know of several variable stars in this faint object.

In the same field lies a true faint external galaxy, NGC 4153, about 13′ south and about 8′ east of NGC 4147. It is not listed in the *Shapley-Ames Catalog,* and hence must be fainter than 13th magnitude. It would be a real feat for an amateur telescope to locate it. I have examined NGC 4147 on half a dozen occasions without noticing the faint galaxy.

In the western part of Virgo, galaxies are crowded like a confusion of silver sands. But eastward from the middle of the constellation, deep-sky objects are more scattered, and amateurs may have difficulty in locating some of them because easily identifiable stars are few.

There is a strip of sky here near declination +02° where several galaxies and a beautiful globular cluster can be readily located by means of a technique that dates back to William Herschel (**Figure 6.8**). The procedure is simple: set your

Figure 6.8
Herschel's 40-foot telescope was a veritable cosmic workhorse, which he used to plow up new deep-sky wonders.

JUNE

telescope on a prearranged starting point, leave it stationary, and watch celestial objects drift through the field according to a timetable.

For this purpose, use a low-power eyepiece with a field not much less than 1° across. To check the field size of an eyepiece, time the drift of an equatorial star centrally across it, and count one minute of arc for every four seconds of time. Once that's completed select a star lying west of the desired galaxy, but having nearly the same declination. The telescope is then left stationary, allowing diurnal motion to carry the object into the center of the field. The drift time required is the same as the difference in right ascension between the star and the galaxy.

Figure 6.9
You can find several galaxies and a globular cluster easily by letting the sky drift over your stationary telescope. Start your scan at the small galaxy NGC 5746.

Begin by picking up the elongated spiral galaxy NGC 5746, about ⅓° west of the 4th-magnitude star 109 Virginis. It is a 10th-magnitude spindle about 6' × 1', less in small telescopes. A more difficult spiral in the same field, NGC 5740, lies about ½° to the southwest of NGC 5746. It is about magnitude 12, and 2' × 1' across.

NGC 5746 is the starting point for our strip (**Figure 6.9**). Center it in the field, clamp the instrument, and be careful not to shift it during the remainder of the experiment. Then relax and watch the stars glide by. Keep track of the time that has elapsed since the starting object was in the middle of the field.

Consult a schedule of times at which some stars and deep-sky objects will pass the midline of the eyepiece field. After a wait of about 13 minutes, the first of a group of galaxies near 110 Virginis will appear at the edge of the field. Be careful not to let the light of this star dazzle your eye when it comes into view about three minutes later.

The brightest of this little group is NGC 5846, a compact elliptical galaxy. Nearly overlapping it is a much fainter but somewhat larger companion, NGC 5850. On an excellent night in 1935, I found it was easy in a 4-inch.

After another gap of about 10 minutes the great globular cluster M5 swims into view (**Figure 6.10**). About ⅕° in diameter, it is one of the better globulars

Figure 6.10
End your scan at M5 — a stunning globular cluster — which seems to float above the stars of Serpens. In between NGC 5746 and M5 lie several faint galaxies.

for small telescopes, because it actually gives the impression of being a cluster rather than an amorphous glow.

The strip method of observing is a handy way of picking up an inaccessible object without using the setting circles on your telescope. From a star atlas select an easily identifiable star having nearly the same declination to the west, and find from their differences in right ascension how long to wait. Also, the strip technique can be used to explore rich Milky Way regions this summer. You can obtain the approximate position of any object of interest by noting how long an interval of time it follows the starting star, and by estimating how far north or south of the center of the field it passes. By the way, on an exceptionally good night, can you see M5 with the naked eye, just north of 5 Serpentis?

Cup and Crow

One point Scotty often repeated in his columns was that when searching for faint objects, it is beneficial, if not imperative, to have a set of star charts with a magnitude limit of at least 9. "On such charts," he said, "there will almost always be one or two stars that appear in the same low-power field as the object of interest. This will help not only in locating but also in identifying objects . . . misidentifying deep-sky objects is a sin common to both novice and experienced observers alike." As a case in point, he referred to a small group of galaxies near the border between Corvus and Crater. One of them, NGC 4038 (the famous Ring-tail Galaxy), he said, deserves special attention, because "several times amateurs have sent descriptions of what they believe is this galaxy, but I'm sure they have mistaken another galaxy for the Ring-tail." The clarification seemed particularly important to Scotty because he had an affinity for this region of sky; it took him almost 60 years to learn to appreciate Crater and Corvus, and he wanted to be sure that his readers got the full benefit of their efforts as they tipped their telescope tubes toward these southern celestial wonders.

There is never a shortage of deep-sky objects. Whatever the season, the sky holds more than enough of these delights to keep you busy all night, every night — if you take the time to search them out with good charts and reference books. Many deep-sky objects have special features to tantalize you. In the June evening sky, almost every direction we look offers something interesting. But near the meridian this month are two constellations that took me almost 60 years to learn to appreciate. Corvus, the Crow, and Crater, the Cup, seem like insignificant asterisms when viewed from the United States. I always found them delicate, with stars that had to be picked out from the background. They are not at all like Cassiopeia, Cygnus, or Delphinus, which almost force themselves on you. My feelings changed when I first saw Corvus and Crater from Central America.

Raise the Glass to the Cup . . .

William Tyler Olcott's classic *A Field Book of the Stars* first made its appearance about the time Halley's Comet swept through the inner solar system in 1910. Olcott's outlines of the constellations were the only ones several generations of young astronomers knew, and many of his patterns are still accepted today. I especially like Crater, the Cup, perched precariously on Hydra and tilted as if to pour a celestial libation of stardust.

As a youth, I found Crater a lucid grouping of stars that was a delight to explore with a 1-inch, 40× refractor. Today, however, my mail indicates that few amateurs pay much attention to this area of the sky. While many of the deep-sky objects here should be better known to amateurs, there are no spectacular ones to draw observers into the area. The light pollution surrounding urban centers leaves this faint constellation little admired.

Burnham's *Celestial Handbook* lists only 15 deep-sky objects down to magnitude 13 in Crater. All but one are from the *New General Catalogue of Nebulae and Clusters of Stars*. There are many fainter *NGC* objects in this part of the sky, as well as a number of others listed in the two *Index Catalogues*. Novice observers may find all this a bit confusing. Several years ago, when a 12-inch was considered a really large amateur telescope, deep-sky objects of magnitude 13 seemed like a realistic limit for most observers. But now, with 17-inch and larger telescopes commonplace, many fainter galaxies that are not to be found in popular observing guides are accessible to amateurs.

A large telescope is not necessary to enjoy the pleasures of Crater, where the paucity of naked-eye stars makes good charts quite necessary; it is almost always indispensable to have an atlas showing stars to at least 9th magnitude.

Figure 6.11 Despite its southerly location, the egg-shaped galaxy NGC 3887 can be spotted in a 4-inch telescope near Zeta (ζ) Crateris.

There are perhaps a score of objects within the reach of an 8-inch under good skies. NGC 3887, an 11th-magnitude spiral galaxy (**Figure 6.11**) not far from 5th-magnitude Zeta (ζ) Crateris, was discovered by William Herschel. Its oval disk is about 2.5′ long and rather easy to find in a region devoid of faint stars. In the mid-1950s the galaxy showed well in a 4-inch f/16 off-axis reflector. In the transparent sky of Kansas, a 10-inch f/8.6 reflector at 60× revealed NGC 3887 to be quite bright and egg-shaped. My old notes also suggest that a trace of the spiral arms was also visible in this instrument. The galaxy is easily picked up in my 5-inch Apogee telescope, but at 20× it is a blur just barely distinguishable from a star.

NGC 3672 is another spiral galaxy discovered by the senior Herschel. It is a little brighter than 12th magnitude, and some 3.5′ long and half as wide. I viewed this galaxy from Madison, Wisconsin, with the 6-inch Clark refractor that once belonged to the famous double-star observer Sherburne W. Burnham. My then novice eye required 150× for the observation. One of this telescope's eyepieces was most unusual for its time; it was of 1½ inches focal length, made by John Brashear, and possibly designed by Charles Hastings. It had a wide apparent field, and gave such unmatched views of deep-sky objects that it played a major role in my growing interest in this branch of observing.

NGC 3637 is a tiny spiral only about 1′ in diameter. With low powers this object will almost always be mistaken for a star. I suggest searching for it with at least 100×. There is a 7th-magnitude star just 3′ southwest of the galaxy. Similar in shape and difficulty is the slightly smaller elliptical galaxy NGC 3732. Again, high magnifications will be a help when searching for this object.

More challenging is the faint spiral galaxy NGC 3865. It was missed by the Herschels, and, not surprisingly, I have never been able to find it with my 4-inch Clark refractor. I have seen it, however, with the 20-inch Clark at Wesleyan University. Its oval disk is about 1.5′ long and of magnitude 13.5 or perhaps a bit fainter.

If you are successful in locating NGC 3865 and want to try something even more challenging, try hunting down the twin galaxies NGC 3634 and NGC 3635, which are nearly in contact with each other. They are almost stellar in appearance and around 15th magnitude. If the sky is perfectly clear and dark, they may be within the reach of a 10-inch telescope at high magnification. A 16-inch should show these galaxies well above the visual threshold.

. . . Then Sail on to Corvus

Many amateurs living at mid-northern latitudes consider the southern constellation Corvus, the Crow, dim and uninspiring. But when I first saw it from Central America I found it far more impressive than Hercules. I often wonder why there is no indication of this star pattern in the Mayan Codex (of Mesoamerican astronomical tables); then again, maybe there is and scholars just haven't recognized it yet. Its four prominent 3rd-magnitude stars, however, do not remind me of a crow. I grew up on the shores of Lake Michigan where sailboats were part of my very essence, and I can see Corvus only as the sail of a small gaff-rigged vessel heading eastward on the back of Hydra.

The first telescopic object I usually turn to is the small but really fine planetary nebula NGC 4361, located in the north central part of the "sail" (**Figure 6.12**). Curiously this object was not included by Smyth in his *Cycle of Celestial Objects,* and William Herschel thought that it was resolvable into stars. This description is all the more strange because even in my 4-inch refractor NGC 4361 looks like a planetary, and in a 12-inch its character is unmistakable. It is not a typical planetary, however, for Lick Observatory photographs show a central nucleus with arms superficially resembling a barred spiral. If NGC 4361 were higher in the northern sky, I can't help but think that it would have ended up in Messier's catalog and thus be considered among the anointed by today's amateurs.

It is wise to use at least 80× to 100× on it. The total visual magnitude is about 10, so the surface brightness is about four times less than M57, the Ring Nebula in Lyra. Its disk appears larger than Jupiter. Its bright, inner core is about ¾′ across, but the outer halo approaches 2′. Careful observers have spotted it with binoculars. The 4-inch Clark refractor at 40× shows this planetary as a circular object filled with mottled light. At 100× the mottling was pronounced and the central star, reported to be of 13th magnitude, was not seen with my left eye. My right eye, however, which had its ultraviolet absorbing lens removed in a cataract operation, showed the star to be at least a magnitude above the visual limit. (This, by the way, is exactly the same experience I had with the central star in the Ring Nebula.)

The hot star at the center of NGC 4361, whose ultraviolet radiation excites the nebula's thin shell of gas to glow, had proved a challenge for me before my cataract operation. I had failed to see it with my 4-inch Clark refractor — a telescope that routinely reaches stars of 15th magnitude. I attribute this to the lower

Figure 6.12
The planetary nebula NGC 4361 in Corvus looks almost like a galaxy — with what looks like a central nucleus and arms like those in a barred spiral.

contrast between the central star and the nebula's light than if the star lay against a dark background. The planetary's nebulous haze reduces contrast between the star and the background, and maximum contrast is essential to achieve optimum performance from a telescope.

Ronald Morales — an experienced observer of planetary nebulae — examined NGC 4361 with an 8-inch f/5 reflector and a range of eyepieces. With a 10.2-mm ocular (100×) the object appeared "bright, round, diffuse with irregular edges, and the central star easily seen." A 16-mm eyepiece (64×) also showed the star, but a 25-mm (41×) revealed the planetary to be a "fuzzy patch with the star only suspected." It is worth remembering that deep-sky objects can often appear quite different when the same telescope is used at various magnifications. The fact that the seemingly small change from 41× to 61× would make the difference in the central star's visibility should serve as a vivid reminder.

When observing faint objects such as NGC 4361, take full advantage of dark adaptation and averted vision. Spending half an hour or more in darkness can add surprisingly to the visibility of faint detail.

Figure 6.13
A firestorm of starbirth activity was triggered by the collision of these two galaxies — NGC 4038 and 4039 — popularly known as the Ring-tail Galaxy or the Antennae.

NGC 4038, with its companion NGC 4039, is popularly known as the Ring-tail Galaxy or the Antennae (**Figure 6.13**). Located along the western edge of Corvus, it is listed as a peculiar galaxy, and that alone makes it a tempting target for amateur and professional astronomers alike. The unusual shape is believed to result from the collision of two galaxies, a theory borne out by computer simulations. Though Burnham devotes considerable space to this object in his *Celestial Handbook*, the visual observer will see only an asymmetrical blur, about 2½′ in diameter and 11th magnitude. NGC 4038-39 is shaped like an apostrophe in my old 10-inch reflector. Photographs made with large telescopes show two separate central masses each with a long curving tail. This is a unique sight that would be a good target at a star party, where it could be viewed in several different telescopes.

Several times amateurs have sent descriptions of what they believe is this galaxy, but I'm sure they have mistaken another galaxy for the Ring-tail. My 5-inch, 20× Apogee refractor shows the pair as a bright blob. An observation made with my 4-inch Clark refractor under the indifferent skies of my old home in Haddam, Connecticut, revealed NGC 4038-39 to be little more than an asymmetrical 11th-magnitude blur. However, at a campsite near Big Sur, California, I viewed a wealth of detail in the Ring-tail with a borrowed 12-inch reflector. Other reports in my files support this. At Omaha, Nebraska, Frank Rolwicz's 10-inch Newtonian easily showed the galaxy's "tail." And viewing with a 10-inch reflector, Morales thought that the galaxies looked like a shrimp.

What they remind me of, however, is heavily inspired by my interests while in grade school. Even before I discovered telescopes, I was crazy about microscopes. I would talk my science teacher into letting me borrow the school's microscope on holidays and weekends, and would bicycle to lakes and ponds near my home in Wisconsin to examine water samples. One creature I found in abundance was the water flea, Daphnia. There was a time when I could instantly identify a dozen species with just a glance in the microscope. So it's understandable why I associated NGC 4038-39's photographic appearance with the grandmother of all water fleas.

About 40' southwest of the Ring-tail is a galaxy of similar size, NGC 4027. I make its visual magnitude to be 11.3, about half a magnitude fainter than NGC 4038. Most telescopes will show both objects in the same low-power field. This spiral is about 2' in diameter and I have seen it in my 4-inch and in the Moonwatch 5-inch Apogee telescope. The background is so uncluttered in this area that faint nebulae are definitely easier to identify than in most other parts of the sky.

A pair of small galaxies in eastern Corvus is probably too faint for a 4-inch telescope, but should be within the grasp of an 8-inch. NGC 4782 and 4783 are almost touching each other. These twin elliptical galaxies are 0.5' across with photographic magnitudes of 12.9, but are a bit brighter visually. They lie about $\frac{1}{2}°$ due north of an 8th-magnitude star, and I suggest searching for them with a magnification of about 100×.

Two more galaxies are nearby. They are very faint, and I suspect they require at least a 12-inch telescope. Some 9' east of the last pair, NGC 4794 is about 14th magnitude and slightly brighter than NGC 4792 to its northwest. Experienced astrophotographers might find these galaxies a challenge. If so, I am sure your results will be worth the effort.

Before leaving the Corvus region, try examining M104, just over the northern border in Virgo. Recall the earlier discussion of Messier's original catalog of star clusters and nebulae, published in *Connaissance des temps*. It had a total of 103 entries, but Messier's personal notes made it clear that he knew of other objects. Through the efforts of 20th-century astronomers who studied the catalog in detail, these objects have been included as entries 104 to 110.

It was the French science popularizer Camille Flammarion who got the ball rolling when, in 1921, he proposed that NGC 4594 be included as M104 because Messier had penned a reference to the "nebula" in his own working copy of his catalog.

Figure 6.14
The Sombrero is one of the most easily recognized extragalactic deep-sky treasures. It lies about 65 million light-years away and is part of the great Virgo Cluster of Galaxies.

Often called the Sombrero Galaxy, M104 is an easy object for even the smallest telescope (**Figure 6.14**). Its bright, 8th-magnitude oval disk is about 8′ long and well within the reach of binoculars. The nearly edge-on galaxy is noted for a dark band slicing across its middle. In their *Observing Handbook and Catalogue of Deep-Sky Objects,* Christian Luginbuhl and Brian Skiff note that "the dark lane can be seen easily" in a 6-inch telescope. On the other hand, John Mallas, the skilled deep-sky observer and coauthor of *The Messier Album,* could not see the lane with a 4-inch refractor.

But on one excellent night back home in Connecticut I saw the lane with a 4-inch Byrne refractor and a 4-inch off-axis reflector. To my surprise the lane was more difficult in the refractor, even though refractors usually scatter less light than reflectors. Scattered light degrades the view by reducing image contrast. When I placed a dark cloth over my head to block light entering my eye from the side, M104's dark lane was definitely easier to see.

Sometimes it's easy to forget just how much the atmospheric debris, which becomes more concentrated as we look near the horizon, affects our view of the sky. This is true whether we use the naked eye or a telescope. For me, many of these southern objects would have much less appeal were it not for my occasional excursions into Central America, where they stand 20° higher than at home. I suspect that whenever eclipse chasers travel to southern locations, they find viewing the night sky just as rewarding as the solar three-ring circus itself.

JUNE OBJECTS

Name	Type	Const.	R. A. h m	Dec. ° ′	Millennium Star Atlas	Uranometria 2000.0	Sky Atlas 2000.0
Collinder 256, Melotte 111	OC	Com	12 25	+26 00	654, 655, 677, 678	107, 108 148, 149	6, 7
M5, NGC 5904	GC	Ser	15 18.6	+02 05	765	244	14
M51, NGC 5194	Gx	CVn	13 29.9	+47 12	589	76	7
M53, NGC 5024	GC	Com	13 12.9	+18 10	699	150, 195	7, 14
M63, NGC 5055	Gx	CVn	13 15.8	+42 02	609	75, 76	7
M94, NGC 4736	Gx	CVn	12 50.9	+41 07	610	75	7
M101, NGC 5457	Gx	UMa	14 03.2	+54 21	570, 571	49	2, 7
M106, NGC 4258	Gx	CVn	12 19.0	+47 18	592	74, 75	6, 7
M108, NGC 3556	Gx	UMa	11 11.5	+55 40	576	46	2, 6
M109, NGC 3992	Gx	UMa	11 57.6	+53 23	574, 575	47	2, 6
NGC 3610	Gx	UMa	11 18.4	+58 47	561	46	2
NGC 3613	Gx	UMa	11 18.6	+58 00	561	46	2
NGC 3619	Gx	UMa	11 19.4	+57 46	561	46	2
NGC 3634	Gx	Crt	11 20.6	−09 01	800, 801, 824, 825	281	—
NGC 3635	Gx	Crt	11 20.6	−09 01	—	281	—
NGC 3637	Gx	Crt	11 20.7	−10 16	824, 825	281	—
NGC 3642	Gx	UMa	11 22.3	+59 05	560, 561	46	2
NGC 3672	Gx	Crt	11 25.0	−09 48	824	281, 282	13
NGC 3683	Gx	UMa	11 27.5	+56 53	560, 561, 575, 576	46	2
NGC 3690	Gx	UMa	11 28.5	+58 33	560, 561	46, 47	2
NGC 3732	Gx	Crt	11 34.2	−09 51	824	282	13
NGC 3738	Gx	UMa	11 35.8	+54 31	575	47	2
NGC 3756	Gx	UMa	11 36.8	+54 18	575	47	2
NGC 3865	Gx	Crt	11 44.9	−09 14	799, 823	282	13
NGC 3887	Gx	Crt	11 47.1	−16 51	847	282, 327	13, 20
NGC 3894	Gx	UMa	11 48.8	+59 25	560	47	2
NGC 3898	Gx	UMa	11 49.2	+56 05	575	47	2
NGC 3945	Gx	UMa	11 53.2	+60 41	560	25, 47	2
NGC 3982	Gx	UMa	11 56.5	+55 08	574, 575	47	2

Ast = Asterism; BN = Bright Nebula; CGx = Cluster of Galaxies; DN = Dark Nebula; GC = Globular Cluster; Gx = Galaxy;
OC = Open Cluster; PN = Planetary Nebula; ✷ = Star; ✷✷ = Double/Multiple Star; Var = Variable Star

JUNE OBJECTS (CONTINUED)

Name	Type	Const.	R. A. h m	Dec. ° ′	Millennium Star Atlas	Uranometria 2000.0	Sky Atlas 2000.0
NGC 4027	Gx	Crv	11 59.5	−19 16	846, 847	327, 328	13, 21
NGC 4036	Gx	UMa	12 01.4	+61 54	559, 560	25, 47	2
NGC 4041	Gx	UMa	12 02.2	+62 08	559, 560	25	2
NGC 4147	GC	Com	12 10.1	+18 33	702	148	7, 14
NGC 4153	Gx	Com	12 10.8	+18 22	—	—	—
NGC 4217	Gx	CVn	12 15.8	+47 06	592	74	2, 6
NGC 4361	PN	Crv	12 24.5	−18 48	845	328	13, 21
NGC 4605	Gx	UMa	12 40.0	+61 37	558	25, 48	2
NGC 4782	Gx	Crv	12 54.6	−12 34	820	284	14
NGC 4783	Gx	Crv	12 54.6	−12 33	820	284	14
NGC 4792	Gx	Crv	12 55.1	−12 30	820	284	—
NGC 4794	Gx	Crv	12 55.2	−12 37	820	284	—
NGC 4814	Gx	UMa	12 55.4	+58 21	558	48	2
NGC 4868	Gx	CVn	12 59.1	+37 19	631	108, 109	7
NGC 4914	Gx	CVn	13 00.7	+37 19	631	108, 109	7
NGC 5005	Gx	CVn	13 10.9	+37 03	630	109	7
NGC 5033	Gx	CVn	13 13.4	+36 36	630	109	7
NGC 5053	GC	Com	13 16.4	+17 42	699	150, 195	7, 14
NGC 5195	Gx	CVn	13 30.0	+47 16	589	76	7
NGC 5198	Gx	CVn	13 30.2	+46 40	589	76	7
NGC 5350	Gx	CVn	13 53.4	+40 22	608	76	7
NGC 5353	Gx	CVn	13 53.5	+40 17	608	76	7
NGC 5354	Gx	CVn	13 53.5	+40 18	608	76	7
NGC 5371	Gx	CVn	13 55.7	+40 28	608	76	7
NGC 5740	Gx	Vir	14 44.4	+01 41	766	243	14
NGC 5746	Gx	Vir	14 44.9	+01 57	766	243	14
NGC 5846	Gx	Vir	15 06.4	+01 36	765	243	14
NGC 5850	Gx	Vir	15 07.1	+01 33	765	243	14
NGC 5866	Gx	Dra	15 06.5	+55 46	568	50	2, 7
Owl Nebula, M97, NGC 3587	PN	UMa	11 14.8	+55 01	576	46	2, 6

Ast = Asterism; BN = Bright Nebula; CGx = Cluster of Galaxies; DN = Dark Nebula; GC = Globular Cluster; Gx = Galaxy;
OC = Open Cluster; PN = Planetary Nebula; ✶ = Star; ✶✶ = Double/Multiple Star; Var = Variable Star

JUNE OBJECTS (CONTINUED)

Name	Type	Const.	R. A. h m	Dec. ° ′	Millennium Star Atlas	Uranometria 2000.0	Sky Atlas 2000.0
Ring-tail Galaxy, NGC 4038–39	Gx	Crv	12 01.9	–18 52	846	327, 328	13, 21
Sombrero Galaxy, M104, NGC 4594	Gx	Vir	12 40.0	–11 37	820, 821	284	14

Ast = Asterism; BN = Bright Nebula; CGx = Cluster of Galaxies; DN = Dark Nebula; GC = Globular Cluster; Gx = Galaxy; OC = Open Cluster; PN = Planetary Nebula; ✶ = Star; ✶✶ = Double/Multiple Star; Var = Variable Star

JULY

CHAPTER 7

Peering into the Cat's Eye

"Skies around the north ecliptic pole in Draco are rather barren," Scotty once wrote, "but a few interesting objects wait for the zealous observer." Take the bright planetary nebula NGC 6543, for example. Scotty believed that pale green planetary nebulae of its caliber were sometimes neglected by amateurs. It's easy to understand why. Some are so small they could be easily mistaken for stars. "When searching for one of these emeralds," Scotty advised, "the observer must first locate the field very carefully, and may have to examine many stars at higher power to discern the nebula's tiny disk." The Cat's Eye is different, however. Through a telescope it appears as a small bluish disk about the angular size of Jupiter, and looks like an out-of-focus 9th-magnitude star. It does offer special challenges to observers with all manner of observing aids. But, as Scotty attests, a glimpse of it through a large-aperture telescope can turn the purest scientist into a hopeless romantic.

"You are all poets," I told a gathering of amateur astronomers at the 1983 annual Texas Star Party. At first they reacted with silence. Then they began to agree. The common thread that binds amateurs together is a love of the grandeur and beauty of the starry deeps. While some may claim it's the science of astronomy that interests them, I believe that deep down it is the ultimate experience of the night sky that holds the real attraction.

After my talk I received a letter from Christine Combs of Colleyville, Texas. "I was introduced to NGC 6543 at the Texas Star Party," she wrote. "The first sight was breathtaking . . . to see the bright inner star surrounded by greenish-blue nebulosity . . . we observe with a 20-inch Dobsonian, usually with 200×. Last night I experienced anew the excitement of seeing this greenish cat eye buried deep in the sky for me to wonder at. My husband, an engineer and long-time amateur, may give a more detailed description, but I'm the poet and always strive for words to describe the feeling aroused by these deep-sky wonders."

The object that captured Combs's imagination is one of the most glorious planetary nebulae in the sky (**Figure 7.1**). Located in Draco, NGC 6543 lies only 10′ from the ecliptic's north pole. The ecliptic is the annual path of the Sun among the

Figure 7.1 NGC 6543, the Cat's Eye (center) in Draco, is estimated to be 1,000 years old. Its complex structure reveals the dynamics and late evolution of a dying star and its companion.

stars — a great circle on the sky, tipped 23½° to the celestial equator. Thus NGC 6543 lies at the center of the circle traced by the Earth's pole during its 26,000-year precessional wobble. Because of this unusual location, precession has little effect on the planetary's right ascension and declination.

Combs's remarks sent me hunting for my earliest observing records, made in grade school more than half a century ago. Sure enough, I had penned: "6543, which *Astronomy with an Opera-Glass* says is a planetary. I see a green star, but much wider than the other stars around it." These notes describe views with a homemade 1-inch refractor at 40×. A larger instrument will easily reveal its nebular nature. My 4-inch Clark refractor shows it as a rather uniform bright disk about 20" in diameter, but high magnification reveals areas of irregular brightness. Photographs show the nebula has a coiled form similar to the much larger Helix Nebula in the constellation Aquarius. Its length is about 22", or two-thirds the separation between the components of the double star Beta (β) Cygni (Albireo).

NGC 6543 made a bit of astronomical history in the 1860s. Before then it was assumed that most nebulous objects were actually clusters of very faint stars, and a failure to see them as such was merely due to lack of a big enough telescope. NGC 6543 was one of these unresolved objects. Then, on August 29, 1864, William Huggins turned an 8-inch refractor equipped with a visual spectroscope toward the nebula. (He was the first person to turn a spectroscope toward a planetary nebula.) Rather than a continuous streak of light spanning the visible spectrum, Huggins discovered that the spectrum of NGC 6543 consisted of three bright lines, the strongest being in the green. Thus, "at a glance" he realized that NGC 6543 was a cloud of luminous gas and not, as some astronomers had thought, a mass of unresolved stars. With this single observation came proof that not all deep-sky objects are made of stars. The mysterious green lines were attributed to a hypothetical element called nebulium. More than half a century would pass before these lines were recognized as due to doubly ionized oxygen.

NGC 6543 offers several challenges for the observer. One concerns the nebula's color, and another the brightness of its central star. Through his obervations, William H. Smyth found the planetary a "very fine pale blue," while other observers have seen it more as green. Smyth seems to have had an unusually acute sensitivity for color; most amateurs today cannot distinguish the multitude of shades he could. I suggest using a magnification of about 100× when making color estimates.

NGC 6543's central star shines at about magnitude 11, but the interior stars of planetaries are more difficult to see than their magnitudes suggest. That the central star is surrounded by bright nebulosity means there exists a contrast effect, which poses a problem when attempting to judge the star's brightness. Modern catalogs generally list 9.5 as the visual magnitude of the star. However, my own estimates tend to favor a fainter value. The central star is also bright when compared with other planetaries. With a 3- or 4-inch telescope at 30×, NGC 6543 may look like little more than a swollen star, but as Combs's letter testifies, with a large instrument the nebula can be very impressive.

I recently hunted up NGC 6543 with my 4-inch Clark refractor. At 120× the view with my normal eye showed the planetary and its central star much as expected. With the other eye, which had its lens removed during a cataract operation, the star was brighter than the nebula! This was due no doubt to much more ultraviolet light from the hot star reaching the retina of my lensless eye. Deep photographs record a faint shell about 4' across surrounding the planetary, but I doubt it can be seen visually.

In 1985 I asked for reports from amateurs who examined NGC 6543 with large telescopes. One such report came from Michael Gardner of Sunnyvale, California. As a member of the Mount Wilson Observatory Association, he had viewed it with the observatory's 60-inch reflector! It was a particularly good night, as fog had rolled into the Los Angeles basin and obscured the city lights. Furthermore, the fine seeing for which Mount Wilson is so noted was even better than usual.

"We were observing at the Cassegrain focus of the 60-inch," Gardner writes. "A 55-mm Plössl produced 450 power and a field of view about 6½' in diameter. NGC 6543 was a stunning blue-green oval. The colors were like a Kodachrome. We saw much more structure in the planetary than the picture [*Sky & Telescope,* July 1985, page 89] shows, and the colors were much more vivid and bluish." Gardner did not see the faint shell about 4' across that surrounds the nebula on deep photographs, but he was not specifically looking for it.

Often planetary nebulae will show best within a very narrow range of magnification for a given telescope. Once you have located one of these glows, spend some time experimenting with different eyepieces (don't forget to try a Barlow lens if you have one). Aperture masks may also improve the view. Careful examination of an object can sometimes reveal details not generally noted in observing handbooks. For example, while viewing NGC 6543 I found the central star a "blazing yellow," due to its strong contrast with the nebula's blue color. I do not recall ever reading about this effect before.

The Crown Jewels

Observing is a continual learning experience. At first, time spent under crystalline starlight has all the flush and romance of a soft embrace on a warm summer's eve. Wasn't that first glimpse through a telescope enchanting? As we grow and become more knowledgeable, we tend to move past the romance and begin to challenge ourselves and our visual limits. We start first with the naked eye and binoculars, then move on to modest-sized instruments, until we find ourselves delving ever deeper into the universe with ever larger telescopes. Once we have years of experience under our belts, we start seeking out celestial treasures that seemed impossible when we first started out. For more than four decades Scotty watched amateur astronomy progress in this way, as if our visible horizons and telescope technology were expanding in sync with the universe. Here Scotty shows how one of the sky's smallest constellations, Corona Borealis, can satisfy both the naked-eye novice and the monster-telescope owner.

The stars are always with us. Night after night the blaze of distant suns stretches from horizon to horizon. For many of us the fascination of a starry sky began even before grade school, as the spectacle of the heavens astonished and excited our imagination. As we tally more and more memorable hours under the night sky, the sensation is cumulative. It makes no difference whether we observe with the naked eye, a 4-inch telescope, or a 36-inch Dobsonian.

With each passing year the parchment of the sky comes to hold more information, more contentment, and more wisdom for us. We know that as we step out under the stars tonight we will be headed for a reunion with some old friends. Nevertheless, as observers we must be prepared to push beyond our accepted frontiers, perhaps even to the point of having to reject some "truths" to which we have devotedly clung.

As darkness falls on July evenings, one frontier I always turn to is Corona Borealis, the Northern Crown. The constellation rides high in the sky, aloof from the fainter stars of Boötes and Hercules. This ancient group is one of the few star patterns that span many cultures in essentially the same form.

Two objects in Corona Borealis are of special interest to naked-eye observers. Both are variable stars, but they have very different behaviors. R Coronae Borealis (**Figure 7.2**) is usually 6th magnitude and just visible to the naked eye under good conditions. It remains relatively constant, sometimes for years; then it can abruptly plunge to around 14th magnitude. Recovery is often slow, and there can be many relapses. Sometimes the star takes several years to regain its initial brightness. I check the naked-eye visibility of R Coronae Borealis every clear night — if it's missing, out comes the telescope.

Some 3° to the southeast is another interesting variable. Unlike R Coronae Borealis, this one attracts the most attention *when* it attains naked-eye visibility. T Coronae Borealis is a recurrent nova and typically hovers around 10th magnitude. But it has exploded to 2nd or 3rd magnitude. This happened in 1866 and 1946, and lesser flare-ups occurred as recently as 1963 and 1975. The next burst will likely be first spotted by an amateur who checks T Coronae Borealis with the naked eye or binoculars.

On the other end of the spectrum, the explosive growth of amateur astronomy has produced an interesting group of specialists — those who delve into the deepest of deep skies. While such people don't lend themselves to easy description, I've pieced together a rough profile based on my mail. They are individuals who observe alone. They use at least a 12-inch telescope, more typically a 17-inch or larger instrument. They are experts at getting the most from their telescopes. They keep the optics as clean as possible to reduce the scattered light that dilutes the contrast between a faint object and the sky background. They also make proficient use of the various nebula filters currently on the market.

Figure 7.2
Why R Coronae Borealis suddenly fades from view is uncertain, though it may be related to a shell of light-absorbing carbon particles surrounding the star. T Coronae Borealis is a nova and brightens periodically.

Objects in the *New General Catalogue of Nebulae and Clusters of Stars* (*NGC*) are small potatoes to these people. The *NGC* objects were discovered visually in the 19th century, so they lack the attraction of fainter quarry — objects that were likely discovered on photographs and many of which have never before been viewed directly! The popular observing guides are therefore of little use. Instead these observers turn to multi-volume catalogues with such alien code names as the *UGC, MCG,* and *MOL**, which are usually found only in professional observatories.

These observers make up only a tiny percentage of amateurs. Most of us prefer a casual stroll through the well-tended gardens of the heavens as opposed to beating new trails through the dense celestial jungle. But every now and then it's fun to take on a challenge — to climb a difficult path and stand for a moment on a mountain peak.

Figure 7.3
The Corona Borealis cluster is a rich swarm of galaxies 940 million light-years distant. Observers who search it out will be rewarded with the sight of dozens of tiny gray spindles scattered across the deeps.

For instance, take veteran deep-sky observer Ronald Morales, who described his quest for the faint Corona Borealis galaxy cluster (**Figure 7.3**) in the May 1990 issue of *Sky & Telescope* (page 563). Morales explained how he keeps a "failed-to-see" list of objects that he returns to on nights when the sky is unusually dark and transparent. By doing this he has located many that he might otherwise have given up hope of finding — objects that have brought him unexpected pleasures. This is a worthwhile practice for every serious deep-sky observer, but I can't recall ever seeing it in print before.

---

* *The* Uppsala General Catalogue of Galaxies, *the* Morphological Catalogue of Galaxies, *and the* Master Optical List of Nonstellar Astronomical Objects, *respectively.*

Morales's name is familiar to many readers. In addition to his own articles, I have mentioned his observations many times in this column. He has an observatory outside Tucson, Arizona, that includes telescopes with apertures up to 17½ inches, but his favorite is a 12½-inch f/7 Cave reflector. Many years back Morales and I spent an evening observing with his 8-inch reflector under the pristine skies of Arizona's Empire Mountains. That night I had my best view ever of the nebulosity surrounding the Pleiades. Morales is also coauthor of the Webb Society *Deep-Sky Observer's Handbook*, Vol. 6: *Anonymous Galaxies*, a source of endless deep-sky challenges.

Dueling Globulars

"The northern winter sky," Scotty wrote, "sparkles with scores of open clusters that dot the Milky Way. Spring brings the realm of galaxies into view, and autumn is for hunting planetaries. But it's the warm summer nights now upon us that are perfect for viewing globular clusters." In fact, if you look at the meridian in mid-July around 9 p.m. with the naked eye and binoculars, you can spot several Messier globulars lined up from north to south, as if clinging to some sinuous celestial vine. In order, they are M92 and M13 in Hercules; M12, M10, and M107 in Ophiuchus; and M80 and M4 in Scorpius. A little tilt of the head east or west will bring others into view: M5 in Serpens; and M14, M19, and M62 in Ophiuchus. A telescope will reveal even more globular wonders. What to do? Which ones to cover? Well, without question, the most popular globular is M13 in Hercules. But Scotty was hesitant to bestow full honors on it, for — as we next read — it appears he preferred another globular over this one.

As you read this, amateurs somewhere are looking at the splendid globular star cluster M13 in Hercules (**Figure 7.4**). Some might be comparing notes on how well various telescopes resolve the cluster into individual stars. Others are using it as a test of eyesight or sky conditions (though with a total magnitude of 5.9, this ball of ancient stars should not be much of a challenge except from places like the fens of southern Connecticut). M13, along with the Orion Nebula, is perhaps the most frequently observed deep-sky object. Indeed, even when Halley's Comet beckoned people to look skyward during 1986, it is probably fair to suggest that more people saw M13 than the comet itself. It is pure coincidence, but nevertheless interesting, that Edmond Halley chanced upon M13 in 1714. The following year he published a paper that was the first to describe a half dozen "nebulous" objects in detail rather than mention them as asides to a star catalog. Of the great globular Halley wrote, "This is but a little Patch, but it shews itself to the naked Eye, when the Sky is serene and the Moon absent."

In June 1986 I mentioned that M13's popularity is derived not only from its large size and naked-eye brightness, but also from its favorable sky location. For observers at mid-northern latitudes the cluster is visible much of the year. On

Figure 7.4
The great globular cluster M13 contains more than 100,000 stars to 21st magnitude. The true stellar population may be close to half a million.

pleasantly warm summer evenings it passes nearly overhead where the view is through the thinnest layer of interfering atmosphere. Furthermore, M13 is conveniently placed on the line between Eta (η) and Zeta (ζ) Herculis, which forms the west side of the Hercules Keystone. With such prominent stars to point the way, it's easy to understand why even novice observers can quickly locate the cluster. M13's association with the Keystone is so interwoven that the relatively small asterism is believed by many to be all there is to the constellation.

Small binoculars or a finder will show M13 as a pale, colorless glow with a diameter as much as half that of the Moon. If encountered accidentally while sweeping, M13 can be quite startling. A 3-inch telescope will just begin to show stars at the cluster's edge, and a 4-inch will add more. The fact that Charles Messier never saw any with his 60× Newtonian of 4½-foot focal length shows just how far telescope making has advanced in two centuries.

By the mid-1800s, W. H. Smyth was extolling M13's appearance in a 5.9-inch refractor. "An extensive and magnificent mass of stars," he wrote in 1844, "with the most compressed part densely compacted and wedged together under unknown laws of aggregation." Smyth had read a similar comment by William Herschel, and this may have influenced his own observations.

Photographs, too, have influenced observers, but not always in a positive way. About 1850, while using Lord Rosse's telescopes in Ireland, Bindon Stoney noted an unusual pattern of three dark rifts (**Figure 7.5**) radiating outward from near the center of M13. The 19th-century observer and author Thomas W. Webb mentions in his classic *Celestial Objects for Common Telescopes* that these lanes were "beautifully seen by Buffham," who used a 9-inch reflector.

Soon afterward photography became the all-powerful tool of astronomy. Mention of Stoney's rifts essentially disappeared from the literature, I suspect because the lanes did not appear on early photographs of M13. I first wrote about the "propeller" in the July 1953 column, but no amateurs of the day reported seeing it. Every few years I would bring it up again, with similar results. It wasn't until 1980, after I asked several more times for amateurs to hunt for them, that John Bortle reported seeing the lanes with his 12½-inch Newtonian reflector at 176×. The cosmic jest surrounding the sighting was that Bortle, tired of reading about the folklore of the lanes in my column, set out to disprove their existence, or so he said in his letter. Sighting the lanes seems to depend upon a careful balance of aperture and magnification. Both Bortle and Dennis di Cicco commented on the importance of magnification. During the Stellafane convention in 1981, di Cicco was surprised by how easily the lanes were seen with the 12-inch f/17 Porter turret telescope at about 180×. However, even knowing their orientation and appearance, he was unable to see them at 95× with a 12-inch reflector that was set up nearby.

Jan Römer of the Delaware Valley Amateur Astronomical Association in Pennsylvania reports that he cannot see the lanes under any conditions with an

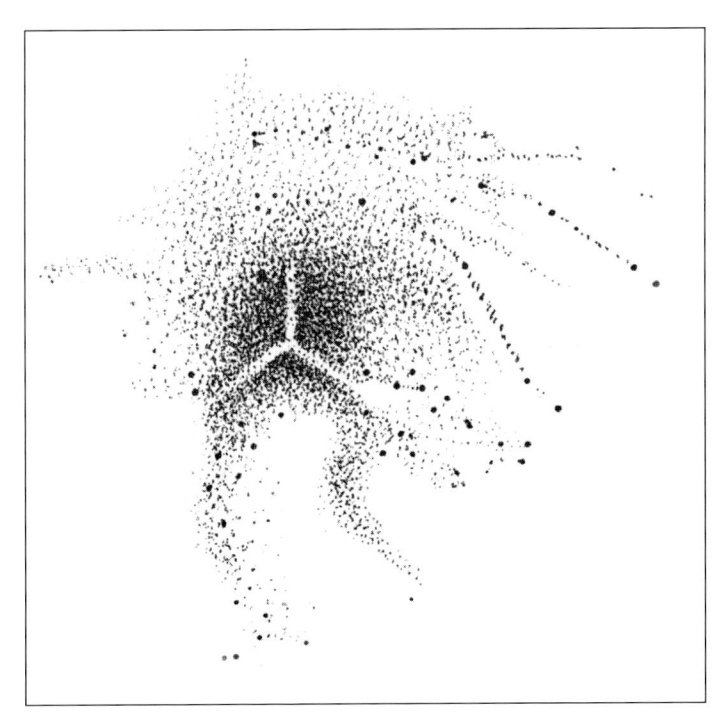

Figure 7.5
In this negative drawing depicting M13 by Lord Rosse, three lines form what looks like a dark propeller near the core of globular cluster M13.

8-inch f/8 reflector at 100×. He does see, however, many star chains crossing the cluster and also several of the star-poor areas mentioned by observer and writer John Mallas. Most observers note that they appear best at magnifications of about 200×. It is exciting to think that these lanes can again be seen after a lapse of nearly a century in which no one reported viewing them. The feature is offset to the southeast edge of the cluster.

Veteran observer Mark K. Stein may hold the record. He writes, "When I lived under the polluted skies of Louisville, I considered some of your descriptions of deep-sky objects the result of a fine, experienced observer having a slightly overactive imagination. But now in the much darker skies of Bloomington, I can plainly see objects I wouldn't have attempted from Louisville." With a 6-inch at his new location Stein has seen the dark lanes in the globular cluster M13. This feat puts him in a special class, for the lanes usually require an 8-inch or, better, a 20-inch. I have never seen them in my 4-inch Clark — but I keep on trying.

On long-exposure photographs the myriad stars in M13 spread out to a diameter of more than 20′, two-thirds of the Moon's diameter. In a 4-inch telescope only the cluster's edge can be resolved. A 10-inch instrument, however, will show more stars than the eye can count, though the center still remains a solid glow. Sharp-sighted amateurs will notice that M13 is not circular as it appears in long-exposure photographs. Early drawings like Lord Rosse's show dark lanes and strings of stars — all still apparent to the visual observer. Of course, the resolution of M13 depends somewhat on magnification. An 8-inch aperture at 30× will not separate the stars because the cluster's core appears too bright. Increasing the power to 300×, however, reveals stars across the entire field. One of my most memorable views occurred when a smog layer had reduced the naked-eye magnitude limit to about 4½. As is often the case, the seeing was very steady then. At 300× the cluster was faint, but the individual stars stood out well. The number of stars attributed to the cluster depends upon which reference you consult, but all agree that it runs well past 100,000.

The Challenger

Every deep-sky observer has a list of favorites: a planetary nebula, a test object for sky transparency, or a special area of the heavens to sweep on nights when the conditions are just right. One of my pet objects is well placed in the July evening skies, near the border between Virgo and Serpens — the remarkable globular cluster M5. Formerly M5 was included in an extension of Libra which curled up into Serpens, and consequently it is catalogued as a Libra object in the older observers' texts, by Smyth, Webb, George F. Chambers, and Garrett P. Serviss. Since the IAU revisions of the constellation boundaries, it has officially been charted in Serpens (Caput).

The German astronomer Gottfried Kirch found M5 in 1702. Messier, with a very small telescope, described it as a nebula. Smyth wrote: "This superb object is a noble mass, refreshing to the senses after searching for faint objects; with outliers in all directions, and a bright central blaze which even exceeds M3 in concentration." It lies at the eastern end of a short chain of three, faint, naked-eye stars: 109 and 110 Virginis and 5 Serpentis. M5 is just 0.4° northwest of 5 Serpentis and is almost ½° in diameter (**Figure 7.6**). I saw it with my naked eye under the clear skies of Arizona, as well as from the summit of Temple I at the Mayan ruins of Tikal in Guatemala.

Telescopically M5 is a treasure house, and every increase in aperture brings new sights. With low power on a small instrument it is very beautiful and bright, but not resolved. Its appearance is comparable to that of M13 in Hercules. At home in Connecticut, my 4-inch Clark refractor shows the cluster with a lacy fringe of stars. Years ago a 10-inch reflector in Kansas revealed some 300 stars, and the background glow hinted at many more just below the telescope's grasp. With 12-inch to 16-inch instruments it appears as one of the most compelling objects on record. John Herschel studied the cluster in the early 1800s with his 18.7-inch speculum-metal mirror. He found M5 to be very compressed at the center, and likened it to a cosmic "snowball." With the 20-inch Clark refractor at Wesleyan University, I would add that M5 seemed more like a starry blizzard.

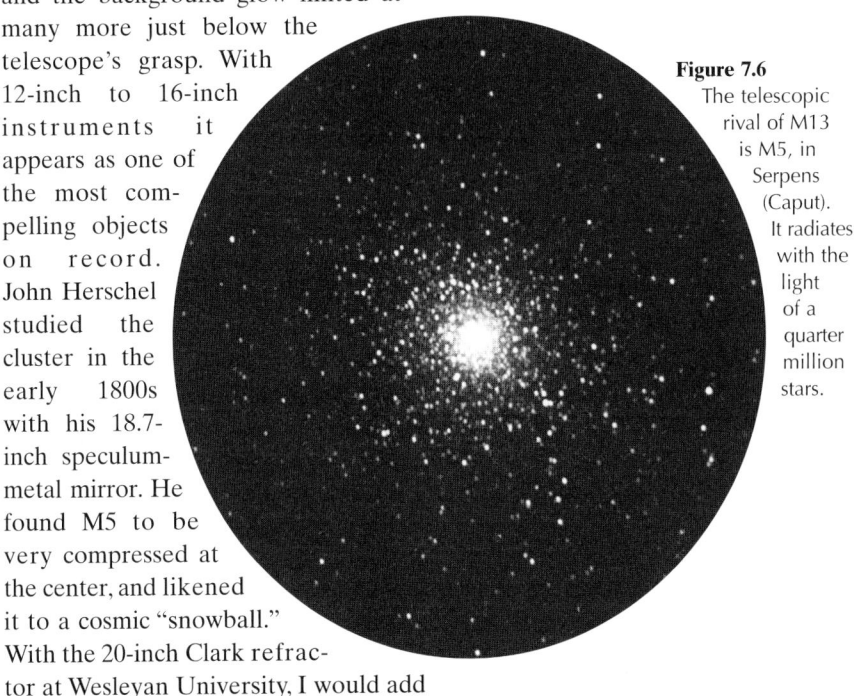

Figure 7.6
The telescopic rival of M13 is M5, in Serpens (Caput). It radiates with the light of a quarter million stars.

The English observer Kenneth Glyn Jones contended that M5 is second only to M13 in the northern hemisphere. But it may come as a surprise to learn that M5 is actually listed as being a tenth of a magnitude brighter (5.75 versus 5.86) in *Sky Catalogue 2000.0*. Both M13 and M5 are of similar size, too. In fact, M5 is the brightest globular cluster in the northern hemisphere of the sky and indeed is surpassed by only two southern globulars, 47 Tucanae and Omega Centauri. Are there times when you can see M13 and not M5 with the naked eye? It would be an interesting project to see what the effects of altitude have on each object. It may be that M13 is the better-known cluster only because it passes nearly overhead for those living at temperate northern latitudes.

The Orphans of Ophiuchus

Warm summer nights in July offer us the richest regions of the Milky Way. Under a dark sky, its smoky trail of fused starlight is dappled with the hazy light of globular clusters. Not surprisingly, Ophiuchus, the sky's 11th-largest constellation (in area) contains a plethora of these objects, including a half-dozen that belong to Messier's catalog: M9, M10, M12, M19, M62, and M107. Naturally, beginners tend to explore the Messier objects first. But Ophiuchus has a host of lesser-known sights worth scouting out. Furthermore, Scotty alerted those with wide-field telescopes to be on the lookout for the sinuous veins of dark nebulosity that empty into black lagoons of obscuring dust near many of the Ophiuchus globulars. "There is something for everyone," Scotty wrote. "Fine objects abound, and the central part of Ophiuchus, lying between the two sections of Serpens, provides a test for the observing capabilities of amateurs with modest-sized telescopes." Indeed, the Serpent Bearer is full of celestial surprises, as we shall see.

Some joys of astronomy are seldom mentioned. There is the pleasure of sifting through a musty 19th-century book with its quaint prose, knowing that it inspired countless people to look at the heavens. Hidden among the pages of another is a fine engraving of an old telescope that, carefully examined, yields ideas for today's telescope makers. Old atlases show interesting constellation figures and star names. While many of us enjoy old books and charts, we also delight in works of recent vintage. It is all as much a part of amateur astronomy as looking through a telescope on a July evening.

Perhaps the best-known contemporary celestial cartographer is Wil Tirion of the Netherlands. In recent years his beautiful star atlases have been widely circulated. Let's consider for a moment the right-hand half of chart 22 in the deluxe version of *Sky Atlas 2000.0*. Scorpius sprawls across the field as it unwinds from brilliant Antares, the Scorpion's heart — to Shaula, the stinger in its tail. For those of us living in the northern United States, the southernmost stars on this chart are, at best, difficult to see. Still, we are better off than amateurs in Canada, England, and even at the latitude of Messier's Paris, since a good portion of the heavens charted here never rises above their southern skyline.

The Milky Way here cascades down the sky as indicated by the shades of blue on the chart. The dust in our own galaxy blocks the view beyond, so the familiar red ovals of external galaxies are lacking from this section of Tirion's chart. However, other interesting features of the sky are apparent. The galactic equator, representing the plane of the Milky Way, rises steeply through the field. To its east is a profusion of open clusters, shown in yellow, but few lie to its west. North of Antares, especially in Ophiuchus, stars are remarkably scarce; in many places it is possible to draw a 2° circle that doesn't contain a single star down to the 8th-magnitude limit of the chart. Yet on the other hand, the star-poor region of Ophiuchus east of Antares contains an abundance of globular clusters. Looking at this atlas page gives us a perspective of the sky that cannot be obtained any other way. The field is much too large for a single view through a telescope, and in photographs covering this area faint stars overwhelm the deep-sky objects.

Wide-field, long-exposure photographs show the area of Ophiuchus and Serpens to be a turmoil of stars mixed with clouds of bright and dark nebulosity. Even good finderscopes may have trouble in this confusion. The chart maker, working more selectively with symbols and colors, can show us at a glance what grandeur the starry heavens hold. Globulars are common, including several which the Herschels missed but which are still within the reach of amateur equipment. Surprises in this part of the sky are common.

One example is the open cluster IC 4756 in the eastern half of Serpens (Cauda). It is one of the largest such objects in the heavens, appearing more than 1° across and just a little smaller than the Beehive cluster, M44, in Cancer. Some 80 stars between 7th and 12th magnitude are evenly scattered across IC 4756's diameter. The Herschels probably missed it because the small fields of their reflectors would have passed right over the group without revealing any concentration of stars. IC 4756 appears as a patch of the Milky Way to the unaided eye. Binoculars or a finder will easily reveal its individual stars. While large reflectors will not show the entire object at once, they will be useful when searching for unusual star chains or dark lanes.

Many years ago Glen Chaple, Jr., of Townsend, Massachusetts, "discovered" a deep-sky object in Ophiuchus; he found about two dozen stars forming a group which could be glimpsed with the naked eye. He likened the cluster to the Praesepe in Cancer, but it was not plotted on his copy of *Norton's Star Atlas*. The object does, however, have a name. It is IC 4665 and is listed in the *Index Catalogue* to the *NGC*. In those days of stargazing, serious deep-sky observers soon learned that no single atlas can suit all their needs. Today, of course, IC 4665 is plotted on most modern star charts.

About 1° northeast of 3rd-magnitude Nu (ν) Ophiuchi is the tiny globular NGC 6517. Only 4.3' across and about 12th magnitude, it usually requires an 8- or 10-inch telescope to positively identify this cluster. However, last year, working with a finder chart that was plotted from the AAVSO *Star Atlas,* I was able to locate NGC 6517 with my 4-inch Clark refractor at 150×. Lower powers made it impossible to distinguish the cluster from a faint field star.

Some 1° northeast of NGC 6517 is the 5th-magnitude star Tau (τ) Ophiuchi. Continuing on a straight line for 1° more brings us to NGC 6539, another globular missed by the Herschels. Of roughly 12th magnitude and 2½' across, it is a challenge for apertures less than 8 inches. My notes contain only sightings made with a 10-inch reflector. And, although I considered NGC 6539 a difficult object 20 years ago, I know that during the last decade many amateurs have observed it with telescopes as small as 8 inches.

Another difficult globular is NGC 6535 in Serpens (Cauda). It, too, was overlooked by the Herschels. Discovered in the 19th century by the English observer

John Russell Hind, it is only about 1½′ in diameter and 11th magnitude. While my notes made years ago in Kansas read "at least 100× needed to establish identity as a globular," in England, D. Branchett, using a 3-inch aperture, found NGC 6535 to be a "faint, elusive object."

Tiny globulars seem to flourish in this part of the sky. Without suitable finder stars it is often extremely difficult to hunt down these challenging objects. One exception is NGC 6366. Locating it is easy, since it is in the same field as 5th-magnitude 47 Ophiuchi, which lies 17′ due west of the cluster. With magnifications of around 100×, the 4-inch Clark shows NGC 6366 to be magnitude 11.5, and some 3′ in diameter. It cannot be recognized as a cluster at 20× in my 5-inch Apogee telescope. At Wesleyan University's Van Vleck Observatory in Middletown, Connecticut, I have used the 20-inch Clark refractor to view NGC 6366. It shows the object as some 5′ across and looking more like a compressed open cluster than a globular.

Figure 7.7
The globular cluster M9 in Ophiuchus lies 30,000 light-years away, close to the center of our Milky Way galaxy. It is accompanied by the inky dark nebula Barnard 64.

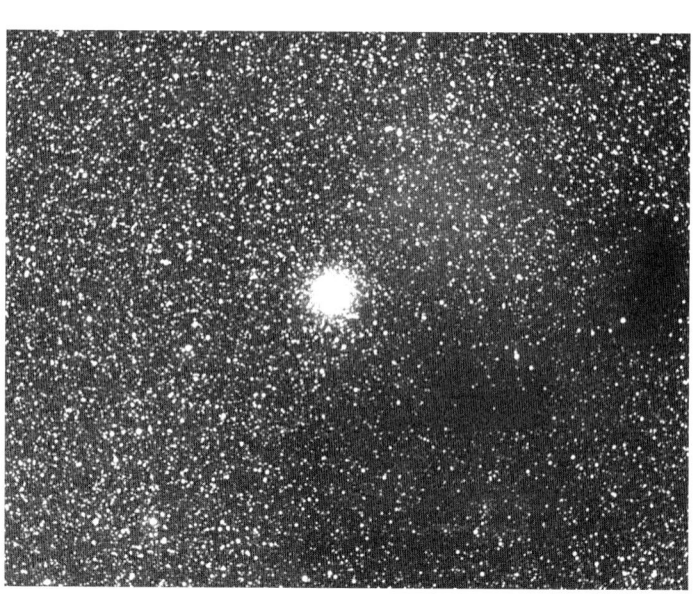

Eastward in Ophiuchus is the globular M9, discovered by Messier in May of 1764. About 6′ across and 8th magnitude, it stands out well against the background stars of the surrounding Milky Way (**Figure 7.7**). Hans Vehrenberg comments in his *Atlas of Deep-Sky Splendors* that with a small telescope M9 can be mistaken for a star. This is worth checking out. The cluster is near the northeast edge of a remarkable dark nebula, Barnard 64, less than ½° west of it. With my 4-inch Clark refractor or 5-inch Apogee telescope the dark nebula is easily seen.

Most dark nebulae are difficult to "see," but here the rich background is rather uniform, and it is interesting to compare the star densities northeast and southeast of M9. Try powers of about 100×. These dark patches really show best on photographs, and it was with photography that E. E. Barnard discovered most of them earlier this century.

There are two globulars only a short hop from M9. About 1½° to the northeast is NGC 6356. Although still in a rich star field, this 9th-magnitude cluster, 2' across, is easy to identify, especially at higher magnifications. The other globular, NGC 6342, lies the same distance south and slightly east of M9. This is more difficult, only 0.5' in diameter and 10th magnitude. However, Canadian Pat Brennan notes NGC 6342 as "faint" but readily seen in a transparent sky. Northeast of NGC 6342 is another dark nebula, Barnard 259. It is not as sharply defined as Barnard 64 but still the stars should appear noticeably thinner here than in surrounding areas.

While checking my files I was amazed to discover that I have never written about the globular cluster M19 in Ophiuchus during the four decades of preparing this column. It lies at about the same declination as brilliant Antares and 7½° east of the Scorpion's heart. The fact that Messier discovered this cluster while scanning the sky from Paris is a tribute to his observing skill. Even from the northern United States, where the cluster climbs nearly 10° higher in the sky than at Paris, M19 is not especially well placed for viewing.

John Mallas, whose location in Southern California was more suitable for examining M19, once remarked that the globular is a miniature of the great Omega Centauri cluster. M19 is 5' in diameter and a bright 6.6 magnitude. Under good skies it is easily seen in most telescope finders. The visual diameters of globular clusters depend to some extent on the size of telescope and the observer's eyesight. Members of an observing group should compare their individual estimates of these objects' sizes, using a micrometer, an ocular grid or reticle, or by timing transits over a crosshair in the eyepiece. The results may well be quite different for telescopes of different sizes. Similarly, apparent magnitudes may differ in large and small instruments.

Two other globulars lie within a short distance of M19. A little less than 2° east and slightly south is NGC 6293. You'll know you're headed in the right direction when you pass a 6th-magnitude star about 1° from M19. Just southwest of this star is the Cepheid variable BF Ophiuchi, which ranges between magnitude 7.5 and 8.5 during a four-day cycle. About 2' across and of magnitude 8.4, NGC 6293 can be seen with only a 2-inch aperture and shows nicely in 4- to 8-inch telescopes. And south and east of NGC 6293 are more great clouds of obscuring material which reveal themselves primarily to the camera. However, those using large-aperture, rich-field telescopes may be able to detect some of the more sharply defined boundaries here. Now return to M19. Just 1½° north-northeast of it is NGC 6284. It is a bit smaller and fainter than NGC 6293, but still suitable for small telescopes. It is a nice cluster for beginners and searching for it provides good experience in hunting out challenging objects. Amateurs who have successfully worked down from the brighter globulars to such faint ones can feel well satisfied with their observing prowess.

More Surprises in the Serpent Bearer

Although Ophiuchus is best known for its profusion of globular clusters, few realize that the Serpent Bearer contains a wealth of planetary nebulae that can both delight and test the mettle of all observers. When summer nights fall upon us we naturally tend to turn our telescopes to the more famous summer planetary nebulae M57 (the Ring) and M27 (the Dumbbell), which are climbing toward the meridian after sunset. But Scotty enjoyed the challenge of dim planetaries, too. He knew how difficult it was to confirm a sighting, especially if the planetary was small and faint, and the atmosphere even slightly unstable. Under such conditions stars and planetaries look the same — like tiny swollen disks. Without access to a special star chart or photograph that shows an object in its precise position relative to nearby faint stars, observers can be at a loss. So how is one to know the difference between the two? For Scotty's readers, one way was to turn to the Deep-Sky Wonders column and find out how other observers worked out this dilemma. Another way was to look at a lot of planetaries. Fortunately, the wide area of Ophiuchus contains a variety of bright ones to help train your eye to recognize these gems.

Ernst J. Hartung's book *Astronomical Objects for Southern Telescopes* contains a useful suggestion for observers of difficult planetary nebulae: hold a small direct-vision spectroscope between your eye and the telescope ocular. The prism spreads each star in the field into a narrow, faint streak. But because the visible light of a planetary nebula consists mainly of a close pair of bright lines in the green part of the spectrum, the nebula's image remains almost entirely unchanged.

In this way, even the tiny faint planetaries can easily be picked out in a crowded star field, and Hartung had much success with objects only an arcsecond or so in diameter. While this powerful technique is not new, few amateurs seem to know of it. It is well worth trying, for the effect is startling. Here is a sample of brighter planetaries in Ophiuchus, some of which are readily recognizable by ordinary viewing, while others require the visual spectroscopic method.

NGC 6369 is one of the easier kind. It is within the reach of all but the smallest telescopes. This 10th-magnitude object lies in a sparse field, facilitating its identification. It is a perfect smoke ring about half the apparent diameter of Jupiter. Several nearby naked-eye stars help guide the way. Ron Morales in Arizona saw NGC 6369 with a 10-inch reflector at 87× as "round, green, and with the edge quite sharp." By using a magnification of 137× he could easily make out the nebula's dark center.

Near the eastern edge of Ophiuchus is a small V-shaped asterism known as Taurus Poniatovii (the Bull of Poniatowski). Named for an 18th-century Polish king, this obsolete constellation contains the famous double star 70 Ophiuchi, a binary with an 88-year orbital period. In 1989, the 4.3 and 6.0-magnitude components were near a minimum separation of 1.5″. A few degrees to the northeast of this group, on the shoulder of the Bull, is the planetary nebula NGC 6572. Wilhelm Struve, who discovered this nebula in 1825, considered it to be one of the "most curious objects in the heavens."

In 15 × 65 binoculars NGC 6572 appears as a star, if one knows where to look. Indeed, the nebula is a mere 16″ in diameter, and could be mistaken for a star if lower magnifications were used even in a telescope. Yet with only a little power it is still a beautiful little gem. Its total light is about equivalent to a 9th-magnitude star, while its surface brightness is a hundred times that of the Ring Nebula in Lyra! With the Clark refractor I have estimated its magnitude to be 9.5. My old 10-inch reflector showed the vivid green color of this object with any power more than 50×. It is interesting to note that older observers have described NGC 6572 as green, while the younger ones tend to call it vivid blue.

The planetary's central star is not difficult. Estimates of its magnitude range from brighter than 10th to less than 12th, but such are the problems when trying to estimate a star's magnitude when it appears against a bright background. There are several reports of the star not being seen with 10-inch instruments. If you search for it, remember to use the highest magnification your telescope and the night will allow. This dims the nebula's glow and improves the contrast between it and the star. A little over 3° north-northeast of NGC 6572 is the planetary PK 38+12.1 (Cannon 3-1), which is plotted on Tirion's charts. Once located it is fairly easy to observe its 6″-diameter, 12th-magnitude disk. It is easy to find with the spectroscopic method (described above).

Certainly within reach of a 5-inch is the planetary NGC 6309. This 11.7-magnitude object is a bit more than 20″ long and only half as wide. Ron Morales calls this the "Box Nebula" and reports a curious gray-green color seen with a 10-inch reflector. The English observer Ed S. Barker's description in the Webb Society handbook for nebulae mentions NGC 6309 as appearing "slightly mottled" in an 8½-inch telescope. He also finds a faint star at each end of the nebula. What do you see?

Finally, I must share with you a letter I received from Canadian amateur Dunstan Pasterfield, who had an interesting idea. During his first nights of observing, he scanned the sky with binoculars and was impressed with the large, very loose open cluster NGC 6633. (Webb's *Celestial Objects for Common Telescopes* lists this cluster in Serpens, but today we find it in Ophiuchus, according to the official constellation borders established in 1930.) When Pasterfield got an 8-inch reflector he immediately turned to NGC 6633. "It's a lovely, great, straggling thing . . . of an absurd shape," he writes. "I keep my eye on it annually. I'm convinced that some day something will happen in 'my' cluster."

It's an interesting concept: adopt a deep-sky object and learn it so well that you would know at a glance if something were unusual, like a nova or stray asteroid or comet in the field. The idea is not new. Robert Evans in Australia has memorized hundreds of galaxies as part of his search for supernovae, and his efforts have paid off. He has become the greatest visual discoverer of supernovae of all time.

But you don't have to learn hundreds of objects. In 1979 Maryland amateur Gus E. Johnson was making his annual inspection of galaxies in Virgo when he noted that M100 didn't look right. The reason was a 12th-magnitude supernova, and the credit for the discovery went to Johnson.

Naked-eye Globular Clusters

July 1994 was a sad month for Deep-Sky Wonders readers, because it was the last time Scotty's words would appear as such in the magazine. Scotty had died the previous December 23rd, while touring Mexico. His final column is reprinted here in its entirety, and revisits some objects discussed earlier in this chapter that were particular favorites. For this installment, Scotty wrote, "Astronomy, as I have known it since about 1922, is full of encouraging surprises. Share them with me." Funny how his last offer could as easily have been his first. His life seemed to have come full circle. And as this book testifies, Scotty's spirit remains with us, and his words will continue to inspire and teach us. Besides, the measure of greatness does not end with one's life, but continues as long as one's words or accomplishments continue to affect the lives of others.

Warm summer nights are a fine time to relax under a dark sky. As you lie back and scan the ghostly band of the Milky Way and its environs, see how many globular clusters you can detect with the unaided eye. If you observe from mid-northern latitudes and can detect 6.5-magnitude stars, there are eight globulars to try for this month in the evening sky: M2 in Aquarius, M3 in Canes Venatici, M4 in Scorpius, M5 in Serpens (Caput), M13 and M92 in Hercules, M15 in Pegasus, and M22 in Sagittarius.

It's not unreasonable to place the naked-eye limit at magnitude 6.5. I make no claim for special powers when my logbook shows I've gone that faint, or on special nights, a magnitude fainter! The rub is that globulars are not stellar points; their light is spread over a tiny area of sky — enough to make them a real visual challenge without optical aid.

I'll assume that 6th-magnitude M13, isolated between the stars marking the western edge of the Keystone of Hercules, is no real contest; many observers regularly use it as a test of the night's clarity. Instead, let's begin with M5, since my records show that amateurs have not paid much attention to this celestial splendor. Shining at magnitude 5.7 and measuring 17.4' across, this globular is easy to find; just look northwest of the star 5 Serpentis, which M5 all but hugs. You'll have trouble, however, if you can't differentiate the cluster from the star.

German astronomer Gottfried Kirch first sighted M5 while hunting for comets in 1702. Charles Messier rediscovered it through the Paris smog six decades later, describing it as a "fine nebula which I am sure contains no star."

The cluster remained unresolved until William Herschel trained his telescope on it in 1791. My 4-inch Clark shows a lacy edge of stars all around its borders. The 20-inch Clark refractor of Van Vleck Observatory in Middletown, Connecticut, transforms it into a dazzling mass of sparkling stardust. It is a shame to let such a brilliant summer object wither on the vine.

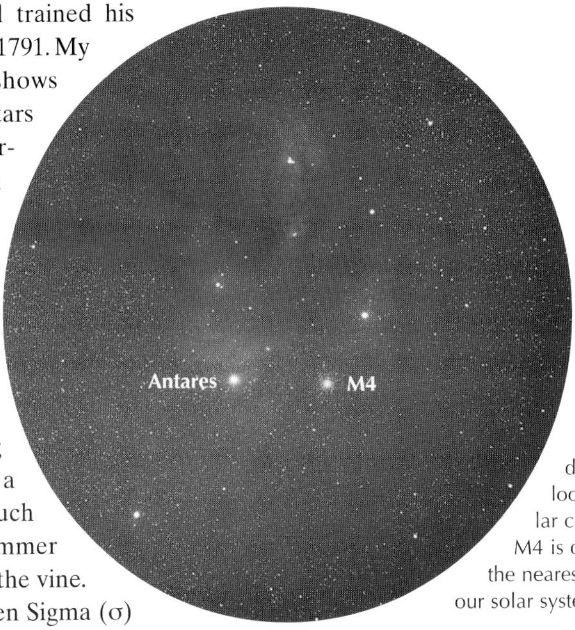

Nestled between Sigma (σ) Scorpii and fiery Antares is 6th-magnitude M4 — a huge globular measuring 26.3′ wide (**Figure 7.8**). While traveling through Central America, I spent many hours looking at M4 with the naked eye, binoculars, and my portable 4-inch rich-field reflector. I have seen it from the wild Petén jungle of Guatemala and at midnight from the top of the pyramids in the old city of Tikal. Always it has been rewarding. To see it with the naked eye, you must filter out the "glare" from brilliant Antares, though this is not difficult from the darkest of sites.

Figure 7.8 Roughly 6,000 light-years distant, the loose globular cluster M4 is one of the nearest to our solar system.

Through a telescope M4 is a dynamic object. Although Webb, in his *Celestial Objects for Common Telescopes,* refers to it as being "rather dim," remember he viewed it from high-northern latitudes. From the Florida Keys, where the Winter Star Party is held, the huge ball of M4 is mighty impressive. While you're in the vicinity, turn to Antares and move your telescope 0.56° to the northwest. There you'll find NGC 6144, a 9th-magnitude globular cluster 9.3′ in diameter. In the open sky this would be a most easy object, but Antares must be out of the field of view if any success is to be expected.

Antares deserves a careful glance as well. It has a 6.5-magnitude companion about 3″ to the west. First noticed in 1819 during an occultation of Antares by the Moon, the companion is said to glow green, but that is an illusion. The companion is easy in a 6-inch telescope, but you'll need a steady atmosphere. If you're having problems with glare from the primary, try using a Lumicon deep-sky filter. Although I haven't tried it myself, I hear it does the job nicely.

Seemingly swimming against a current in the Milky Way, the 5.1-magnitude globular M22 lies about 2° northeast of Lambda (λ) Sagittarii, the top of the

Teapot. Abraham Ihle of Germany apparently discovered it in 1665 while following the motion of Saturn. The problem in detecting this cluster is not with its brightness — which is nearly a full magnitude brighter than M13 — but with its proximity to the horizon and summer haze. It's nicely positioned, though, being just east of 24 Sagittarii.

In 1992 Brian Skiff spied 6.4-magnitude M15 from the Texas Star Party. It's not terribly difficult to see with the unaided eye. The trick is to separate it from a star of comparable magnitude immediately to its east. By the way, if you own a large telescope, see if you can spot the dark patch with two dustlike lanes near the cluster's center, which Webb first reported.

Two globulars are more demanding. Shining at magnitude 6.3, M3 lies in a very star-sparse region of sky. Furthermore, it nearly abuts a 5th-magnitude star, so you'll have to resolve the two to make a positive sighting. Overshadowed by its stunning neighbor M13, the 6.5-magnitude globular M92 in Hercules is right at our magnitude limit. It is not only challengingly faint but also isolated northeast of the Keystone.

Finally there is M2, which culminates late on July nights. At magnitude 6.5 it, too, is at our preselected limit. But I've had no problem sighting it. I have seen it repeatedly from the bayous of Louisiana during my stint as a celestial navigator for the Air Corps.

JULY OBJECTS

Name	Type	Const.	R. A. h m	Dec. ° '	Millennium Star Atlas	Uranometria 2000.0	Sky Atlas 2000.0
70 Ophiuchi	**	Oph	18 05.5	+02 30	1272, 1296	249	15, 16, A1
Antares, Alpha (α) Scorpii	Var	Sco	16 29.4	−26 25	1397, 1419	336	22
Barnard 64	DN	Oph	17 17.2	−18 32	1371	337, 338	15, 22
Barnard 259	DN	Oph	17 22.0	−19 19	1370, 1371	337, 338	15, 22
BF Ophiuchi	Var	Oph	17 06.1	−26 35	1395, 1417, 1418	337	22
Cannon 3-1, PK 38+12.1	PN	Oph	18 17.6	+10 09	1248	204	15, 16
Cat's-Eye Nebula, NGC 6543	PN	Dra	17 58.6	+66 38	1065, 1066	30	3
Corona Borealis Cluster, Abell 2065	CGx	CrB	15 22.7	+27 43	646	112, 154	—
IC 4665	OC	Oph	17 46.3	+05 43	1273	203, 248	15
IC 4756	OC	Ser	18 39.0	+05 27	1270, 1271	205, 250	15, 16
M2, NGC 7089	GC	Aqr	21 33.5	−00 49	1286	255, 256	16, 17
M3, NGC 5272	GC	CVn	13 42.2	+28 23	651	109, 110	7
M4, NGC 6121	GC	Sco	16 23.6	−26 32	1397, 1419, 1420	336	22
M5, NGC 5904	GC	Ser	15 18.6	+02 05	765	244	14, 15
M9, NGC 6333	GC	Oph	17 19.2	−18 31	1370, 1371	337, 338	15, 22
M13, NGC 6205	GC	Her	16 41.7	+36 28	1158, 1159	114	8
M15, NGC 7078	GC	Peg	21 30.0	+12 10	1238	210	16, 17
M19, NGC 6273	GC	Oph	17 02.6	−26 16	1395	337	22
M22, NGC 6656	GC	Sgr	18 36.4	−23 54	1391	340	22
M92, NGC 6341	GC	Her	17 17.1	+43 08	1135	81	8
NGC 6144	GC	Sco	16 27.3	−26 02	1397	336	22
NGC 6284	GC	Oph	17 04.5	−24 46	1395	337	22
NGC 6293	GC	Oph	17 10.2	−26 35	1395, 1417	337	22
NGC 6309	PN	Oph	17 14.1	−12 55	1347	292	15
NGC 6342	GC	Oph	17 21.2	−19 35	1370, 1371	337, 338	15, 22
NGC 6356	GC	Oph	17 23.6	−17 49	1370	292, 293, 337, 338	15, 22

Ast = Asterism; BN = Bright Nebula; CGx = Cluster of Galaxies; DN = Dark Nebula; GC = Globular Cluster; Gx = Galaxy; OC = Open Cluster; PN = Planetary Nebula; * = Star; ** = Double/Multiple Star; Var = Variable Star

JULY OBJECTS (CONTINUED)

Name	Type	Const.	R. A. h m	Dec. ° '	Millennium Star Atlas	Uranometria 2000.0	Sky Atlas 2000.0
NGC 6366	GC	Oph	17 27.7	−05 05	1322	248, 293	15
NGC 6369	PN	Oph	17 29.3	−23 46	1394	338	22
NGC 6517	GC	Oph	18 01.8	−08 58	1320, 1344	294	15, 16
NGC 6535	GC	Ser	18 03.8	−00 18	1296	249	15, 16
NGC 6539	GC	Ser	18 04.8	−07 35	1320	294	15, 16
NGC 6572	PN	Oph	18 12.1	+06 51	1272	204	15, 16
NGC 6633	OC	Oph	18 27.7	+06 34	1271	204, 205	15, 16
R Coronae Borealis	Var	CrB	15 48.6	+28 09	645	113, 155	7, 8
T Coronae Borealis	Var	CrB	15 59.5	+25 55	667, 1206	155	7, 8

Ast = Asterism; BN = Bright Nebula; CGx = Cluster of Galaxies; DN = Dark Nebula; GC = Globular Cluster; Gx = Galaxy; OC = Open Cluster; PN = Planetary Nebula; ✶ = Star; ✶✶ = Double/Multiple Star; Var = Variable Star

AUGUST

CHAPTER 8

Scanning the August Pole

What is the most neglected area of the entire northern sky? The north polar region, of course — especially for anyone who owns a German equatorial mount. It's a shame because, as Scotty reminded us, one advantage of living in the north temperate zone is that the polar region of the sky is available every clear night of the year. But Scotty had a way to solve this problem so that we could turn our sights to this rich but overlooked area. "Observers using equatorially mounted telescopes may find it difficult to sweep the sky so close to the pole," he wrote. "In this case, try turning the mount about 90° so the polar axis points east or west. Dobsonian and other altazimuth telescopes have no problem working this part of the heavens; their Achilles' heel is the area near the zenith." That done, observers should have no problems scanning the heavens around Polaris with the ease of an altazimuth mount. Such simplicity of thought was a well-known Scotty trademark.

The regions near the north celestial pole are usually neglected by amateurs, who seem more attracted to the spectacular sights farther south. But sometimes we overlook the obvious. Polaris, for example, is a variable star. In fact, it is the brightest Cepheid in the sky. *Sky Catalogue 2000.0* gives its range as 0.15 magnitude over a 4-day period, but studies done during the 1980s show that the range is decreasing, leading some astronomers to speculate that the star may cease to vary altogether. Currently Polaris varies by only a few hundredths of a magnitude and is thus well below the range detectable by the eye.

Polarissima is the name John Herschel gave to the "nebula" he found closest to the north celestial pole during his great sky surveys of the early 19th century (**Figure 8.1**). More commonly called NGC 3172, it is a small spiral galaxy about 1' across and magnitude 13.5. Although some have seen it with telescopes as small as 6 inches, under average sky conditions NGC 3172 will be a challenge for an 8-inch. Here is an object that is better seen with high magnification. Try also for a faint pair of galaxies just a few degrees from Polaris, in the far northeastern corner of Cepheus. NGC 2276 and 2300 are only 6' apart. Both are 11th magnitude, but NGC 2300 will be spotted first. It is round and 1' across, while its neighbor to

Figure 8.1
The spiral galaxy NGC 3172 is often called Polarissima because of its proximity to Polaris, the bright star at left.

the west is twice as big and does not stand out as well in the telescopic field. This pair stays at nearly the same altitude all year, providing a convenient test for sky transparency.

While near the pole, try for another tiny galaxy. NGC 6217 in Ursa Minor is 11th magnitude, and I estimate it to be about 1' long. (Catalog entries, often based on photographic images, list it as somewhat larger.) This spiral should be within the reach of any telescope 8 inches or larger, but on good nights I have found it with a 4-inch. This open-armed barred spiral is visible from north-temperate latitudes throughout the year and provides a good test for sky transparency.

Near the extreme western edge of Cepheus is the open cluster NGC 6939. Through a small telescope it appears about 5' in diameter and shines with a total light equivalent to a 10th-magnitude star. In their book *Revue des constellations*, R. Sagot and J. Texereau describe NGC 6939 as: "Not very notable in a 3-inch 18-power refractor; a round milky spot with very faint stars in a 3¾-inch at 45×." In the catalogs, this open cluster is listed as being fainter than NGC 188, but it is far easier to pick up because the stars are concentrated in a smaller area of sky. From Springfield, Vermont, my 5-inch Apogee telescope held it in view even when stopped down to 2¾ inches. A 10-inch shows a sprinkling of stars. And only about 1° to the southeast, just over the border into Cygnus, is the fine galaxy

Figure 8.2
While scanning Cepheus, look for the star cluster NGC 7380. (North is to the upper right.)

NGC 6946. This spiral is easily seen in even small telescopes. It shines at 9th magnitude and is roughly 10′ across.

Tucked in the southeastern corner of Cepheus is the bright cluster NGC 7380. Its 50 stars (**Figure 8.2**) are scattered over an area about 10′ in diameter. Having a total visual magnitude of 6.4, NGC 7380 can be readily seen in binoculars. Small telescopes will reveal a conspicuous double star, magnitudes 7.6 and 8.6, at the southwestern edge.

The most northern galactic cluster in the sky, NGC 188, is also one of the oldest known, 14 to 16 billion years[*]. It is located just 4° south of Polaris and 1° south of 2 Ursae Minoris — a bright star that was engulfed by Cepheus when the constellation boundaries were redefined by Eugene Delporte in 1930. NGC 188 is 15′ across, so use low power. It contains some 150 stars, most of which are fainter than 13th magnitude. Despite a total magnitude of 9, it is virtually invisible with poor transparency, or in too feeble a telescope. On fine nights I see it as a ghostly glow

---

[*] *Currently the age of the universe is estimated at 12–14 billion years, while the ages of some of the oldest stars in globular clusters are estimated to be around 13–15 billion years. As astronomers work on a more accurate determination of the ages of stars and the universe, this seeming paradox of stars older than the universe is being resolved.*

AUGUST

in the 4-inch Clark refractor. In his *Celestial Handbook,* Robert Burnham, Jr., describes the appearance of NGC 188 in a low-power 6- or 8-inch telescope: "A large but dimly luminous spot with only a few of the brighter members showing individually." Although it shows nicely in an 8-inch telescope, you will have to search it out from a few bright stars nearby that some people mistake for the cluster. Back in Kansas, on a night when the naked eye reached magnitude 7.5, I counted several dozen stars in NGC 188 with my 10-inch reflector at 86×.

Inside the Cepheus pentagon is a more challenging open cluster, NGC 7142. John Herschel described it accurately as "a large, rich, loose cluster of stars of magnitude 10 or 11." To me it seemed an evenly spread layer of small stars. Look for some faint stars in a patch 10′ across. Finding them can be a problem, for the brightest single member (except for two obviously foreground stars) has been measured as visual magnitude 12.4. Hence, NGC 7142 does not show in a 2-inch finder, and I generally search with the main telescope after plotting the cluster on a detailed chart. High magnification helps after this cluster has been found. One fine night in Connecticut, the 4-inch Clark at 40× showed an even glow in which a dozen stars twinkled. Since 100× revealed more, I put on a Barlow lens to double the power again, and saw a host of minute stars.

Open cluster NGC 7510 in Cepheus is about 9th magnitude and 3′ across. Of it, Canadian amateur Dunstan Pasterfield writes, "Very easy to find, an attractive object framed by surrounding stars. It has an unusual shape, like a very thin arrowhead that is slightly bent at the tip. A hint of nebulosity. About 7-10 stars — delightful thing." Have others seen any nebulosity here? Could the impression of a glow come from faint stars in the cluster not seen directly?

More Sights in Cepheus

Though the north celestial pole lies in the constellation Ursa Minor, Cepheus reaches up to declination 88½°. Most of its bright stars are 20° or 30° from the pole, in the Milky Way where it crosses the southern part of the constellation. "Here one might expect many clusters and planetary nebulae," Scotty wrote, "especially since neighboring Cassiopeia has a profusion of them." But Cepheus is relatively poor in traditional deep-sky wonders. It has no dramatically bright galaxies (only one is plotted on *Sky Atlas 2000.0*) or any globulars. And, surprisingly, rich open clusters are not as profuse as one would expect from a Milky Way region. In all, one could say that Cepheus has few good objects. But "good" is a relative description, as Scotty reveals here. Remember, even a faint object can be a good one if it presents a challenge.

Thanks to their never setting from mid-northern latitudes, some of my favorite double stars within 15° or so of the pole can be inspected whenever the mood strikes, or shown to a friend who drops in. The circumpolar doubles range from difficult binaries to glorious objects at low power.

Polaris itself is a wide optical pair, first seen as double by William Herschel in 1779. Its two stars, magnitudes 2.1 and 8.9, are 18″ apart. Argument still goes on about the smallest aperture needed to show the companion. Long ago, William Dawes suggested it as a test for a 2-inch, but it has been seen in smaller telescopes. Why not make some systematic tests with a graduated set of cardboard diaphragms on your own telescope?

An easy pair that can be split even in big tripod-mounted binoculars is Struve 1694 (also written as Σ 1694) in Camelopardalis. It consists of a 5.3-magnitude star and a 5.8-magnitude companion, separated by about 22″ and moving in parallel paths in space. The "Σ" prefix (often seen in star catalogs) means that a double star is listed in Wilhelm Struve's famous catalog.

Look in Cepheus for Struve 2923, an unequal 9″ pair. Its components, of magnitudes 6.3 and 9.4, should be resolved by 2- or 3-inch telescopes. This, too, is a common proper motion pair. Half a degree away is a difficult binary that makes an interesting object for larger amateur instruments: Struve 2924. This system of 6.6 and 7.0 magnitude stars has an orbital period of 226 years, according to the calculations of Wulff D. Heintz. The true orbit is actually nearly circular, but the apparent orbit is a highly foreshortened ellipse. Struve discovered this pair about 1830, when the separation was 0.8″. In the 1930s it had closed to less than 0.2″, and I've seen it widened to 0.5″.

Figure 8.3
Binocular and small-telescope users will find the nebula NGC 7023 to be an interesting view.

NGC 7023 is an object that *Sky Catalogue 2000.0* calls "one of the brightest reflection nebulae," an encouraging description to say the least. It is centered on a 6th-magnitude star that is easily seen in binoculars (**Figure 8.3**). From my earliest days as an observer I have notes that refer to the nebula as "real bright," and question why it is not plotted in *Norton's Star Atlas*. I suspect it is an easy object for big binoculars.

Figure 8.4
Planetary nebula NGC 40 in Cepheus measures about one light-year across. It has an extended halo (not seen here) probably formed by the mass ejection of gas from its atmosphere.

Two rather difficult nebulae, on the other hand, are NGC 40 and IC 1470. NGC 40 is a 10th-magnitude planetary about 0.6′ in diameter (**Figure 8.4**). Though located in a region devoid of bright guide stars, it was spied with my 5-inch Apogee telescope. IC 1470 lies on the galactic equator near the Cassiopeia border. On a good night it should be an easy object for an 8-inch telescope. Tom Reiland, of Glenshaw, Pennsylvania, was observing IC 1470 with an 8-inch f/5.3 Newtonian when he happened upon a smaller and fainter object about 12′ south and slightly west of it. At first he thought it might be a comet, but increasing the power from 54× to 130× revealed half a dozen stars spread across 30″ with a hint of nebulosity. IC 1470 is now regarded as a diffuse nebula, though at one time it was thought to be a planetary. Using a pair of 5-inch binoculars, I searched quite a while before locating this faint object, which is only 1′ across. But once found, it was relatively easy.

Malcolm J. Thomson observes with a 16½-inch f/5 reflector from Santa Barbara, California. A recent letter tells of his surprise when his observations of a gaseous nebula in Cepheus showed more detail than the famous Sir John Herschel saw. This object is NGC 7129. Herschel described it as "a very coarse triple star involved in a nebulous atmosphere; a curious object. The nebula is extremely faint and graduates away." The stellar triangle is about 2' in extent. Thomson, however, has seen five stars. He writes: "At 70×, three stars are visible in the nebulosity, A in the south following part, B in the south preceding. The third, fainter star, C, is northeast of A. With 160×, a fourth star was visible slightly southwest of C. Using 222×, I could glimpse a fifth star just southeast of A. All five components were readily visible with 333×."

The 18¾-inch speculum-metal mirror used by John Herschel for his observations in 1829 was perhaps equivalent to a modern 12-inch aluminized mirror. One indication of Thomson's observing skill is his independent rediscovery of the very faint galaxy NGC 5296 in Canes Venatici, which was missed by both William and John Herschel and was first noted by Lord Rosse with his 72-inch in 1850.

I looked at NGC 7129 with my 4-inch Clark refractor on a night so clear that the Triangulum Galaxy M33 was intermittently visible to the naked eye. At 120×, the nebula and John Herschel's three stars were clearly seen, but even with 310× there was no trace of the other two stars.

The Great Planetaries of Summer

"They are ephemeral spheres that shine in pale hues of blue and green and float amid the golden star currents of our galaxy." What Scotty was describing, of course, are the planetary nebulae, which, he extolled, "are often the delight but also the bane of amateur astronomers." Bright planetary nebulae are uncommon, and planetaries of all shapes are rare because their lifetimes are short. The Messier catalog contains only four of them — M27, M57, M76, and M97. To date, only about a thousand are known, and perhaps only a hundred of them are suitable for amateur telescopes. The most famous planetary in the heavens is unquestionably M57, the Ring Nebula in Lyra.

To some people, the ethereal gas bubbles of planetaries have a compelling pull all their own. They float on the foam of the Milky Way like the balloons of our childhood dreams, so delicate they appear. If you want to stop the world and get off, the lovely planetaries sail by to welcome you home. The Ring Nebula, M57 in Lyra, is one of the best-known objects in the summer sky (**Figure 8.5**). It was discovered by the French astronomer Antoine Darquier in 1779 while comet hunting with a 3-inch refractor. He described it as "a very dull nebula, but perfectly outlined; as large as Jupiter and looks like a fading planet."

Figure 8.5 M57 in Lyra appears ring-like because we are looking down a barrel of gas cast off by a dying star thousands of years ago.

By Herschel's time its "smoke-ring" form was known and justly admired. William Herschel thought it to be a ring of stars just beyond the resolution of his telescopes. His son John first called attention to the fainter nebulosity which fills the interior of the ring, likening it to gauze stretched over a hoop. In Ireland, Lord Rosse used his 6-foot speculum-metal reflector in the 1840s to detect structure in M57's ring, but curiously he does not mention the central star first seen with a smaller telescope by the German Friedrich von Hahn around the year 1800. (Rosse does say that he never really viewed M57 under excellent conditions when his 6-foot reflector was working at its best.) All this reminds us of the superior optics which today's amateurs enjoy, for under good conditions the central nebulosity and star can be glimpsed in apertures as small as 6 and 10 inches respectively.

The Ring Nebula differs from most planetaries by the almost perfect sharpness of its outlines, and the completeness of the ring form, in contrast to such objects as the Dumbbell Nebula, where visual scrutiny would never suggest the typical planetary construction. However, long-exposure photographs of M57 show a second ring outside the first, of a fragmentary and curdled appearance. Also designated NGC 6720, M57 is easily found halfway between Beta (β) and Gamma (γ) Lyrae. It is oval, 80″ by 60″, and is of the 9th magnitude visually.

For early observers, the trick was to see it as a smoke ring with a dark core. As the average amateur's telescope became larger and eyepieces better made, the ring was no longer a challenge, especially since this object bears magnification well. Appearing a bit more than 1′ across, M57 looks like a 9th-magnitude star in finders. The Apogee telescope shows the ring as very bright, but no other detail is visible. At powers of 250× and up, a curious effect takes place. The oval outline of M57 takes on a lemon shape with the ends of the oval appearing rather pointed. They also appear more diffuse and wispy. A power of 600×, however, is none too great if there is sufficient aperture to support it. Even at high magnification, the interior of the nebula retains a thin film of haze that can show some structure.

I have probably looked at the Ring Nebula with a greater variety of telescopes than I have at any other heavenly subject. One of my best views came with the 12-inch f/17 Porter turret telescope on Breezy Hill in Springfield, Vermont, the site of the annual Stellafane convention. At 200× the ring was bright, slightly elongated, and of uniform luminosity. An increase to 600× changed the picture dramatically. No longer was the smoke ring evenly bright. Instead, two sides of the ring were made up of curved and twisted streamers. The oval now had pointed ends, and the central region was full of turbulent detail.

The sky above Breezy Hill that night was superb, with the naked-eye magnitude limit approaching 7. M33 could even be glimpsed without optical aid (a favorite test of mine for those rare nights of really excellent observing conditions).

So what does it hold for the explorers of the wilds? In 1874, Edward S. Holden studied it with the 26-inch refractor at the U. S. Naval Observatory in Washington, D. C. He commented that the interior of the ring appeared to be filled with "glistening points" of light. This report remained unique for about a century. In 1979, I was looking at the Ring with the Porter turret telescope using 600×. To my surprise I, too, saw a scattering of faint stars across the center of the nebula and against the brighter parts of the ring. However, these stars were not visible in the 20-inch Clark refractor at Connecticut's Wesleyan University. But that telescope is plagued by light pollution. The problem today is to see the gossamer bands of nebulosity that cross the planetary's core. They are visible in the Porter telescope. This instrument is particularly noteworthy for its high-contrast views of the planets and deep-sky objects.

On a top-class night, a 12- or even 10-inch telescope can show the planetary's central star. In moments of exceptional atmospheric conditions a 12-inch or larger instrument may reveal a scattering of stars across the central vacancy and even amid the ring itself. However, stars do not spread beyond the outer edge. I've checked several photographic atlases and noticed a definite lack of faint stars around M57. From my own dark-site observations I know that the edge of the Milky Way does cross this region. Could there be a dark halo surrounding the glowing ring that blocks the view beyond? If so, why do we see stars inside the ring?

For those of you interested in viewing the central star of M57, it is worth noting that over the years, estimates of its magnitude have ranged from 14.5 to less than 16. Although some have suspected the star as being variable, the data are far from conclusive, as anyone who has tried to estimate the magnitude of a star embedded in nebulosity knows. Needless to say, the central star is very difficult to glimpse. It should be a routine object for a 12-inch telescope, but often cannot be seen with a 17½-inch. The usual technique of increasing magnification to darken the sky behind a faint star does not work as well as expected. This is due to the faint nebulosity filling the interior of the ring and reducing the star's contrast with its background. A telescope's focal ratio may play some part in rendering the star visible. At f/17 the 12-inch Porter turret telescope usually makes easy work of M57's central star, while significantly larger apertures have difficulty.

Try also looking for three galaxies that are near M57; one is even in the same field of view. The brightest of the three, NGC 6713, lies just a bit more than 1° northwest of the Ring. Albert Marth discovered it with the 48-inch speculum-metal reflector that William Lassell constructed in the early 1860s on the Mediterranean island of Malta. I wouldn't be surprised if a modern 16-inch reflector, with its high-efficiency optical coatings, had a total light-gathering power similar to that of the speculum-metal mirrors in Lassell's telescope. The *NGC 2000.0* catalog lists NGC 6713 as photographic magnitude 14, so visually it might appear a little brighter — say 13.2. Another galaxy is NGC 6700, located about 1½° southwest of M57. Édouard Stephan discovered this one with the 31-inch silver-on-glass reflector at Marseilles Observatory in France. *NGC 2000.0* lists it as photographic magnitude 14 as well, so it should appear similarly brighter visually.

Much more challenging is a tiny 15th-magnitude barred spiral discovered by E. E. Barnard. IC 1296 lies just 5′ northwest of the ring. I do not know of any amateur sightings of this galaxy.

The Dumbbell's Many Faces

Next to M57, the Dumbbell Nebula (M27) is the sky's second most sought-after planetary nebula. Its ghostly green orb is really the glow of dimly lit shells of gas blown off an aging star shining weakly at the nebula's core. As the shells expand, they fade, and only a few tens of thousands of years pass before their ghostly shapes disperse into space. The extreme limit of visibility of the expanding shell and of its tiny illuminating source, a white dwarf star, intrigued Scotty. Naturally, he loved to share the visual mysteries of planetaries with his readers. And he couldn't resist tossing out a planetary nebula challenge or two to those who would pursue them. The Dumbbell was not lacking in challenges for Scotty or his readers.

The August sky contains many delightful planetary nebulae — ephemeral spheres of blue and green gas that float amid the pearly star currents of the Milky Way. Certainly one of the most observed is the Dumbbell Nebula, M27, in Vulpecula (**Figure 8.6**). Binoculars will show it, but no instrument ever exhausts the additional detail that may be seen as larger telescopes are turned to it. Thirty years ago my mail was filled with complaints from amateurs who had difficulty finding M27, but I haven't heard such remarks for the past 10 years. This is a tribute to the growing observing skills among the fraternity of today's amateurs. If you're having trouble locating it, here's one method that works well for me. Set your finder on Gamma (γ) Sagittae, the head of the celestial arrow. Sweep about 5° north and you should see an M-shaped pattern of stars composed of 12, 13, 14, 16, and 17 Vulpeculae; this group is more conspicuous to the eye than most star charts lead you to believe. M27 is just ½° south of the M's central star.

The noted English observer Thomas W. Webb saw M27 as "two hazy patches in contact," and it is this appearance that gives rise to the nebula's popular name. However, I don't believe that many observers see the dumbbell shape without having a bad wrench to the imagination. John Mallas, who observed with a 4-inch refractor, drew M27 as a rectangle twice as long as it is wide. With my 4-inch Clark refractor, a quick look reveals the planetary as two cones with their apexes in contact. After finding the best eyepiece for the evening's sky conditions, and by using averted vision, I usually see the faint nebulosity between the brighter parts of the cones. By gently rocking the telescope back and forth — which sets the planetary in motion and helps the eye capture the faintest extensions of light — I found the end result is a full circle of light, just as one would expect of a planetary. The luminosity of the interior varies greatly, but the circular outline remains firmly fixed.

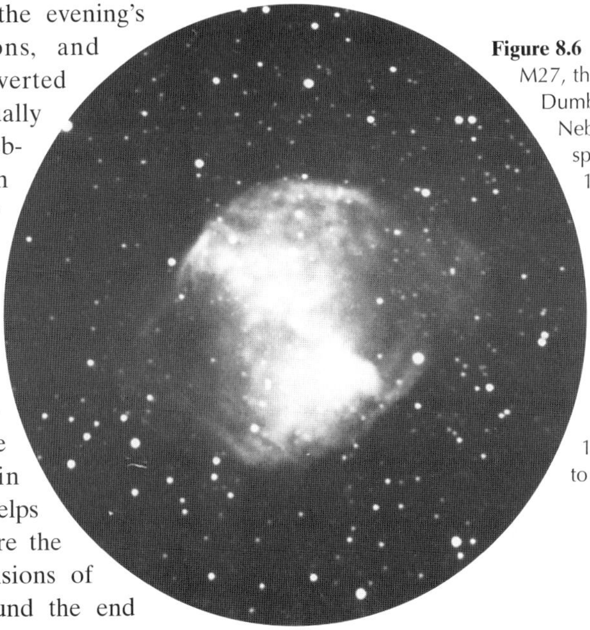

Figure 8.6 M27, the famous Dumbbell Nebula, was spotted in 1764 by Messier and was the first planetary nebula discovered. Its light takes about 1,000 years to reach us.

It is hard to assign a "best" type of telescope for viewing M27. My 5-inch Apogee telescope with a fixed power of 20× shows it as a bright sphere with the dumbbell shape rather mild. My 10-inch, f/8.6 reflector shows M27 much better at 300× by means of a Barlow lens than at the same power with a short-focus eyepiece. The latter left the sky gray, and contrast with the nebula was poor. The Porter turret telescope reveals intricate inner detail and a strong edge all the way around the nebula. M27 is often cataloged as having dimensions of 8' × 4', but as mentioned above, it really appears round and about 8' across, especially on photographs. Its total brightness is equal to that of a 7.6-magnitude star, and there is a 12th-magnitude central star that is a difficult object for most amateur telescopes.

The fact is, this fickle planetary has as many shapes as there are observers; its appearance can even vary from minute to minute as observing conditions change. Several years ago I had a good lesson in just how critical a role the atmosphere plays in observing faint, nebulous objects. During May 1983, I was in western Pennsylvania when Comet IRAS-Araki-Alcock raced across the sky from Draco to Cancer. It moved so swiftly that within minutes I could detect a change in its position with the naked eye. What really interested me, however, was the coma's diameter. At times it was 2° or 3° across, while at other times it was fully 6°. This must have been due to atmospheric changes that otherwise would have gone unnoticed.

According to Burnham's *Celestial Handbook*, M27 appears to be expanding at the rate of 1″ per century. If this rate has remained constant, then the nebula is perhaps 48,000 years old — if it has slowed over the years then the object could be much younger.

So the test with M27 is not in seeing it — the glowing bubble, as I mentioned, should be easy in finders if the sky is dark — but rather in what is seen. Can you trace the faint wisps of nebulosity from the ends of the bar to where they meet above and below it?

Houston's Uncertainty Principle

The next time you're out under a very dark sky, look up at the Milky Way. How wide is it? The answer is not as simple as it seems. The answer depends not only on where you look but how you look. It also depends on the conditions and state of the atmosphere and on your own physical condition and eyesight — all factors that vary with time. Thus, the "shores" of this Galactic River can be as hard to define as the extended coma of a comet. Even star charts do a poor job of tracing the visual breadth of this, the sky's largest deep-sky wonder. But the problem doesn't end there. The same difficulty we experience in this simple exercise with the naked eye applies also to the telescopic observing of nebulae, especially of the planetaries, whose expanding shells gradually fade until they blend with the sky background. Scotty was excited about this visual conundrum, which he explains here.

This may sound foolish, but all my life I've wanted to find the edges of the Milky Way. Obviously our galaxy can't go on forever. It must stop at some point. But where? On any clear night I can clean my glasses and wander out to a dark corner of my yard with a ruby-colored flashlight and star charts in hand. After letting my eyes dark-adapt, I can carefully plot the apparent edges of the Milky Way (**Figure 8.7**). The rub is, however, if I repeat the process two hours later, I usually get a border that is nowhere near my original one.

Books and atlases don't seem to offer much help. *Norton's Star Atlas* plots the boundaries of the Milky Way in great detail, but they're not what I see. The atlas's preface concedes that the boundaries are difficult to define, and that the outline on the charts is based on observations by the 19th-century astronomy popularizer Richard A. Proctor. Thus, my 1957 edition of *Norton's* uses data about a century old!

Wil Tirion's *Sky Atlas 2000.0* might seem like a better bet since it was first published in 1981. The Tirion charts show more twists and turns at the edge of the Milky Way, but here again they are often quite different from what I see. The introduction to this atlas notes that the outline shown is one that the Dutch astronomer Antonie Pannekoek prepared in the 1920s. I'm not sure how he derived it, but it is probably a combination of photographic and visual observations, both of which have their limitations.

Over the years my conviction has been growing that the atmosphere is really what determines the limits visible to the eye, and the atmosphere can change on a time scale of minutes. In late 1985 I set out to watch Halley's Comet with my 5-inch 20× Apogee telescope when it was near the Pleiades. At 11:30 p.m. sweeping did not turn up the comet; I had to revert to charts to find it. The cosmic

Figure 8.7 The Milky Way is an awe-inspiring sight on a cloudless dark night. Here, Deneb (upper left), Vega (right) and Altair (lower left) seem to frame the galaxy's star clouds.

visitor looked like a 7th-magnitude star with barely a trace of a blur at its edges. At midnight I looked again and found a whole different object before me. The comet was easily swept up and the bright coma appeared about 1° across. By 12:20 a.m Halley had returned to being a slightly nebulous star.

All this is worth remembering by the deep-sky hunter in search of faint galaxies, supernova remnants, stray H II regions, and diffuse planetary nebulae. You can use every trick in the book to press your telescope to its limit, but the atmosphere is still going to be a highly variable "filter." I call it Houston's Uncertainty Principle, and it confounds your ability to predict the magnitude limit of a given telescope to no better than one full magnitude.

In my flirtations with the boundary of the Milky Way, I have spent a lot of time looking at the region around Lyra. The Tirion charts show the Milky Way's edge crossing the lower section of the lyre, including the whole area between Beta (β) and Gamma (γ) Lyrae and its resident deep-sky showpiece M57, the Ring Nebula.

Édouard Stephan also discovered several star clusters here, including one involving the bright, naked-eye double star Delta (δ) Lyrae. This group is bright but sparse, having only a dozen or so stars in a 20′-wide area. It is easy to see in binoculars. The bright double may lead the eye astray and be the reason so few observers seem to be aware of the rest of this cluster.

From Delta Lyrae it's only a short skip to the nearly equilateral triangle of stars Epsilon (ε), Zeta (ζ), and brilliant Alpha (α), or Vega. Few amateurs seem to know that all three, and not just Epsilon, are double stars. Epsilon (ε) of course, is the famous Double Double, whose bright components are separated by 3½′ and create a test for naked-eye acuity, much more so than Alcor and Mizar in the handle of the Big Dipper. As a youth I could split them easily, but I have not been able to do this for several years. Each component is itself a pair — the separations are currently 2.6″ and 2.3″ — well within the range of a 3-inch telescope.

I've never met an amateur who claims to have resolved Vega. Perhaps most are not aware that it has a companion. Brilliant Sirius and Antares are well known as binaries, but unlike them Vega is just an optical pair with no physical connection between the components. At present the 9.5-magnitude companion lies about 1′ south of the magnitude 0 primary. The glare will make this a difficult pair at best. Don't be confused by another 9.5-magnitude star 2′ northeast of Vega.

The third member of the triangle, Zeta, is an easy pair. The 4.3-magnitude primary has a 5.9-magnitude companion ¾′ to the southeast. But a real challenge would be to look for a 15th-magnitude star cataloged as being just 25″ northeast of the primary. It would be interesting to know the accounts of your adventures in looking at these stars.

Telescopic Delights in Delphinus

The tiny diamond-shaped form of Delphinus is popular among summer stargazers. (My wife, Donna O'Meara, loves to spot it whenever she's looking up on a warm summer's eve.) Like the Pleiades, this small but tight grouping of moderately bright suns has the power to draw attention to itself. "Unlike many of the large constellations that date to antiquity," Scotty wrote, "most smaller asterisms are the handiwork of celestial cartographers from the 17th and 18th centuries. There are exceptions, however, and one of my favorites is Delphinus, the Dolphin." Scotty explained that his love for Delphinus was a long-lived one. "I became well acquainted with Delphinus when I joined the American Association of Variable Star Observers in the 1930s and started doing 'serious' astronomy," he wrote. Indeed, the first variable star he would log in his record book would be Z Delphini. The constellation's 189 square degrees also offers a fine blend of stellar and deep-sky wonders, making this delightful constellation a naked-eye, binocular, and telescopic attraction.

Summer lies hot and tranquil on the land. The gigantic storms of winter and the turbulent atmosphere that accompanies them are only memories now. At this time of year the seeing is steady all night.

West of the meridian in late evening lie the great star fields dancing with the brilliance of Sagittarius, Scorpius, and Scutum. The eastern sky, however, is a virtual desert of bright stars. The Great Square of Pegasus has little to offer the naked-eye observer, and Equuleus is likewise dim. On nights when a bright Moon floods the heavens with its golden light, the eastern sky appears almost devoid of stars. Near the meridian, however, is the small constellation Delphinus the Dolphin (**Figure 8.8**). To the eye it appears as little more than an overgrown asterism, but it contains some compelling objects for anyone with a telescope.

Delphinus has a distinctive diamond of stars forming the creature's head. While these have never received official cluster status, there is some question as

Figure 8.8
The constellation Delphinus looks like a diamond with a tail and can be found about 12° northeast of brilliant Altair, the brightest star in Aquila, the Eagle.

to whether several of them are physically related to one other or to fainter stars in the immediate area. The constellation is thinly covered with the outliers of the Milky Way and galactic dust as well. This blocks our view of the universe beyond, so the area lacks the profusion of galaxies that are visible farther to the east.

Nevertheless, I developed a fondness for Delphinus years ago when I began making observations for the American Association of Variable Star Observers. The area was rich in variables, and I could make half a dozen estimates in less time than it took to search out a single star in a less crowded field.

In my youth I knew Delphinus as Job's Coffin, a moniker whose origin even such star-name experts as Richard H. Allen acknowledged as being lost in the sands of time. Although my library is limited, the term appears in a copy of Elijah H. Burritt's *Geography of the Heavens,* published in the 1830s. I suspect that the name was just one of those fashionable tags that caught on among sky-watchers. Today we're surrounded with similar fads drawn from popular music, television commercials, and adolescent attire. (Can anyone pinpoint the source of ripped jeans being a fashion statement?)

On a similar note, at one Stellafane convention in Vermont I heard a 20-year-old amateur refer to the star Betelgeuse as "beetle-juice." While many readers are familiar with this term, I never encountered the pronunciation before World War II. In fact, I wonder if it originated with an Army Air Corps cadet at a military base where I taught celestial navigation during the war. One of the cadets was having trouble saying Betelgeuse, so he changed it. Before long even the experienced navigators had picked up the altered pronunciation.

Delphinus has several noteworthy double stars. Gamma (γ), at the northeast corner of the diamond, is a beautiful pair with a 4.5-magnitude primary and a 5.5-magnitude companion some 12″ to its west. Although the pair is physically connected, the orbital period is so slow that the position angle of the two stars has barely changed since it was first measured by Wilhelm Struve in 1830. The pair has long been known for its contrasting colors, but there is some question as to exactly what the colors are — different observers report different combinations. All agree, however, that Gamma Delphini is a splendid sight in small telescopes.

In the 1930s I was at the University of Wisconsin and had access to Washburn Observatory's 6-inch Clark refractor, used by Sherburne W. Burnham to observe double stars. One of the many doubles he discovered with that telescope was Beta (β) Delphini, at the southwest corner of the diamond. The components have magnitudes of 4.0 and 4.9, and, though an extremely challenging object for amateur telescopes (the pair's separation varies between 0.2″ and 0.65″), Beta Delphini has an orbital period of only 26.6 years. For about half of the orbital period of the pair, the stars are separated by about 0.30″. By the turn of the century the stars will be nearing maximum separation; the most likely reason Burnham succeeded where other great double-star observers had failed is that he chanced upon the pair when it was near its maximum separation in August 1873.

In 1950 I examined the star with my newly completed 10-inch reflector. Then the separation was near a maximum of 0.6″ with the companion due north of the primary. My first attempts to split the pair failed because the companion was lost in the diffraction spike caused by the telescope's secondary mirror holder. Success came only after rotating the tube 45° in its cradle to shift the position of the spike. Ten years later the companion had moved roughly due east of the primary, and the separation had shrunk to a minimum of 0.2″. I was unable to detect even a slight elongation of the double star's image. Try experimenting with masks to determine the smallest aperture which will show the pair.

Alpha (α) Delphini is a challenging double for a different reason. With a separation of almost 30″ (two-thirds the apparent diameter of Jupiter), the pair would seem to be within reach of the smallest telescope. But at magnitude 13.3, the secondary is 9½ magnitudes (more than 6,000 times) fainter than the primary. It's located to the southwest. An old trick is to place a bright star just outside the field of an eyepiece when searching for faint companions.

Figure 8.9
Distant, shimmering NGC 7006 is one of the most remote globular clusters that appear to swarm around the Milky Way Galaxy.

The interesting globular cluster NGC 7006 lies about 3½° due east of Gamma (γ) Delphini. If you have a motorized drive, switch it off, set your telescope on Gamma, and 15 minutes later the globular will be in view (**Figure 8.9**). Though readily detectable in a 3-inch instrument, it is small — 1.1′ in diameter — and so concentrated that it may at first be mistaken for a 10th-magnitude star. Although a relatively easy object for a 6-inch telescope, this cluster is among the most distant globulars known, located more than 110,000 light-years from Earth. Large telescopes may show it with a clumpy appearance, but I doubt it can be resolved in any amateur instrument. NGC 7006 affords an interesting contrast with the nearby great globular cluster M15, in Pegasus, which can be seen with the naked eye.

Somewhat easier is the globular NGC 6934. More than 5' in diameter and shining with the total light of a 9th-magnitude star, the cluster can be glimpsed in binoculars. Because of its setting, I find it a particularly pretty object for rich-field telescopes. (**Figure 8.10**)

Figure 8.10 Another globular cluster easily seen in small telescopes or binoculars is NGC 6934. Through larger telescopes, individual stars can be seen around the edges and near the concentrated knot of stars at the center.

NGC 6905 is a planetary nebula situated in a coarse clustering of stars in the very northwest corner of Delphinus. William Herschel discovered it in 1782, and his son John Herschel speculated on a possible connection between the nebula and the many faint stars around it. Its 12th magnitude elliptical disk is about 45" in diameter (roughly the same size as Jupiter's disk) and is relatively easy for a 6-inch telescope under good skies. On one of those exceptionally rare nights when the Connecticut atmosphere forgot that its main function is to depress amateur astronomers, I fished out NGC 6905 with my 4-inch Clark refractor. At least a 10-inch telescope is needed to make out the 14th-magnitude central star. Barbara Wilson of Houston, Texas, reports seeing a lot of internal structure in NGC 6905 with a 20-inch reflector. The central star has an extremely high surface temperature of about 100,000° Kelvin.

NGC 6891 is a small object about 12" in diameter, roughly three times smaller than NGC 6905. At low magnifications it looks almost stellar. The planetary is listed in *Sky Catalogue 2000.0* as being of photographic magnitude 11.7, but it seems brighter visually (other sources list it as between 9th and 10th magnitude). I have seen it from Connecticut with a 4-inch off-axis reflector. Observing with a 16-inch reflector from the Sonoran desert outside Tucson, Arizona, Ron Morales notes that NGC 6891 resembles an unresolved globular star cluster.

Unlike many authors, T. W. Webb took great pleasure in writing about rich star fields. These are not the discrete objects like star clusters and galaxies that interest most deep-sky observers, but rather wonderful starry vistas that are nothing more than chance alignments of random suns. Webb wrote about them simply because they were delightful to look at.

Philip Harrington has a similar attitude in his book *Touring the Universe Through Binoculars*. Accordingly, he lists several asterisms that are new to amateur observing guides. One such group involves 6th-magnitude Theta (θ) Delphini. Harrington notes that there are about two dozen stars here of 9th magnitude and brighter that are visible in binoculars.

About 1½° southeast of Theta lies NGC 6956, a small galaxy discovered by William Herschel. It is about 2' across and, by my estimate, magnitude 13.2. While Herschel considered it "very faint" as seen with his 18¾-inch reflector, perhaps the telescope's speculum-metal mirror was tarnished or the sky conditions were unfavorable on that particular night. Today NGC 6956 can be held easily in a 4-inch refractor.

Although galaxies are sparse in Delphinus, NGC 6928 is the brightest member of a foursome located 1½° south of Epsilon (ε) Delphini. Its cigar-shaped disk is about 2' long and magnitude 12½. An 8-inch telescope should easily snare it on a good night, but a larger aperture will probably be needed to see its companions. They all lie within ¼° of NGC 6928.

AUGUST OBJECTS

Name	Type	Const.	R. A. h　m	Dec. °　'	Millennium Star Atlas	Uranometria 2000.0	Sky Atlas 2000.0
Alpha (α) Delphini	✶✶	Del	20　39.6	+15　55	1216, 1217	209	9, 16
Beta (β) Delphini	✶✶	Del	20　37.5	+14　36	1217, 1241	208, 209	9, 16
Delta (δ) Lyrae	✶✶	Lyr	18　53.7	+36　58	1152, 1153	117, 118	3, 8
Dumbbell Nebula, M27, NGC 6853	PN	Vul	19　59.6	+22　43	1194, 1195	162, 163	8, 9
Epsilon (ε) Lyrae	✶✶	Lyr	18　44.3	+39　40	1132, 1153	82, 83, 117	3, 8
Gamma (γ) Delphini	✶✶	Del	20　46.7	+16　07	1216	164, 209	9, 16
IC 1296	Gx	Lyr	18　53.3	+33　04	1152, 1153, 1174, 1175	—	—
IC 1470	BN	Cep	23　05.2	+60　15	1070	34, 58	3
NGC 40	PN	Cep	00　13.0	+72　32	24	3, 15	1, 3
NGC 188	OC	Cep	00　44.0	+85　20	2, 5, 6, 1035	1, 2	1, 3
NGC 2276	Gx	Cep	07　27.0	+85　45	1, 3, 4, 518, 522	1	1
NGC 2300	Gx	Cep	07　32.0	+85　43	1, 3, 4, 518, 522	1	1
NGC 6217	Gx	UMi	16　32.6	+78　12	1046	11	2
NGC 6700	Gx	Lyr	18　46.0	+32　17	1175	117	—
NGC 6713	Gx	Lyr	18　50.7	+33　57	1153	117	—
NGC 6891	PN	Del	20　15.2	+12　42	1242	208	16
NGC 6905	PN	Del	20　22.4	+20　07	1217	163	9, 16
NGC 6928	Gx	Del	20　32.8	+09　56	1241	208, 209	—
NGC 6934	GC	Del	20　34.2	+07　24	1265	208, 209	16
NGC 6939	OC	Cep	20　31.4	+60　38	1074, 1075	32, 55, 56	3
NGC 6946	Gx	Cep	20　34.8	+60　09	1074, 1075	32, 56	3
NGC 6956	Gx	Del	20　44.0	+12　31	1240	209	—
NGC 7006	GC	Del	21　01.5	+16　11	1215	164, 165, 209, 210	9, 16, 17
NGC 7023	OC+BN	Cep	21　00.5	+68　10	1061	32, 33	3
NGC 7129	OC+BN	Cep	21　41.3	+66　06	1060	33	3
NGC 7142	OC	Cep	21　45.9	+65　48	1060	33	—
NGC 7380	OC+BN	Cep	22　47.0	+58　06	1071	58	3
NGC 7510	OC	Cep	23　11.5	+60　34	1070	34, 58	3

Ast = Asterism; BN = Bright Nebula; CGx = Cluster of Galaxies; DN = Dark Nebula; GC = Globular Cluster; Gx = Galaxy;
OC = Open Cluster; PN = Planetary Nebula; ✶ = Star; ✶✶ = Double/Multiple Star; Var = Variable Star

AUGUST OBJECTS (CONTINUED)

Name	Type	Const.	R. A. h m	Dec. ° '	Millennium Star Atlas	Uranometria 2000.0	Sky Atlas 2000.0
Polaris, Alpha (α) Ursae Minoris	**	UMi	02 31.8	+89 15	1, 2, 517, 518, 1033	1, 2	1, 2, 3
Polarissima, NGC 3172	Gx	UMi	11 50.0	+89 07	1, 517, 518, 1034	1, 2	—
Ring Nebula, M57, NGC 6720	PN	Lyr	18 53.6	+33 02	1152, 1153, 1174, 1175	117	8
Struve 1694	**	Cam	12 49.2	+83 25	520	9	2
Struve 2923	**	Cep	22 33.3	+70 22	1048	34	3
Struve 2924	**	Cep	22 33.0	+69 55	1048	34	3
Theta (θ) Delphini	*	Del	20 38.7	+13 18	1241	209	16
Vega, Alpha (α) Lyrae	**	Lyr	18 36.9	+38 47	1132, 1153	82, 117	3, 8
Zeta (ζ) Lyrae	**	Lyr	18 44.8	+37 36	1153	117	3, 8

Ast = Asterism; BN = Bright Nebula; CGx = Cluster of Galaxies; DN = Dark Nebula; GC = Globular Cluster; Gx = Galaxy; OC = Open Cluster; PN = Planetary Nebula; * = Star; ** = Double/Multiple Star; Var = Variable Star

SEPTEMBER

CHAPTER 9

Wandering Through Lacerta, the Lizard

"No region in the heavens is barren," Scotty wrote. "No constellation is fruitless for the observer. Even a lifetime of exploring the celestial display cannot exhaust the surprises that dance so beautifully before the amateur astronomer." His words ring true because some regions of sky remain frequently overlooked by backyard observers. Consider Lacerta, the Lizard. In the September 1972 Deep-Sky Wonders, Scotty told how James P. Brown of Kingsport, Tennessee, pointed out to him that, for many years, the column had curiously omitted that constellation. Tucked between Cygnus and Andromeda, Lacerta extends almost a full hour in right ascension and about 20° in declination, but its brightest star shines at a dim 4th magnitude. "Inconspicuous as Lacerta, the Lizard, may be," Scotty proffered, "it nevertheless offers some good open clusters for September viewing."

As the last glow of evening twilight drains from the western sky during September evenings, the small constellation Lacerta crawls high overhead. This celestial lizard was created in 1687 by Johann Hevelius to fill the void between Cygnus and Andromeda. Some eight years earlier, the French astronomer Augustin Royer used the handful of naked-eye stars in this area to portray a scepter and hand of justice commemorating King Louis XIV of France. And, almost a century after Hevelius's figure, the German Johann Bode used these same stars to form Frederici Honores as a tribute to his sovereign Frederick the Great. But just as the rule of a monarch is temporary, so too was the acceptance of these asterisms, and today only the lowly lizard occupies this space.

Lacerta is a weak little constellation tucked alongside the galactic equator and much overshadowed by its vivid neighbors Cygnus and Cassiopeia. Because Lacerta's naked-eye stars are all inconspicuous, they provide an opportunity for observers to test their skills at finding their way far from bright guidepost stars. Its brightest star, Alpha (α) Lacertae, is only magnitude 3.8, so from a light-polluted urban location the celestial lizard may be a bit difficult to find. Because of a lack of bright guide stars to help point the way, this area is a good place for the novice telescope user to practice hunting with a finder.

There is a good reason why amateurs should get familiar with Lacerta. Since 1910 three novae have blazed out in this area, and more will certainly come. Because there are no stars brighter than 4th magnitude to distract the observer, a relatively faint interloper should be evident.

Though it has no spectacular deep-sky objects, it does contain several bright clusters that can be reached with modest instruments. As many observers know, areas of the sky that appear empty to the naked eye often hold interesting goodies for a small telescope. Sentiment often drives me to take a look at two multiple stars in Lacerta that I learned of in my youth, while surveying the sky with a copy of the venerable *Norton's Star Atlas*. Struve 2894 consists of a pair of 6.1- and 8.3-magnitude stars separated by 16″. This puts them well within the grasp of a 2-inch telescope. The second, 8 Lacertae, is a pair of 5.7- and 6.5-magnitude stars about 22″ apart. Two other stars of 9th and 10th magnitude lie 82″ and 49″ from the primary, respectively, and may be part of the same physical system.

NGC 7243 is the only cluster in Lacerta mentioned in the classic observing guides by William H. Smyth and Thomas W. Webb (the 1881 revision of Smyth's *Cycle of Celestial Objects* by George F. Chambers added another open cluster, NGC 7209). Smyth notes that the surrounding area is very rich, especially to the north of the group. To find NGC 7243 look for a small keystone of stars, including Alpha, Beta (β), and 4 Lacertae, which should fit within the field of any finder. NGC 7243 is along the southwest edge of this keystone, about 2½° west and a

Figure 9.1
The open cluster NGC 7243 is a fine sight in large binoculars and small telescopes.

little south of Alpha. It is a splashy coarse star grouping about 20′ across (**Figure 9.1**) with a total magnitude of about 7.5, making it a good object for deep-sky binoculars with 80-mm objectives. *Sky Catalogue 2000.0*, however, lists the group as having a total magnitude of 6.4, which might make it visible to the naked eye under the best observing conditions.

The cluster stands out especially well from the stellar background when I stop my 4-inch Clark refractor down to 1.8 inches. According to *Revue des constella-*

tions by R. Sagot and Jean Texereau, NGC 7243 in a 4-inch at about 50× is a rich triangular cluster of many stars between 9th and 11th-magnitude. The number of stars increases from about 15 in a 2-inch to 60 in an 8-inch. I found no definite shape in a 12-inch recently, but counted at least 80 stars within a ⅓° area. Look for a wide double star at the cluster's center, particularly if you have a 6-inch or larger telescope. Try using different magnifications on this cluster. Often certain features of an object are apparent only at specific magnifications. This is also an interesting object to try sketching. First rack your eyepiece out of focus to the point where only the brightest stars are seen. Add these and continue the process until sharp focus reveals the faintest stars.

Figure 9.2
The large and loose open cluster NGC 7209 in Lacerta.

About 4° southwest of NGC 7243 is NGC 7209. Its rather scattered bright stars (**Figure 9.2**) cover an area roughly 20′ in diameter and are visible in my 2-inch finder. Try star-hopping from the first cluster to the second one by following the small semicircle of naked-eye stars 4, 5, and 2 Lacertae. NGC 7209 is slightly smaller and fainter than NGC 7243. It stands out nicely from the Milky Way background of star dust, lying within a rough pentagon of five brighter stars that can all be seen if the field of view is ¾° or larger. As many of the cluster's stars are between 9th and 10th magnitude, NGC 7209 is excellent in my 4-inch refractor. It shows more than 50 stars in a 6-inch. Some amateurs call this object triangular too, but I wonder whether they have been looking at NGC 7243 instead. Readers who record their impressions of this cluster should note instrument size, magnification, and sky condition. Look for an orange 6th-magnitude star 15' northeast of NGC 7209.

For those who want a challenge, the small open cluster NGC 7296 is 40' east of 4th-magnitude Beta Lacertae. This 10th-magnitude cluster lies on a rich background, which might be a hindrance since NGC 7296 is only 4' in diameter. At 20× my 4-inch Clark refractor had a difficult time with this object. Glare from 4th-magnitude Beta and the rich Milky Way background made the cluster hard to pick up. Yet NGC 7296 can be found with a little patience. By increasing the refractor's magnification to 80×, NGC 7296 was easily found, and the view at 150× was even better. Sweeping for the cluster at these higher magnifications is made easier by bright Beta Lacertae — if you get lost, return to the star and begin the search again. After examining the cluster at 150×, try sweeping nearby, or just letting the stars drift. This is an excellent way to discover many curious sights never seen during low-power sweeps.

Near the northern border of Lacerta, just west of the halfway point on a line between Beta Lacertae and Epsilon (ε) Cephei, is a rather difficult open cluster, NGC 7245. It is only 5' across and of magnitude 11.5. NGC 7245 can be identified in 20 × 65 binoculars but is much more satisfactory in a 10-inch telescope. Recently, I could not locate it with a 5-inch at 20×, but when I switched to my 4-inch refractor at 100× it was clearly seen. There is a triangle of three stars between 8th and 10th magnitude here. More than a dozen stars can be seen with a 12-inch telescope, but perhaps only half this many are within the grasp of an 8-inch instrument.

Somewhat richer than NGC 7245 is the open cluster IC 1434. Generally, objects in the *Index Catalogue* supplement to the *NGC* are faint and hard to locate, but the cluster IC 1434 is easily found. Continuing on a straight line from 9 Lacertae to Beta Lacertae for about the same distance that separates these stars will put you in the cluster's vicinity. IC 1434 is an 8' grouping of perhaps 30 stars. Of 10th magnitude, it is a dull little cluster, but one that is improved considerably by running up to 150× or 200× and using averted vision.

As viewed from Connecticut with a 6-inch, IC 1434 appears as a compact ball of faint stars, about 6' to 8' in diameter; I have the feeling that this object might be a very pretty sight in a 16-inch telescope.

The group lacks solidarity and appears as if several clusters have been jumbled together. The English amateur Guy Hurst, writing in the Webb Society *Deep-Sky Observer's Handbook*, Vol. 3, *Open and Globular Clusters*, mentions that the cluster seems backed by "considerable haze" (by the way, this handbook incorrectly states the cluster is in Cygnus). The surrounding sky is quite rich, and I recommend that you spend some time just sweeping the nearby area.

A cluster missing from most catalogs and charts is NGC 7394. Canadian observer Pat Brennan describes it as "a coarse grouping about 10' by 3' across with a bright star at the southeast end." In addition to the bright star at its southeast end, there are about 10 fainter ones. Since the field is not particularly rich, the cluster can be easily recognized! When John Herschel swept up this grouping in 1829, he noted in his record book that the measured position referred to "a double star, the last of a poor cluster of about a dozen stars."

More challenging than any of these clusters is the 12th-magnitude planetary nebula IC 5217. It is located just slightly northeast of center in the stellar keystone

Figure 9.3
The challenging 12th-magnitude planetary nebula IC 5217 in Lacerta is roughly 8,000 light-years distant.

I mentioned earlier. The planetary is slightly oval and about 8" long (**Figure 9.3**). Although it should be visible in a 6-inch telescope, larger apertures will make the task of locating this object easier. It will show as a disk at magnifications of 100× or more. There is a central star of magnitude 14.6, but I know of no amateur sightings of it. With a Lumicon O III filter I have seen IC 5217 with the 4-inch Clark.

There are no bright Lacerta galaxies, but NGC 7331 lies just over the line in Pegasus. It is a 10th-magnitude spindle about 10' × 2' in extent. A 2-inch finder reveals it with effort, but it shows well in my 5-inch Japanese binoculars. NGC 7331 is of current astronomical interest because of a possible link with Stephan's

Quintet (see *Sky & Telescope* for March 1977, page 170). The latter is a group of much fainter galaxies, very difficult for amateurs, lying 30′ to the south-southwest.

Cruising Through Cygnus

"If one constellation dominates September evening skies," Scotty said, "it is the one known to the ancients as Cygnus, the Swan, and to less imaginative moderns as the Northern Cross." Regardless of how we see the constellation, one thing is for sure: few of us can resist sweeping our binoculars and rich-field telescopes across the dark and bright expanses of Milky Way that flow through Cygnus like a snow-banked river at night. A main attraction of this constellation is its many open clusters, especially the bright Messier objects M29 and M39. But Cygnus has much more to offer. In fact, Cygnus is one constellation that has it all: double stars, variables, dark nebulae, gaseous nebulae, galaxies, planetary nebulae, and even a supernova remnant — and, of course, plenty of telescopic challenges.

Deep-sky observers are discovering Cygnus. Does this sound peculiar? After all, the constellation is a landmark of the late-summer sky and is awash with bright lanes of the Milky Way. Even a naked-eye view suggests that Cygnus should be a deep-sky wonderland. But only recently have amateurs pursued the individual delights that the constellation has to offer.

Most deep-sky objects in Cygnus have been overlooked because they were never mentioned by Smyth or Webb in their observing guides. Early observers paid surprisingly little heed to these objects. Smyth mentions only M29 and M39 in his famous (but now scarce) *Cycle of Celestial Objects* (written in 1844). Webb is almost as brief in his *Celestial Objects for Common Telescopes* (first written in 1859). But he says of Cygnus: "I had at one time projected a survey of the wonders of this region with a sweeping power; but want of leisure, an unsuitable mounting, and the astonishing profusion of magnificence, combined to render a task hopeless for me which, I trust, may be carried through by some future observer."

In addition to the two open star clusters Messier cataloged in Cygnus, Smyth also called attention to the planetary nebula NGC 6826. Webb did slightly better by adding a wisp of the Veil Nebula near 52 Cygni, three clusters, and two interesting star fields (one containing the cluster NGC 6871, which he did not mention per se).

Both authors went on at length about the variable and multiple stars in the constellation, but it seems they did not spend much time at the eyepiece exploring Cygnus themselves. If they had, certainly they would have written about some of the splashy open clusters scattered through this corner of our galaxy.

Later authors followed in the footsteps of those before them, and amateurs weren't directed to the lesser-known sights in Cygnus until the late 1970s when Burnham's *Celestial Handbook* became widely available. This work includes 32 deep-sky objects in the Swan, and there are even more plotted on Tirion's *Sky Atlas 2000.0*.

Today's amateurs, however, are not content to use just the popular guides. Frequently they roam the sky on their own, and I regularly receive mail from observers who have "discovered" a star cluster. Most of these objects are listed in the *New General Catalogue of Nebulae and Clusters of Stars*. When I write back saying their cluster is in the *NGC* they reply, "If it's known, why isn't it plotted on the star atlases?" The answer is that the charts include only a representative sample of deep-sky objects, and most of them have been selected from the observing guides mentioned. The Tirion charts, for example, show some 2,500 deep-sky objects, which is less than one quarter of those listed in the *NGC* and its *Index Catalogue (IC)* supplement.

A superficial inspection of these catalogs suggests that only about half of the objects listed in Cygnus are plotted on the Tirion atlas. Alister Ling of Montreal has been sweeping Cygnus in search of objects not usually mentioned in observing guides. Writing in the newsletter *Betelgeuse*, he says that in crowded star fields he looks for hazy blobs in his viewfinder that often turn out to be magnificent clusters when seen with the main telescope. This search technique is especially useful for locating large open clusters. Ling mentions three clusters, NGC 6910, 6997, and 6871, that have never been described in this column. All are plotted on *Sky Atlas 2000.0*, so why not hunt them down yourself?

Ling has also been using a UHC filter on his 12-inch Dobsonian at 56×. The North America Nebula (NGC 7000) "jumped out" and detail never before visible was evident. Furthermore, while casually sweeping nearby he stumbled upon the elusive Pelican Nebula (IC 5070). On another night he viewed the

Figure 9.4
The glorious North America Nebula in Cygnus may be illuminated by an inconspicuous star that is obscured by dark clouds near the nebula's "Atlantic coast." The hard-to-spot Pelican Nebula flies alongside to the west.

section of the emission nebula IC 1318 east of Gamma (γ) Cygni, the center star of the Northern Cross. It appeared as a "huge sprawling region of nebulosity split in two."

One of the most controversial objects among amateur observers, the North America Nebula (**Figure 9.4**) is familiar from photographs as a large diffuse glow about 3° east and 1° south of Deneb. There has been remarkable diversity of opinion on how small a telescope can show this object visually. Some argue that the nebula cannot be observed readily without photography, but the region is so rich that it is well worth sweeping with anything from binoculars to large telescopes. Early telescopic observers all missed NGC 7000 because of its large size (relative to the fields of their instruments) despite its brightness. Only in the last generation or two, with the rise of rich-field designs, has the nebula become generally accessible to telescopes.

If observing conditions are very good, and you know what size and shape to expect, the North America Nebula can be made out easily with the naked eye. An opera glass makes it more apparent, but this is not true when more powerful binoculars are used. When I was in grade school I had a small folding opera glass, bought for a dollar at a carnival, and it showed the nebula! Actually, NGC 7000 is difficult to see in most telescopes. With a 5-inch Moonwatch Apogee telescope you should know beforehand what it looks like, and the nebula is downright challenging in a 6-inch f/4. However, a few years ago it was brilliant when I saw it in Edgar Everhart's 11.4-inch Wright telescope, which was then located in Connecticut.

In northern Cygnus there is a sparse cluster that may have been spared the indignity of a "nonexistent" label because it carries the protection of a Messier number. The open cluster M39 has a distinct triangular shape. The group is 32′ in diameter and 4.6 magnitude. Hence it's within reach of the naked eye. Discovery is credited to the French observer Jean-Baptiste Le Gentil in 1750, who described it as "very dim" but visible without a telescope. In a small, low-power telescope M39 shows as a nice sparkling cluster. Smyth called it a "splashy galaxy field of stars." Recent printings of Webb's observing guide call M39 a "grand open cluster," but the original never mentioned the group. It appears to have been added to the guide during a revision of the sixth edition by Thomas E. Espin in 1917.

Le Gentil refers to the triangular shape of M39 but orients it differently than most observers today see it. It is not clear to me, however, if his observation of the shape was made with the naked eye. Will someone try examining M39 with a range of instruments (don't forget binoculars and the unaided eye)?

Several years ago, while observing from the Southern California desert, I glimpsed a dark streak running about 5° east-southeastward from M39. There is no doubt about a vacancy some 6′ across on the east side of M39. From it a narrow, dark lane can be traced eastward with difficulty for perhaps 2½° until it merges with a much wider and pronounced stream that leads to the Cocoon

Nebula, IC 5146. The obvious dark lane is Barnard 168, which is shown without a label on *Uranometria 2000.0* charts.

The detection of dark nebulosity depends on many factors. I lean toward using long-focus instruments because my experience has shown that they tend to scatter less light and provide a higher-contrast image than do rich-field telescopes. I have had some dramatic views of dark objects with my old 10-inch f/8.5 Newtonian reflector and the 12-inch f/17 Porter turret telescope in Springfield, Vermont.

About 6° south of the mysterious open cluster NGC 6811 (discussed on page 208) is NGC 6819. I found this grouping early in my high-school days. It was barely detectable in my homemade 40× 1-inch refractor. According to information in Vol. 2 of *Sky Catalogue 2000.0*, NGC 6819 is a whopping 3½ billion years old. The cluster is 7th magnitude, 5' across, and contains perhaps 20 stars of magnitude 12 and fainter. The Webb Society handbook on clusters mentions several reports of nebulosity involved with the group. Have others seen this?

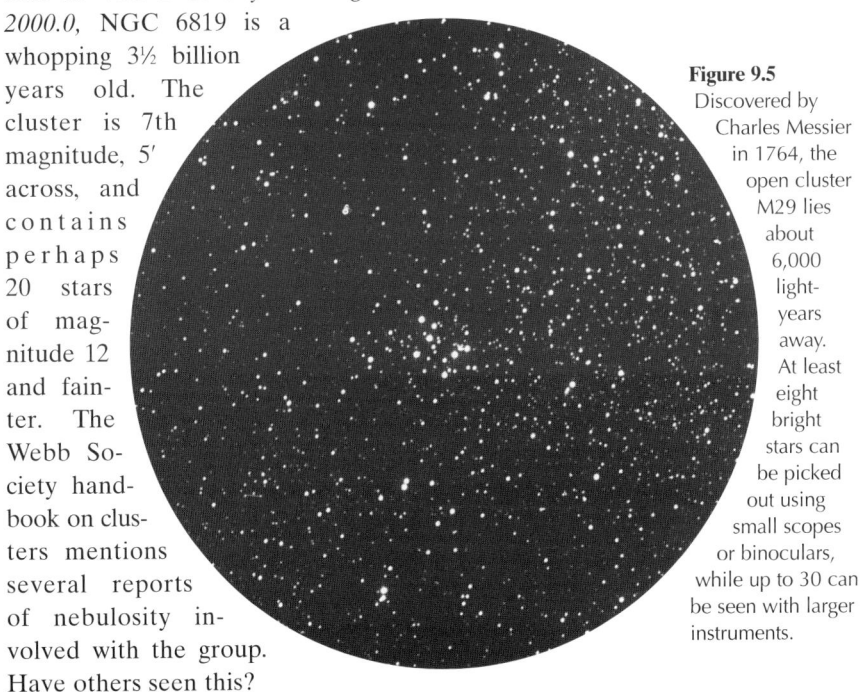

Figure 9.5 Discovered by Charles Messier in 1764, the open cluster M29 lies about 6,000 light-years away. At least eight bright stars can be picked out using small scopes or binoculars, while up to 30 can be seen with larger instruments.

In contrast with the age of NGC 6819 is M29, estimated to be only 10 million years old (**Figure 9.5**). While the former is nearly as old as our solar system, the latter formed not long before humans appeared on Earth. M29 is slightly larger than NGC 6819 and shines with a total light equal to that of a 6.6-magnitude star. I doubt the cluster can be picked out from the rich Milky Way background with the naked eye.

NGC 6866 is about the same size as M29 and a magnitude fainter. Catalogs list 80 stars here. I have no reports of NGC 6866's appearance in a 16-inch or larger telescope, but a 6-inch captures a nebulous glow that is probably caused by unresolved stars. In a 10-inch at 150×, the field of view is filled with sparkling starry excitement.

NGC 6894 is a planetary nebula in Cygnus, which photographs show as a faint diamond 44″ in diameter. Though it has been cataloged as magnitude 14.4, amateurs can see it, and Agnes Wolfe writes that it seems brighter than the catalog values. James Corn of Phoenix, Arizona, observed it on July 16, 1952, as magnitude 12 and large.

Another object this month is the planetary nebula NGC 6826 in Cygnus. Although not well known today, I found a reference to it while looking for earlier mention of the Veil Nebula in an 1856 edition of Elijah Burritt's *Geography of the Heavens*. With a total light equal to a star of magnitude 8.8, in modest apertures this planetary shows a fine greenish disk about 25″ across.

NGC 6826 is plotted in *Norton's* with the Herschel number 73[4]. Since the planetary's 11th-magnitude central star was easily seen by William and John Herschel, they considered this nebula a transition object between regular planetaries (whose central stars were too faint to be seen) and stars involved in diffuse nebulae. At 300×, a 10-inch reflector I had in Kansas showed NGC 6826 to have uniform light across its center. The planetary takes magnification well and is a fine object for even small apertures.

Also try finding the spiral galaxy NGC 6946 in extreme northern Cygnus. This is only 10° from the plane of our Milky Way, where it is unusual to find galaxies since the interstellar dust concentrated there obscures the light from these distant objects. Though listed as about 9′ × 7′ in size, this galaxy seems even less round to me. Despite the glittering foreground star field, NGC 6946 stands out well.

Unveiling the Veil

The most popular deep-sky object in Cygnus today is the Veil Nebula, though it is difficult to observe. This visually fragmented bubble of glowing gas is so large that most telescopes cannot contain it in one field. The challenge, of course, is seeing it at all. Over the years amateurs provided Scotty with conflicting reports as to its visibility. "While some have missed it even with large telescopes," he wrote, "Burnham has seen it with binoculars. My opinion is that the Veil is more a test of observing skill than optical parameters." Indeed, Scotty witnessed the Veil Nebula evolve from something that amateurs knew only on photographs to one they routinely studied with small telescopes. Conversations at star parties, he said, once swirled around how to view such "difficult" objects. Today, those conversations are part of amateur astronomical history, an intriguing fireside tale to share with novices, and the Veil is yet another of amateur astronomy's ever-growing list of "ex-test" objects.

The most famous celestial sights have been passed down to us throughout the centuries. The Orion Nebula is one. Another is the Andromeda Galaxy, M31, first noted in writing by the Persian al-Sūfī in A.D. 964. It seems unlikely that any observing guide published after Messier's 1764 discovery of M27 fails to mention this beautiful planetary nebula in Vulpecula, and amateurs teach each other to find it.

All things considered, amateurs might think that the deep-sky objects sought out at today's star parties have been scrutinized and described over and over again by past observers. However, this is not the case. There are some objects which, in a sense, belong to the modern amateur. They deserve careful attention, for published visual descriptions are surprisingly few.

One such object is the Veil Nebula in Cygnus (**Figure 9.6**), a broken bubble of luminous gas some 2° in diameter. Although ignored by generations of telescope users, in the last 30 years the Veil has progressed from a difficult test object to a

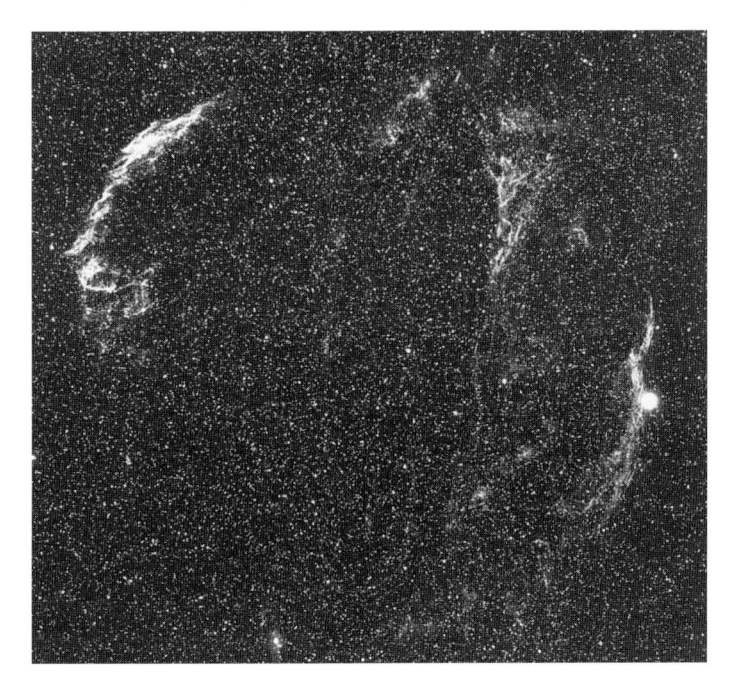

Figure 9.6
The Veil Nebula in Cygnus is believed to be the remnant of a supernova explosion that occurred at least 15,000 years ago. The filamentary remains are expanding at nearly 30 miles per second.

reasonable target for anything from binoculars to the largest amateur telescopes. It is an excellent nebula for training the eye, perhaps the most important observing "accessory," to help us get the most out of the telescope we are using.

Astronomers believe the Veil comprises the remains of an ancient supernova. When I was young the Veil was usually listed as a planetary nebula, since the concept of supernovae and their remnants didn't come until the 1930s when I was in college. Back then the Veil was considered a unique object. Today, however, we know of several others like it, including S147 in Taurus.

William Herschel discovered the brighter parts of the nebula as he swept the sky with his 18¾-inch speculum-metal reflector in 1784. He included the east and west sections of the loop with his class V objects (very large nebulae). In *Norton's Star Atlas* the westerly section carries his designation 15[5], and it also bears the name NGC 6960.

The 4th-magnitude double star 52 Cygni appears to be involved with NGC 6960 but is actually a foreground object not related to the nebula. Incidentally, the *NGC* mistakenly identifies the star as Kappa Cygni, and other catalogs have sometimes copied the error. Even in small telescopes, 52 Cygni is an excellent double; its orange and blue components, magnitudes 4.5 and 9.5, are separated by 6½″.

The eastern loop of the Veil is NGC 6992–95 (14^5 in Norton's). Like the western part, it is a strip of sculptured light about 1° long. Between the outer parts of the loop are several fainter wisps of nebulosity, which show well in long-exposure photographs. The larger triangular piece, discovered photographically, carries no designation number. Just to its east is a smaller patch of nebulosity, NGC 6979. Herschel again was its discoverer, and placed it with his class II objects (faint nebulae).

Why has the Veil Nebula been overlooked for so long? It certainly could have been seen with the reflectors and comet-seeking refractors trained on the sky back in the 19th century. I suspect there were two reasons for omitting the Veil from earlier observing guides. First, Herschel called NGC 6960 only "pretty bright" and NGC 6992–95 "very faint," although the latter is in reality the more easily visible. Smyth may have been more responsible for diverting subsequent amateur attention from the Veil. In his *Cycle of Celestial Objects* he mentions only NGC 6960, and then only as an afterthought in his section on 52 Cygni. He unenthusiastically noted that the field of view in his 6-inch refractor required "considerable attention" before he was able to make out any of the nebulosity.

Other popular observing guides by T. W. Webb and Charles Barns make no mention of the Veil, although by the time of the publication of Barns's work in the 1920s, photography had revealed the splendor of the nebula's twisted filaments.

In the 1940s and 50s, amateurs considered the Veil a test object. Often my mail contained bitter reports of failure to see the Veil's faint glow. However, today it is more typical to receive detailed descriptions from confident observers. My own experience has followed a similar pattern, which I attribute to the growing education of the eye, and to knowing the Veil can be seen visually.

Richard Wilds of Topeka, Kansas, has glimpsed the three brightest sections of the Veil (those with *NGC* numbers) with a 2-inch refractor. In his *Celestial Handbook,* Burnham mentions seeing NGC 6992–95 in large binoculars. Indeed, both the east and west arms of the loop are easy in my 20 × 125 Japanese military binoculars. In a 12-inch f/5 telescope the Veil Nebula in Cygnus is beautiful, and so bright that one notes it even when sweeping. But in a 5-inch f/5 the Veil is visible only with difficulty to keen eyes. Why is this so, when the surface brightness of extended objects depends on f-ratio rather than on aperture?

The seeming contradiction is removed when the eye is considered along with the optics of the telescope. Physiologists have shown that the eye can discern a spot 6° in apparent diameter that is only 6 percent brighter than the background, while a 2° spot must be 11 percent above background. Remember, a galaxy with a diameter of 3′ when magnified 100 times has an apparent diameter of 5.0°.

Hence, the greater magnification usually obtained with larger telescopes (because of their longer focal lengths) may help to reveal nebulae near the visual threshold. This is contrary to popular recommendations, but I have successfully used high powers, especially on planetaries. Of course, when magnification is increased the surface brightness rapidly becomes less. Hence one should not use high powers on objects that already have large angular diameters.

My most spectacular view of the Veil was in the mid-1960s from Connecticut, where, at best, the skies are usually undistinguished. I was sweeping with Edgar Everhart's 12-inch Wright reflector when the giant bubble-like nebula was picked up. The Wright reflector has a 12-inch primary mirror and an 11½-inch corrector plate. This visual instrument works at f/4, and is actually a variation of the Schmidt camera. Everhart, who discovered Comet 1964h, constructed this telescope especially for comet seeking, and mounted it atop a 35-foot tower to obtain a clear view of the horizon. With an Erfle eyepiece giving about 50×, it provided some unusual views of deep-sky objects.

While discussing the instrument's capabilities, we wondered if it would show the two arcs of the Veil Nebula in Cygnus. We tried to find NGC 6960 (the western segment) first, because the 4th-magnitude star 52 Cygni lies on the middle of the arc. Since Everhart's telescope is an altazimuth (for comet sweeping), he pointed the instrument by sighting with the 3× finder while I watched through the main eyepiece.

As we swept over 52 Cygni, the nebula was seen without straining. It appeared to stream across the field. In the Wright telescope the nebulosity did not look as it does in photographs, but this was still one of the most fascinating sights I have seen in a decade or more.

Later we tried for NGC 6992–5 (the eastern arc); there are no bright stars near, so we had to proceed carefully. Sweeping casually, we might have missed this object, but once it was definitely in the field it could be seen easily. Though not as bright as NGC 6960, it was sharply defined; the structure was less evident than in the companion nebula. In all, it was about as noticeable as the faint outer parts of the Orion Nebula.

New nebula filters have apparently helped many locate the Veil. A typical sighting is that by John Bartels. He set up a 4½-inch f/8 surplus telephoto lens with a 32-mm Plössl eyepiece and Henzl 300 nebula filter at the light-polluted airfield of Travis Air Force Base. NGC 6992-5 could be seen without the filter, but was easier with it. NGC 6960 was located only with the filter, but after knowing where and what to look for he was able to glimpse this western loop without it.

How times have changed. Now I often receive reports of the Veil being seen in binoculars. Even novice observers have seen it well enough to record the eastern portion of this giant 2°-diameter broken loop as brighter than the western strand passing near 52 Cygni. At the 1984 Texas Star Party, Lee Cain with his 17-inch Dobsonian binoculars saw the Veil so bright that he almost "feared it would ruin my dark adaptation." Bryce Heartwell of Alberta, Canada, considers the Veil "totally beyond description" as seen with his 17-inch reflector. When he added a Lumicon UHC filter to the system, "the Veil showed as much detail as photographs."

My Japanese 5-inch binoculars, though very heavy, originally had only a shaky tripod. I remounted them on a 3-inch pipe held in concrete down to the bedrock that is Connecticut. A well-greased pipe flange allows motion in azimuth, while the altitude motion is provided by the binoculars' built-in trunions. Though makeshift, the mounting is granite steady and turns smoothly. To test the pros and cons of binocular vision for astronomical observing, I first tried using both eyes and then only one. The sense of reality and impact of the scene were clearly greater with two eyes. Furthermore, it was less tiring for protracted gazing. With the crescent Moon in the sky I selected the portion of the Veil nearest Epsilon Cygni as a difficult test. Yet the binoculars promptly revealed these wisps of nebulosity, and with averted vision I could even glimpse structure.

I have also heard from experienced observers that the Veil can be seen with the naked eye through a UHC nebula filter. This is amazing when you consider that only a few years ago most amateurs thought of the Veil as one of the sky's great telescopic challenges. Dozens of observers have written me enthusiastic letters about the Veil's appearance in telescopes equipped with UHC filters. Both the eastern (NGC 6992–95) and western (NGC 6960) sections of the nebula show delicate filamentary structure.

The following report from Joanne Konst of Kenton, Ohio, will be of interest to everyone who plies the heavens through city lights. Her house lies just two blocks from the center of town on a corner lot bathed in the glow of street lights. From her second-floor balcony she scans the sky with a 10.1-inch Dobsonian reflector and a battery of Lumicon filters. "The Veil Nebula in Cygnus reveals many features," she writes. "NGC 6960 on the western edge was difficult without the filter due to glare from 4th-magnitude 52 Cygni. The eastern section of the loop, NGC 6992, is very detailed. While I can see it without a filter, using one enhances the view and brings out faint details. The edges look like ragged cotton, and I can see many delicate filaments. NGC 6979, as well as several other patches within the confines of the loop, are visible with a filter." Perhaps her report will inspire others to try their telescopes from less-than-perfect observing locations.

The Mystery of NGC 6811

Scotty claimed that in spite of the electronic revolution sweeping amateur astronomy, we should remember that the eye itself performs extensive image processing. As an example, he introduced us to the open cluster NGC 6811, a small group of stars in Cygnus's northwest wing. At a glance, the cluster appears rather innocuous, containing about 70 stars and shining near 7th magnitude. Indeed, when Scotty looked at it in his youth he recalled it to be "an unimpressive, nondescript group." But a letter he received in the mid-1980s from amateur observer Tommy Christensen not only changed his view of the cluster, but set him hot on the trail of a new and wonderful mystery. Yet Scotty was magnanimous; he resisted the temptation to go out on his own and solve the mystery. Instead, he turned the case over to his readers.

Several years ago I received a letter from Tommy Christensen, who lives in Odensa, Denmark, and observes with a 3½-inch refractor. Along with a description of M33 and the Veil Nebula was a brief note about the open star cluster NGC 6811 in Cygnus (**Figure 9.7**). He called it one of the most beautiful clusters he has seen and mentioned "a dark band about 5' thick running through the middle of the cluster, not completely without stars, but nevertheless conspicuously dark." He also likened it to a "smoke ring" of stars.

Figure 9.7
Does the open cluster NGC 6811 in Cygnus appear to have a dark center as seen through your telescope?

My recollection of NGC 6811 was of an unimpressive, nondescript group of stars near Delta (δ) Cygni. A quick look through my notes showed no reference to a ring of stars. Furthermore, I checked observing guides and my file of reports from amateurs. Again, nothing unusual was mentioned. A 6-inch instrument will show some 75 faint stars shimmering inside an area 15' across. It reminds some amateurs of an opera-glass view of the Beehive Cluster in Cancer.

A print from the photographic Ross-Calvert atlas did, however, suggest that NGC 6811 has a dark center. But the cluster is not well resolved on the print, and it's difficult to compare a blue-light photograph with what is seen visually. I could have waltzed out to my telescope and settled the matter, but I thought it better to solicit comments from readers; this I did in my September 1985 column. I was purposely cryptic, asking only if anyone had seen something "unusual" in NGC 6811.

Observers took the bait! In poured scores of letters. It's quite impossible to tell just how many people reported observations. Some letters contained comments from a single observer, while others offered the consensus of as many as 20 amateurs at a star party.

The reports came equally divided in thought. One group assumed the cluster formed some kind of figure. There were letters extolling bells, butterflies, dinosaurs, and a pair of fighting peacocks, and one delightful comment from a person who saw "Nefertiti's head-piece." Several of those who wrote saw three-leaf clovers, and one amateur went whole hog for a four-leaf clover (his friend, however, saw a frog!). Using several 17-inch telescopes at the 1985 Astrofest in Wisconsin, I saw the cluster as either a butterfly or rotifer outline.

The other camp comprised observers who saw a dark center in NGC 6811. They also usually mentioned dark lanes. One even used Christensen's term "smoke ring." A young observer from Ohio simply said "stars in a ring."

This is a beautiful, albeit minor, example of how people see things differently. Everyone was looking at the same cluster, but because of experience, conviction, or psychological factors, each saw it in a different way.

I expected some correlation between telescope aperture and visual description. To some extent this was true. The largest telescope used to report a "dark center" was an 11-inch reflector. The smaller the telescope, however, the more pronounced the feature appeared. With apertures of 16 inches and larger, the dark center just doesn't seem to show. So how can we say what is the "real" description of the cluster?

Then I got a letter from Marton Konecny of Czechoslovakia. He observes with a 2½-inch Zeiss refractor and magnifications from 14× to 140×. While Konecny comments that Americans might think this a small instrument, it is about the standard amateur telescope in his country. He too saw NGC 6811 as a clear cut ring of stars with no hint of the butterfly pattern I had once suggested in an earlier column. I only hope he gets a chance to view through a 16-inch someday! Konecny sent a drawing of NGC 6811, which is unusual in that it records a glow around some of the stars. Have others noted this?

Hunting Cosmic Pearls in Aquila

Aquila, the Eagle, soars highest on cool September evenings. Its brilliant ice-blue gem, Altair — the southern star of the Summer Triangle — adds a flash of sparkle to the lacy boundaries of the Milky Way that brush past this celestial raptor. Despite its prominent placement, Aquila contains some interesting objects, but surprisingly few are truly spectacular. Aquila does have a bounty of planetary nebulae, however, which are within the reach of amateur-sized telescopes. Until Scotty began introducing them to his readers, these cosmic pearls, as he referred to them, were largely ignored in the popular literature. Here Scotty introduces us to several planetaries in the neighborhood of Altair. Many of them have gone unseen, because they are not plotted on some popular sky atlases, such as Wil Tirion's *Sky Atlas 2000.0*. Furthermore, published magnitudes for faint planetary nebulae are usually dismally underestimated, causing many amateurs to shy away from them. But times, equipment, and knowledge continually change, and Scotty was right on top of new observing trends, as we shall see.

September brings a slack tide of celestial treasures. As twilight gives way to dark, the golden sands of Sagittarius remain in the south and the Northern Cross in Cygnus still floats overhead. But nothing very new or dramatic is coming up in the east. Those who observe late into the night often have their first taste of the crisp autumn air that will follow as days grow shorter. And just before dawn the luminous cone of zodiacal light rises from the eastern skyline. It is one of the most pleasant times of the year to be out under a starry sky.

The evening meridian is dominated by the three brilliant stars of the Summer Triangle. At its southern apex is 1st-magnitude Altair, a celestial neighbor located only 16 light-years from Earth. In my mind's eye I see Altair wrapped in a galactic blizzard of tiny luminous spheres. We know them as planetary nebulae, pearls on the grand scale of the galaxy.

Until a decade or two ago most deep-sky observers concerned themselves with only the few bright planetaries large enough to show some detail in small telescopes. Objects like the Ring Nebula in Lyra and the Dumbbell in Vulpecula were standard fare at summer star parties. Although a handful of amateurs observed others, hardly anyone dreamed of searching for the tiny glowing gas bubbles scattered around Altair. Indeed, few even knew they were there!

Times have changed. Today's observers have access to larger telescopes. They have better catalogs of deep-sky objects, and, more important, better star charts. They also have nebula filters. Today's observers eagerly venture into cosmic depths that their counterparts of a generation ago hardly knew existed.

Perhaps the best place to begin our search for Aquila's planetaries is on the pages of *Uranometria 2000.0*. I consider this atlas among the best ever made for observers. The charts have a generous scale, with 1° equal to about ¾ inch. This is roughly the same scale as the finder charts that have endured through more than 80 years of use by the American Association of Variable Star Observers. Plotting a position on the *Uranometria 2000.0* charts is easy because epoch 2000.0 grid lines are shown for every degree of declination and, except near the celestial poles, for every four minutes of right ascension. Stars are shown to about magnitude 9.5, the average limit for a 2-inch finder, and more than 10,000 deep-sky objects are plotted.

But there's more to deep-sky observing than just knowing the location of an object. Without such basic information as its size and brightness, we could sweep over our target and not even know it. A 10th-magnitude planetary that's 5' in diameter is going to look a lot different from one only 5" in diameter. To that end, observers will welcome *Uranometria 2000.0*'s companion volume, *The*

Deep-Sky Field Guide by Murray Cragin, James Lucyk, and Barry Rappaport. It gives basic information about all the deep-sky objects plotted in the atlas. It is arranged by chart number, making it particularly useful in the field, since information about all the objects on one chart can be found on one page (or several adjacent pages).

Altair is on chart 207 of *Uranometria 2000.0*. The corresponding page of *The Deep-Sky Field Guide* lists a dozen planetary nebulae. Of these, only three are from the *NGC*; the rest are from more modern compilations. Since almost all the *NGC* objects were discovered visually, they are generally the easiest to see in amateur telescopes.

One of the objects to which Smyth calls attention in his classic *A Cycle of Celestial Objects* is NGC 6804. It is small, only about 1' in diameter (**Figure 9.8**). He comments that in "very powerful" telescopes NGC 6804 appears fan-shaped. William Herschel, who discovered the object, thought he could resolve it into

Figure 9.8
Seemingly lost among the multitude of stars in the Aquila Milky Way is NGC 6804, a planetary nebula some 4,700 light-years distant.

component stars and thus placed it in his class VI (very compressed cluster of stars). *Norton's Star Atlas* once carried the Herschel number 38[6] for this object, although we now know it to be a planetary nebula.

Careful examination of NGC 6804 with different apertures would make an interesting project at a star party of serious observers. It is about 30" in diameter (the minimum size of Jupiter) and 12th-magnitude. Its visibility depends heavily on sky conditions. I have seen it on several occasions here in Connecticut with a 4-inch Clark refractor, but not in recent years with our increasingly light-polluted skies. My observing notes from the 1950s in Kansas always call the planetary "easy" in a 10-inch reflector.

One of the great pleasures of deep-sky observing is the individuality that certain objects acquire in the eyepiece. I'm always delighted to learn that someone sees an object in a new perspective. One such example comes from Robert Moseley of Coventry, England, who tracked down NGC 6804 while testing a new 10-inch f/6 reflector. His best view was at 120×. He writes, "It gives the impression of a highly condensed but partially resolved cluster. It is a faintish oval nebulosity with a 12th-magnitude star toward its northeast edge. With averted vision at least one other star could be seen superimposed on it." Moseley questioned the 13th magnitude I had given for NGC 6804 in an earlier column. Published magnitudes for planetary nebulae cause many disagreements, and I believe it is best to slightly mistrust all of them and to record your own magnitude estimates with your notes.

Just 50' to the north is another planetary. NGC 6803 is only 6" across and 11th magnitude, which should give it a much higher surface brightness than NGC 6804. Oddly enough, however, my observing notes suggest that I have had a more difficult time with this object.

Most of the other planetaries on *Uranometria 2000.0* chart 207 are less than 10" in diameter. As such, they appear starlike at the low magnifications most observers use for hunting down objects. But planetaries emit most of their light at a few discrete wavelengths, especially two emission lines of doubly ionized oxygen near 5000 Å. Some enterprising amateurs have attached small, narrow prisms to telescope eyepieces to distinguish between faint stars (which have their light drawn into short streaks) and the tiny planetaries (which remain starlike).

Another trick is to use a nebula filter (such as Lumicon's O III filter) that isolates the oxygen emissions. The narrow passband of this filter dims surrounding stars because it blocks most of their light while leaving the planetary's brightness relatively unchanged. Repeatedly flicking the filter between the eye and eyepiece causes the planetary to blink, making it easy to identify which "star" is the nebula.

If you are prepared for the challenge, here are three planetaries that will test your abilities. I have never seen any of them. Furthermore, because the light of a planetary is confined to a few wavelengths, I question how its brightness determined by conventional methods (usually involving photography) compares with what the eye sees.

PK 52–2.2 (the designation is from the *Catalogue of Galactic Planetary Nebulae* by Lubos Perek and Lubos Kohoutek) is less than 10" across and 12th magnitude. The *Deep-Sky Field Guide* notes that it has a "ring structure." If you can locate this object, its appearance will serve as a reference for the other small planetaries. It has a 14th-magnitude central star.

PK 45–2.1 is less than 5" across. Although it is listed as visual magnitude 12.7, its smaller size should give it about the same surface brightness as PK 52–2.2 if you use very high power. It is listed as having a "stellar image."

I suspect that the most difficult of the three is PK 52–4.1, which is less than 10" in diameter and magnitude 13. If you succeed here, you can test your skill on several even fainter planetaries listed in the *Deep-Sky Field Guide*.

SEPTEMBER OBJECTS

Name	Type	Const.	R. A. h m	Dec. ° ′	Millennium Star Atlas	Uranometria 2000.0	Sky Atlas 2000.0
8 Lacertae	✶✶	Lac	22 35.9	+39 38	1122, 1142	87, 123	9
52 Cygni	✶✶	Cyg	20 45.7	+30 43	1169	120	9
Barnard 168	DN	Cyg	21 53.2	+47 12	1104, 1105	86	9
Cocoon Nebula, IC 5146	OC+BN	Cyg	21 53.4	+47 16	1104	86	9
IC 1318	BN	Cyg	20 22.2	+40 15	1127, 1128, 1148, 1149	84, 85, 119, 120	8, 9
IC 1434	OC	Lac	22 10.5	+52 50	1086	57	—
IC 5217	PN	Lac	22 23.9	+50 58	1086, 1102, 1103	57, 87	9
M29, NGC 6913	OC	Cyg	20 23.9	+38 32	1127, 1128, 1148, 1149	84, 85, 119, 120	9
M39, NGC 7092	OC	Cyg	21 32.2	+48 26	1104, 1105	86	9
NGC 6803	PN	Aql	19 31.3	+10 03	1244	206, 207	16
NGC 6804	PN	Aql	19 31.6	+09 13	1244, 1268	206, 207	16
NGC 6811	OC	Cyg	19 38.2	+46 34	1109, 1110	83, 84	8, 9
NGC 6819	OC	Cyg	19 41.3	+40 11	1129	84	8, 9
NGC 6826	PN	Cyg	19 44.8	+50 31	1091, 1109	55, 84	3, 8, 9
NGC 6866	OC	Cyg	20 03.7	+44 00	1128	84	8, 9
NGC 6871	OC	Cyg	20 05.9	+35 47	1149	119	8, 9
NGC 6894	PN	Cyg	20 16.4	+30 34	1171	119, 120	—
NGC 6910	OC	Cyg	20 23.1	+40 47	1127, 1128	84, 85	8, 9
NGC 6946	Gx	Cep	20 34.8	+60 09	1074, 1075	32, 56	3
NGC 6979	BN	Cyg	20 51.0	+32 09	1169	120	9
NGC 6997	OC	Cyg	20 56.5	+44 38	—	—	—
North America Nebula, NGC 7000	BN	Cyg	20 58.8	+44 20	1106, 1126	85	9
NGC 7209	OC	Lac	22 05.2	+46 30	1103	87	9
NGC 7243	OC	Lac	22 15.3	+49 53	1103	57, 87	3, 9
NGC 7245	OC	Lac	22 15.3	+54 20	1086	57	—
NGC 7296	OC	Lac	22 28.2	+52 17	1085, 1086	57, 58	—
NGC 7331	Gx	Peg	22 37.1	+34 25	1142	123	9

Ast = Asterism; BN = Bright Nebula; CGx = Cluster of Galaxies; DN = Dark Nebula; GC = Globular Cluster; Gx = Galaxy; OC = Open Cluster; PN = Planetary Nebula; ✶ = Star; ✶✶ = Double/Multiple Star; Var = Variable Star

SEPTEMBER OBJECTS (CONTINUED)

Name	Type	Const.	R. A. h m	Dec. ° '	Millennium Star Atlas	Uranometria 2000.0	Sky Atlas 2000.0
NGC 7394	OC	Lac	22 50.6	+52 10	1085	—	—
Pelican Nebula, IC 5070	BN	Cyg	20 50.8	+44 21	1106, 1107, 1126, 1127	85	8, 9
PK 45–2.1	PN	Aql	19 24.4	+09 53	1244	206, 207	16
PK 52–2.2	PN	Aql	19 39.2	+15 56	1219, 1220	207	8, 16
PK 52–4.1	PN	Aql	19 42.3	+15 09	1219, 1243	207	—
Struve 2894	**	Lac	22 18.9	+37 46	1143	122, 123	9
Veil Nebula (east), NGC 6992–95	BN	Cyg	20 56.4	+31 43	1169	120	9
Veil Nebula (west), NGC 6960	BN	Cyg	20 45.7	+30 43	1169	120	9

Ast = Asterism; BN = Bright Nebula; CGx = Cluster of Galaxies; DN = Dark Nebula; GC = Globular Cluster; Gx = Galaxy; OC = Open Cluster; PN = Planetary Nebula; * = Star; ** = Double/Multiple Star; Var = Variable Star

OCTOBER

CHAPTER 10

The Great Square of Pegasus

Compared with the brilliant, star-studded sky of winter, the autumn sky is but a vast celestial prairie of feeble suns. Even Pegasus, whose Great Square is the season's hallmark asterism, appears drained of starlight to the naked eye. And though the Square is one of the most familiar of all star patterns in the sky, it is also one of the more difficult for beginners to find, because of its large size. On a star chart, the Square appears so small. But Pegasus is enormous. It is the seventh-largest constellation, covering 1,121 square degrees of sky. Yet, despite this size, Pegasus is not a cornucopia of conspicuous deep-sky objects. In fact, it has only one Messier object, the globular M15. Scotty, however, was a digger, and you could leave it to him to find any telescopic nuggets. "Whether you use a 2.6-inch refractor or a 29-inch reflector," he wrote of Pegasus, "there is plenty to delight your eye."

October is a most auspicious month for amateur astronomers. The summer haze and humidity have given way to cooler days and crisp, clear skies at night. Darkness comes earlier, dewing of a telescope's optics is generally less of a problem, and the sky is not so jammed with star clouds that confusion rules.

The Milky Way stretches from east to west across the northern star patterns, but here we are looking in the direction approximately away from the center of the galaxy. Star swarms marking the galaxy's plane are thinner, and it is easy to star hop and make finder searches for objects embedded within them. Some of the most beautiful sights for small telescopes are in and around this corner of the Milky Way.

Overhead the Square of Pegasus dominates. As a small boy I thought of the Great Square as a baseball diamond where the Norse gods Odin and Loki pitted their mythical teams against one another — stray meteors were pop flies and clouds meant the game was called on account of rain. To my father's dismay, I always rooted for Loki!

Only schoolchildren and a few ROTC veterans of the U. S. Army and Navy are familiar with the old *Star Atlas and Workbook of the Heavens*, published by American Education Publications. Although long out of print, the 32-page booklet was used heavily in schools during the late 1960s and early '70s. It contained a novel method for learning the easily recognized constellations. Rather than having the Big Dipper as a starting point, it had a key constellation for each season: Leo, Scorpius, Pegasus, and Orion for spring, summer, fall, and winter, respectively. The reason was simple. As the Dipper's position around the pole changes with the season and time of night, its pattern sometimes becomes difficult for beginners to locate.

Tests of pattern recognition showed that difficulty can arise if a star chart is rotated just 15° with respect to the sky. Furthermore, the concept of seasonal key constellations grew out of research done during World War II at the celestial navigation school at Selman Field in Monroe, Louisiana; this system produced the best constellation learning among the 500 cadets tested. The key asterism for fall was the Great Square of Pegasus (**Figure 10.1**). While none of the Square's stars are 1st magnitude, it dominates October's star-poor evening sky. Counting the stars within the Square is a good indication of the sky's limiting magnitude. For example, if you see 13 stars you are reaching magnitude 6.0.

Figure 10.1
The Great Square of Pegasus is the starting point for many autumn star-hops.

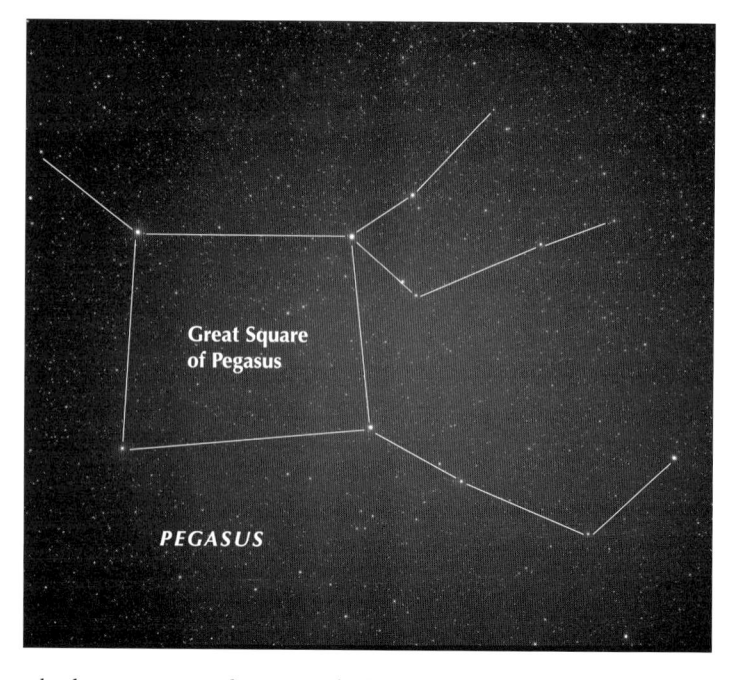

How many naked-eye stars can be seen within the Great Square of Pegasus? A quick look at the Skalnate Pleso *Atlas of the Heavens* suggests more than 100, but many of the stars plotted are below the traditional 6th-magnitude naked-eye limit. However, with dark skies and good transparency, many observers report glimpsing stars as faint as 7th magnitude. And in 1901, the Lick Observatory astronomer Heber D. Curtis found that, by shielding off the light of the sky and taking other precautions, he could just detect stars of magnitude 8.3.

One answer to the Pegasus question comes from John Bartels, who counted 38. He mentions, however, that some high cirrus clouds may have interfered. This suggests his limit was magnitude 6.6. If you push your limit to magnitude 7.0, then 70 stars should be visible.

How many deep-sky objects are visible in Pegasus? The present-day list in Robert Burnham's magnum opus, *Burnham's Celestial Handbook*, carries 23. Moreover, in the text there is mention of the famous "Stephan's Quintet," which effectively adds five more galaxies to the list. During the years of writing this column I have discussed 16 objects in Pegasus. Despite the apparent paucity of deep-sky targets, the number of objects given in the various handbooks is appropriate for typical amateur instruments in use at the time the books were published. However, in the years that have passed since Burnham's complete work became available, the size of amateur telescopes has increased greatly. The rise of the Dobsonian reflector has been a major reason.

Telescopes of 17-inch aperture are now off-the-shelf items of modest cost. There are a dozen or more amateur groups in the United States that either now have or are completing instruments with apertures of 24 inches or more. Such light-gathering power brings within reach of the backyard observer virtually every deep-sky object in the *NGC* and *IC* compilations. Thus the Great Square of Pegasus alone contains more than 100 suitable objects.

Not all the objects are easily located. So, if you do your searching with a finder, it is important to know the size of its field of view. Use a star atlas to determine the distance from an easily identified star to the object of interest and then offset from the star with the finder. You can try this method on NGC 16, a tiny galaxy 1½° south of the bright star Alpha (α) Andromedae at the northeast corner of the Square of Pegasus. (Although the star is officially in Andromeda, it is sometimes marked on charts as Delta (δ) Pegasi, which refers back to when some stars were shared by the two constellations.)

If your finder has a 3° field, and you place Alpha Andromedae at its northern edge, NGC 16 should be just about centered. This elliptical galaxy is only 1′ in diameter and 13th magnitude. With a 6- or 8-inch aperture, scan a 30× or 40× field with your eye for an object that is a bit too big to be a star. It is best not to sweep the telescope for this galaxy since an object this faint will probably be rendered invisible if the field is moving. Once you suspect you have the galaxy, higher magnification can be used to confirm it.

Another good star from which to begin a search for galaxies is Alpha (α) Pegasi at the Square's southeast corner. Just about 3° to its south is NGC 7479, a barred spiral that appears just over 3′ long and a bit brighter than 12th magnitude (**Figure 10.2**). If you have a motorized mount, you can also set on Xi (ξ) Pegasi and wait 18 minutes for this object. If your eye is properly dark-adapted, the galaxy should be visible in even a 3-inch telescope, but a 6-inch is better. A cloth over your head and the eyepiece gives good protection from stray light. I have seen it easily with my 4-inch Clark refractor, but with this small an instrument it is not possible to see any detail. On the other hand, the 12-inch f/17 Porter turret telescope at Stellafane

Figure 10.2 Just about 3° south of Alpha (α) Pegasi is NGC 7479, a very asymmetric spiral galaxy with a bright, long bar and looping spiral arms. Star formation occurs over the entire visible portion.

in Springfield, Vermont, offers a more interesting view. At 300× the central bar is obvious, and there is a hint of a spiral arm at one end.

A 12-inch f/5 reflector set up near the Porter telescope did not offer as good a view of NGC 7479 even though I thought the mirror was good. It may have had something to do with the longer focal length of the Porter telescope, or a better eyepiece. The importance of fine-quality eyepieces has been overlooked by many amateurs. Recently several types of expensive but very high-quality eyepieces have come on the market. Judging from my mail, it seems that these oculars distinctly improve the deep-sky performance of a telescope. Objects once considered only within the reach of large amateur instruments are being seen in smaller telescopes equipped with fine eyepieces.

Several other small galaxies lie near Alpha Pegasi. NGC 7448, an 11.8-magnitude spiral located 1½° northwest of the star, is about 2′ long and half as wide. Like NGC 7479, it is readily seen with the 4-inch Clark. However, increasing the aperture does not improve the view other than to make the object appear brighter. My old 10-inch reflector in Kansas simply showed it as a featureless glow.

NGC 7454 is a challenging galaxy, located about ½° northeast of NGC 7448. Although more than a magnitude fainter, its surface brightness appears nearly the same owing to a diameter of just 0.5′. At low powers NGC 7454 is easily mistaken for a faint field star. It, too, appears quite featureless even with a large telescope. A few years ago I had a peek at it with a 16-inch reflector, and though it was easily seen, no detail was visible. For an instrument of this size, there exist nearly a dozen more galaxies in the immediate area — all below the 13th-magnitude limit for galaxies plotted in Tirion's *Sky Atlas 2000.0* but listed in the *Revised New General Catalogue of Nonstellar Astronomical Objects* (*RNGC*).

The magnitudes listed for galaxies are often deceptive to the amateur observer, for the visibility of one of these dim glows depends on its size and shape, in addition to its overall brightness. A 4-inch telescope can sometimes reach 12th-magnitude stars, but this does not mean that it can show 12th-magnitude galaxies. Two facts might encourage the amateur. Almost all the entries in the *NGC* were discovered visually. In addition, long practice improves the ability to detect faint objects.

Here are some test galaxies of differing difficulty, located in Pegasus. Much of the difficulty in finding a faint object is avoided when one knows just where to look among the field stars. First look for NGC 7678, a 1.7′ × 1.1′ spiral; I saw it with a 10-inch in Kansas and it is easy in the 20-inch refractor of Van Vleck Observatory. NGC 7469 is a 1.3′ × 1′ spiral and slightly fainter. I could not find it with a 6-inch refractor but years back I viewed it in a 13½-inch reflector. For a very demanding test, try NGC 7619 and 7626, a pair of tiny elliptical galaxies of about the 12th magnitude. In the 20-inch they could be seen with some difficulty, but a 4-inch refractor was insufficient on a good night. What is the smallest aperture with which you can fish out this pair?

Can someone explain NGC 7772, which *Norton's* atlas shows inside the Great Square of Pegasus? It is missing from the *Atlas of the Heavens*, but the *NGC* describes it as "a cluster of scattered stars of about 10th magnitude." However, I see no particular concentration of stars in its place. The original discovery seems to have been by Sir John Herschel.

Finally, a bit of observing trivia: who knows how close on the sky the first and last entries in the *NGC* lie? NGC 1 is a 12.8-magnitude (my estimate with a 10-inch reflector) galaxy in Pegasus. The last entry in the original 1888 catalog is NGC 7840, but a 1973 revision by Arizona astronomers Jack Sulentic and William Tifft states that the object was not found on photographs made with the 48-inch Schmidt telescope in California. Thus, the last real object appears to be NGC 7839 — also in Pegasus and at essentially the same declination as NGC 1. The positions in the original catalog place NGC 7839 only 5.6′ southwest of NGC 1. However, NGC 7839 is very faint (it does not even have a listed magnitude in the revised *NGC*) and may be well beyond the reach of most amateur telescopes.

Using the list of objects in Vol. 2 of *Sky Catalogue 2000.0,* which is well suited for amateur instruments, I find the closest readily observable object to NGC 1 is NGC 7819, a faint galaxy in Pegasus. It is about 3.8° north of NGC 1.

Thus, NGC 1 and NGC 7839 are the first and last valid entries in the *NGC,* and it so happens that these galaxies are separated by less than 6′ on the sky. NGC 1 is about 1′ across and, by my estimate, about magnitude 12.8. It forms a pair with NGC 2, which is more than a magnitude fainter and 2′ to the southeast. NGC 7839 is even fainter and 5½′ southwest of NGC 1.

Although NGC 7840 is technically the last entry in the *NGC*, it apparently doesn't exist even though it was reported by such an outstanding observer as Albert Marth, who worked with William Lassell's 48-inch reflector on the island of Malta.

Two Spectacular Autumn Globulars

"Some deep-sky objects are remembered for their beauty, others are cherished for their scientific importance," Scotty wrote. "Some are sought out as test objects to establish the excellence of a telescope, observer, or atmospheric condition, and there are those best known for their unusual features." Finding any one object to meet all these criteria would be hard, but two autumn objects fit many of the categories — the splendid globular clusters M15 in Pegasus and M2 in Aquarius. For many observers, globular star clusters encapsulate all that is stunning and test-worthy. For instance, both M2 and M15 hover at the limit of naked-eye visibility, presenting an enjoyable challenge for dark-sky observers. Resolving the cores of these globulars with small telescopes is equally challenging, no matter where one lives. And their visual appearance has sparked controversies that have persisted for decades. For instance, are their disks colored? Are they marred by dark lanes and patches? Scotty believed that observers had overlooked such features for nearly a century.

The archetype of modern observing handbooks for amateurs is William H. Smyth's *Bedford Catalogue*. When this English observer prepared his classic as the second volume of *A Cycle of Celestial Objects* in the early 1840s, he mentioned only two deep-sky objects in Pegasus. This is unusual considering that it is the seventh-largest constellation. There is little doubt that Smyth was acquainted with more "nebulae" in Pegasus, for he had access to earlier catalogs, especially the great works of the Herschels. However, he chose only those that he felt amateurs would be interested in viewing.

In 1859 Thomas W. Webb published *Celestial Objects for Common Telescopes*. In many respects it relied heavily on Smyth's work. Webb originally included only one deep-sky object for Pegasus. He added a second in later editions, but it was different from Smyth's selection. Other observing guides added more.

All handbooks from the time of Smyth agree on including one object in Pegasus — M15 (**Figure 10.3**). Although Messier first saw it in 1764, it had been discovered by Jean-Dominique Maraldi in 1746. (Incidentally, some sources incorrectly state that Maraldi discovered M15 while searching for De Chéseaux's comet of 1746, when it was actually M2 that he found then.)

M15 (also called NGC 7078) is easily located about 4° northwest of Epsilon (ε) Pegasi. It is one of the few Messier objects easily picked up by sweeping. Make the first sweep northward from Epsilon by turning the declination axle, and stop after going about 5°. Now move one field to the west and move south by the same amount. Repetition of these steps, as if mowing a lawn, should bring the telescope across the field of M15 in a short while. This globular is unmistakable because there is a 6th-magnitude star beside it in the same low-power field. According to *Sky Catalogue 2000.0,* M15 is about 12″ across, with a V magnitude of 6.35. This magnitude, derived from photoelectric measurements, is close to what the eye sees. Therefore, M15 should be visible to the naked eye. Remember, though, that eyes vary from person to person, not to mention over time (I see red stars brighter now than I did years ago), so the actual perceived brightness of an object is never completely fixed. One way to improve the chance of seeing it without optical aid is to look through a long cardboard tube painted black on the inside. This will cut down the often unappreciated flood of sky brightness and improve your magnitude limit.

The view of M15 is impressive with anything from binoculars to the largest telescope. Telescopes of 4-inch aperture and less will not resolve the core of M15. My 4-inch Clark refractor at 40× shows M15 as a slightly oval disk, more luminous in the center, with edges just beginning to break up into individual stars. Increasing the magnification enhances the view, and at 200× stars at the center

Figure 10.3
The core of globular cluster M15 in Pegasus is extremely densely packed, suggesting that a sudden, runaway collapse due to the gravitational attraction of many stars in a small region of space may be occurring.

of the cluster start to be resolved. I once succeeded in resolving the core with my old 10-inch f/8.6 reflector and a low-power eyepiece coupled with a Barlow lens yielding 200×. However, the same ocular combination on a 12-inch f/5 instrument did not break the center up into individual stars completely. I have always recommended using a Barlow for boosting magnification. Eye relief is better, and dust on the ocular scatters less light than if an equivalent high-power eyepiece is used alone.

I had an astounding view of the cluster with a 17-inch reflector and 9-mm Nagler eyepiece. M15's stars nearly filled the field. I also remember vividly a breathtaking view of this globular cluster many years ago with the 36-inch reflector at Steward Observatory in Arizona, which showed stars splashed all over the field of view. Stars literally erupted in the eyepiece field, and with averted vision the cluster's stellar population seemed to double.

The outer reaches of the cluster are so frayed that the eye cannot tell where it ends. Smyth commented on star chains radiating outward from the cluster. I have confirmed this appearance with the 4-inch Clark refractor at my Joseph Meek Observatory. But I find more interesting a remark by T. W. Webb, who wrote, "Buffham, with 9-in. spec. finds a dark patch near the middle, with 2 faint dark 'lanes' or rifts, like those in M13, unnoticed by h. or D`A." Are we to believe that a 9-inch speculum-metal mirror could show features that were missed by the skilled eyes of John Herschel (h.) and Heinrich d'Arrest (D'A.), or are these dark markings just an illusion? My notes also do not mention any of the dark patches suspected by John Mallas when he observed with a 4-inch refractor. Perhaps some amateurs would like to check on the existence of such features.

For years, dark lanes reported in the Hercules globular M13 were also thought to be illusions, but several years ago they were rediscovered by amateurs and are now common fare at star parties. Visibility of the M13 markings depends on telescope aperture and magnification. Perhaps those in M15 are similar. October is prime time for amateurs to give M15 a good working over with different telescopes under a wide range of conditions. Although I encouraged amateurs with 16-inch and larger instruments to follow up these reports, I've never had the patch or lanes confirmed. It's worth pursuing because professional astronomers have found dust clouds in the southern globular NGC 362. Although their work is based on CCD observations and image processing, it seems reasonable that some dust clouds could be visible to the eye.

In the past I have mentioned a planetary nebula within M15. Known as Pease 1 and located on the cluster's northeast side, it was discovered in 1927 on photographs made with the 100-inch reflector at Mount Wilson Observatory. I had always considered this tiny (1″) 14th-magnitude object probably beyond the reach of any amateur telescope, but recently three observers sent interesting reports. All used the Lumicon O III filter to locate the planetary with the blinking method. By flicking the filter in and out between the eye and eyepiece, the planetary appeared to blink relative to the stars.

On to M2!

The distance to M15, 40,000 light-years, is similar to the 50,000 value listed for M2, our other spectacular autumn globular (**Figure 10.4**). One way to pick up M2 is to aim your telescope a quarter of the way from 3rd-magnitude Beta (β) Aquarii to 2nd-magnitude Epsilon (ε) Pegasi. You will find a great glowing heap of stars, with the brighter ones sprinkled like stardust over a disk 10′ or 12′ across. M2 has a total visual magnitude of 6.3. Thus one would expect it to be visible to the naked eye under excellent observing conditions, but such sightings are uncommon.

The famous variable star-observer and comet discoverer Leslie Peltier finds M2 a more difficult object for the unaided eye than M33, the large spiral galaxy in Triangulum. In the clear dark skies over the Yucatán Peninsula in Central America I could view M33 directly, but M2 required averted vision before it could be glimpsed directly. But I have seen M2 often with the naked eye in Kansas, Missouri, Arizona, and even from the bayous of Louisiana. M2 is one of those happy objects that are a delight in any size instrument. Binoculars give enough detail to keep the amateur interested, while the view I once had with Wesleyan University's 20-inch Clark refractor was spellbinding.

Figure 10.4 Fully 37 percent of the light from the globular cluster M2 in Aquarius comes from the stars populating the cluster's central square arcminute.

M2 was discovered by Maraldi in September 1746. Because he could not resolve the glow into stars, he initially mistook it for a comet. Indeed, when Charles Messier first saw M2 some 14 years later, he did not resolve it into stars either and likened the cluster to the nucleus of a comet. Interestingly, in today's instruments, 13th-magnitude stars on the fringes of M2 can be glimpsed in an 8-inch, but in a 4-inch the cluster is generally unresolvable and condensed like a comet. In fact when a number of observers independently discovered Comet Kobayashi-Berger-Milon (1975h) in the same field as M2, it was not possible to tell the globular from the comet until the comet's motion tagged it.

Smyth mentions that an observer named Samuel Vince viewed M2 with William Herschel's 40-foot reflector in 1779 and saw tiny stars right to the center of the cluster. Today, a good 12-inch telescope can resolve M2 even to its core, as I have seen it with the Porter turret telescope at Stellafane. Using this instrument I find John Herschel's description of M2 most suitable, for he called the cluster "a heap of fine sand."

Two observations of M2 in recent times warrant attention by amateurs. Kenneth Glyn Jones in his *Messier's Nebulae and Star Clusters* mentions a faint greenish-blue glow around the cluster under certain observing conditions. John Mallas, who used a 4-inch refractor, reported seeing a dark lane crossing the northeast corner of the cluster. Perhaps some astrophotographer could investigate Mallas's dark lane by taking a set of different exposures. By the way, Steve Coe of Glendale, Arizona, writes that M2 has "lovely chains of stars meandering outward from the core, and several dark lanes are visible."

Sweeping Through Sagitta

After darkness falls in mid-October, look halfway up the sky, just west of the meridian and just north of scintillating Altair. There you will see four close-knit stars marking out Sagitta, the celestial Arrow. Sagitta is one of the best known of the very small constellations, mainly because amateurs use it as a guide to the famous Dumbbell Nebula in dim Vulpecula. Although Sagitta is small (its official boundaries have varied over time), it now covers an area of 80 square degrees, making it the third-smallest constellation. The constellation lies in a rich section of the Milky Way and hosts a bounty of celestial riches, including the once mysterious and ambiguous star cluster M71 and a neglected open cluster known as H 20. "So sparse is this cluster," Scotty wrote, "that without careful positioning of the telescope it is easy to overlook."

As a child I would stand outside on autumn evenings and fantasize about the constellations. I would watch as the horse-archer Sagittarius shot a golden arrow at Scutum (Sobieskii's Shield). The arrow would strike the top of the shield, tearing a great hole in it, and the fragments would fall back together as the arrow-shaped open cluster M11. The arrow would then soar upward into the star clouds, where it would hang poised for another target in the Milky Way or perhaps another galaxy or even some imaginary other universe.

Now, over half a century later, the arrow Sagitta (**Figure 10.5**) still hovers where I saw it as a child. When first described by Eratosthenes, this little constellation consisted only of a group of stars running 4° from east to west. By the 19th century, map makers had enlarged its area to 10°, and when the International Astronomical Union reorganized constellation boundaries in the 1920s Sagitta's territory grew to 20° across, where it will probably remain.

As old as it is, however, Sagitta seems to be little observed. It lies in a bright Milky Way star cloud, so the field is swamped with faint stars. The *RNGC* describes numerous *NGC* objects in this part of the sky as "nonexistent."

However, this usually means that the object, though recorded by earlier observers, was not found on photographs examined by the authors of the *RNGC*. As many amateurs know, a group of stars that stands out as a cluster in a telescope can become lost on a photograph, especially if the field is full of faint background stars. An interesting project, though one with little promise of fame, would be to search visually using, say, a 12-inch telescope, for the objects listed as nonexistent or "not found" in the *RNGC*.

Figure 10.5
The constellation Sagitta straddles the Milky Way and points to several deep-sky wonders. South of it lies the constellation Aquila, with the bright star Altair.

One cluster not easily overlooked in Sagitta is M71. Indeed, it was visible with my childhood 1-inch 40× refractor. You will find it south of the center of a line running between Gamma (γ) and Delta (δ) Sagittae, about 20′ northeast of 6th-magnitude 9 Sagittae in the shaft of the arrow. I have always wondered why Messier saw so little in the cluster (**Figure 10.6**). He described it as very faint with no stars and wrote, "The least light extinguishes it." Maybe it was a poor night when he examined M71, which was brought to his attention by Pierre Méchain in 1780 (after being discovered by earlier observers). On fainter objects, Messier usually saw more detail.

With medium magnification, my 4-inch Clark refractor shows individual stars rather uniformly distributed across the 8th-magnitude cluster. M71 has been

Figure 10.6 Once an object of contention, M71 in Sagitta is now recognized as a loose globular cluster rather than a rich and remote galactic cluster.

called both an open and a globular cluster. The early 20th-century dean of visual observers, E. E. Barnard, noted that M71 looked like a globular cluster in his 6-inch refractor. And the French observing handbook *Revue des constellations* gives this description: "In 10 × 50 binoculars and a 2½-inch refractor at 20×, well seen, large and diffuse. Globular appearance, detached in a rich field, unresolved, with 3¼-inch refractors at 30× to 45×. Fine cluster 6′ by 5′, some stars visible with an 8-inch reflector (150×); about 20 stars of magnitude 12 and fainter, on an irregular milky background, in a 12½-inch at 80×."

My old 10-inch f/8.6 reflector, which, with its ¾-inch-thick plate-glass mirror, was essentially a forerunner of today's Dobsonians, gave a magnificent view of M71 at 100×. Stars were visible across the entire disk, and the object looked decidedly like an open cluster. The 20-inch Clark refractor at Wesleyan University's Van Vleck Observatory in Connecticut shows something more globular. Volume 2 of *Sky Catalogue 2000.0* terms M71 "globular" without mention of any earlier controversy.

Many observers have commented on M71's arrow-shaped appearance, but in photographs it is usually round. Its image in Hans Vehrenberg's *Atlas of Deep-Sky Splendors* is quite triangular. What do you see? Most observers find the object uniformly illuminated, with no central core. Yet in the *Messier Album*, John Mallas drew the cluster with a pronounced bright edge, rather like a boomerang. I find no mention of such a feature in my observing notes, but the

Webb Society *Deep-Sky Observer's Handbook* for open and globular clusters mentions a "nebulosity concentrated in the western part." What do you see? Remember to record the size and magnification of your telescope.

About ½° south-southwest of M71 is the scattered group H 20, a very sparse bunching of about 15 stars in an area 7' across. They have a total magnitude of 7.7, which makes the cluster brighter than M71. However, H 20 is much more difficult to pick out from the background sheen of the Milky Way, as compared with the tightly packed but fainter stars comprising M71. Here is a clear case of not being able to judge the conspicuousness of a cluster from its published magnitude alone. Experienced observers know not to draw a mental image of an object by its listing in a catalog, regardless of the amount of data given. So sparse is this cluster that without careful positioning of the telescope it is easy to overlook. Often, sweeping the area with a 4- or 6-inch telescope will fail to locate it. So inconspicuous is this object that it was not included in the *NGC* compilation.

To the west and a little north is a delicate splash of faint celestial fire known as NGC 6802. This small cluster is rather strongly elongated north-south. As a guide, there is a small grouping of stars to the west called Brocchi's cluster, and NGC 6802 is at the east end of a 2°-long string of stars easily seen in any telescope.

The eastern end of Sagitta contains three very challenging planetary nebulae. *Sky Catalogue 2000.0* lists NGC 6886 as 4" across and of photographic magnitude 12.2. In the early 1900s Heber D. Curtis made a composite drawing of this planetary based on photographs exposed from 10 seconds to an hour. In addition to calling the object 10th magnitude, he sketched it as 6" across. He also noted two projections that make the planetary look like a low-resolution photograph of Saturn. To date these projections have escaped my visual searches even with the 20-inch Clark. Perhaps someone can look for these features with one of the 29-inch Dobsonians now turning up across the United States.

The main disk of NGC 6886 is within the reach of a 4-inch telescope, but a word of warning is in order: Under poor seeing conditions the planetary may be difficult to distinguish from its surrounding stars. Todd Hansen of Potter Valley, California, used a 10-inch f/5.6 Newtonian to view the nebula. He notes it as "extremely small, bright, and slightly oval; at 180× sometimes like a fuzzy star; sometimes almost seems double."

Another small planetary is NGC 6879. It is listed as 5" in diameter and photographic magnitude 13.0. Hansen could not make out the planetary's disk. In the Webb Society *Handbook* (Vol. 2) Pat Brennan reports that NGC 6879 appears stellar in an 8-inch reflector. He identified the planetary by using a small prism at the eyepiece of his telescope. This simple accessory spreads the light of stars out into tiny spectra, while the planetary remains as a dot since its light is essentially from a single emission line at one wavelength.

Another planetary best detected with a prism is IC 4997. Its 2" disk is listed as photographic magnitude 11.6, and it has a central star perhaps two magnitudes fainter. Hansen, probably on an especially good night, thought he could distinguish the star at 360× but failed to do so on another night. English amateur Ed

Barker saw a slight indication of the planetary's disk at 308×. At the time he was using an 8½-inch reflector.

Although American amateurs have lagged behind their European counterparts in using a prism to locate small planetary nebulae, they have been quick to use nebula filters. These dim the field stars while leaving the nebula's brightness relatively unchanged. The filter can be flipped in and out between the eyepiece and eye, making the nebula appear to blink compared to the stars.

Most of the many variable stars in Sagitta are faint, but S Sagittae is an interesting object for observation with binoculars. It is of the Cepheid type, varying between the visual magnitude limits 5.3 and 6.1 in an eight-day period.

Before leaving this rich area, double-star observers with small telescopes will want to view Zeta (ζ) Sagittae, an easy pair of 5th- and 9th-magnitude stars separated by 8″. (The brighter component is a very close binary, beyond the reach of amateur instruments.) Theta (θ) Sagittae is another easy double, magnitudes 6 and 9, separation 12″.

Unraveling the Helix

One of Scotty's greatest and most respected traits was that he did not write to promote himself but to report on the progress of amateur astronomy. He made it his job to involve amateurs in that revolution; of course, he was often the one to get a fire going. Scotty was probably most satisfied when he received reports from amateurs who wanted to share their opinions on a particular subject of controversy. One great example of how Scotty promoted this sharing of knowledge can be found in his articles on the Helix Nebula in Aquarius. The Helix is a tough celestial nut to crack. Its pale hazy disk is often missed by amateurs who are accustomed to seeking much smaller objects. For advice, he would say, "Averted vision is needed, and the eyepiece field should be at least ½° in order to surround the nebula with some contrasting dark sky." As you will see here, readers of Deep-Sky Wonders had widely varied opinions on the object's visibility, and Scotty respected each and every one of them. His column was a forum for the amateur's voice to be heard.

The planetary nebula NGC 7293, also known as the Helix Nebula (**Figure 10.7**), lies in Aquarius about a third of the way from Upsilon (υ) Aquarii to 47. It has a total magnitude of about 6, but its large apparent diameter — nearly half that of the Moon — spreads the light out and makes it a difficult object visually. The Herschels overlooked this nebula with their large reflectors. It is best to view this planetary with a rich-field telescope. With exceptional skies, an experienced eye will sometimes see traces of the nebular structure so vividly recorded in photographs.

I recently saw the Helix Nebula with the 4-inch Clark refractor, and was certain that it was glimpsed in a 2-inch finder. Burnham notes in his *Celestial Handbook* that it can be spied in binoculars. Years ago I suggested that readers send me their observations of the Helix Nebula, which I sometimes call the "Sunflower." Over 200 letters were received, giving a good idea of its visibility in many instruments and at a variety of magnifications.

Harry Cochran of Brentwood, Texas, found the Helix difficult in a 12½-inch at 67×, though the view was better at 117×. On the other hand, to Leonard P. Farrar of Rialto, California, the planetary appeared much like its photograph. He used a 10-inch mirror by Alika Herring in a mountain sky so clear that a flashlight beam was invisible.

Small telescopes in relatively poor skies seldom revealed color. Ted Komorowski told of a gray disk easily visible in his 8-inch f/7.5 at 56×. Yet Ray Lima of Jacksonville, Florida, readily saw blue-green in his telescope of the same aperture. When an object is near the visual threshold, color is not usually seen, only gray. To obtain maximum color perception, use the lowest magnification available.

The nebula's central hole was sighted by only a few observers, who included Michael Pleinis, Aberdeen, South Dakota (4- and 6-inch telescopes), and Mark

Figure 10.7
The Helix Nebula is the planetary nebula nearest to our Sun. At 15' in apparent diameter, it is also the largest.

Grunwald, Mishawaka, Indiana. Among others, Tom Burton of Santa Cruz, California, and N. Taylor in New Zealand could not see the central hole. Taylor noted that at higher powers NGC 7293 filled the field and details were lost.

The performance of instruments similar to Moonwatch Apogee telescopes varied. James H. McMahon, China Lake, California, glimpsed NGC 7293 at the limit of visibility, after he had failed to see it a year before. Buddy Tempest, Columbus, Indiana, found it without using averted vision, while William O'Brian of Gary, Indiana, easily saw the planetary at 16× and 25×, though it was invisible in 7 × 50 binoculars.

The bulk of my correspondents indicated that the Helix was more readily seen in binoculars and finders than in telescopes. Yet some observers, such as Billy Perkins of South Boston, Virginia, had the opposite experience.

A few amateurs compared the Helix with familiar objects. For example, Fred Lossing of Ottawa, Canada, thought the binocular appearance of NGC 7293 was similar to that of M33 in his 8-inch. Jan Finkelstein, Brooklyn, New York, saw a resemblance to M57 in his 2.4-inch refractor.

Atmospheric clarity obviously played a major role for seekers of the Helix. Observers on mountains did much better than smog fighters, though the latter had easier viewing than one would have anticipated from Hans Vehrenberg's statement in his *Atlas of Deep-Sky Splendors:* "Even on dark and extremely clear nights it is barely distinguishable as a very faint patch."

Grunwald's excellent results came after a cold front had just passed, making the sky so clear that M33 was visible to the unaided eye. Such favorable conditions should always prompt searches for the toughest deep-sky objects.

Most observers agreed that NGC 7293 was captivating. Edward Stockton of Lithia, Florida, recalled: "It was so faint that it seemed like a figment of the imagination, but its shape was unmistakable."

In 1983 I again asked for observers' comments on the Helix Nebula, and a number of people replied.

Jim Meketa of Newton Center, Massachusetts, easily viewed the Helix with 7 × 35 binoculars. (Binocular vision often shows a fainter object than could be detected in a view with one eye alone. You can experiment by viewing a faint object with binoculars and covering one of the objectives. Chances are the object will disappear from the field.) Meketa's binoculars showed the Helix as a "small ghostly doughnut," but a 4½-inch finder on his 18-inch reflector showed some detail. The ring appeared unbroken, and there was an "unmistakable" tenuous glow inside it. There was no hint of the helical structure seen on photographs that give rise to the nebula's popular name. Meketa believes that the Helix might be seen with the naked eye under excellent conditions.

In Georgia, David Riddle observed NGC 7293 with a wide range of instruments. It was "visible" in a 6 × 30 finder and "easy" with 8 × 40 binoculars. The central star was seen with a 6-inch f/4 Newtonian, and his best view came with an 8-inch f/6 reflector at 80×.

From Riverside, California, Stephan Karnes could not see the Helix with a 14-inch reflector until a nebula filter was added to the eyepiece to cut down the urban light pollution. Like Meketa, he noted faint nebulosity filling the center of the ring. Another Californian, Steve Gottlieb of El Cerrito, had trouble viewing the Helix with a 13-inch reflector at his home, while it was seen easily from high in the Sierras with a 6 × 30 finder and nebula filter. His conclusion is that the visibility of the Helix is controlled more by sky conditions than telescope size.

One of the best reports came from Joanne Konst of Kenton, Ohio: "NGC 7293, the Helix Nebula, is invisible without a UHC filter," she writes. "But with the filter I see a round glow about 15' across, and a dark center is obvious. Three stars are seen against the nebulosity, but otherwise there is no detail."

A "Field Day" in the South

Occasionally a great celestial event forces masses of Northern Hemisphere observers to pack their gear and head south. Take, for instance, the 1986 return of Halley's Comet. Countless droves of amateurs armed with telescopes of various sizes descended upon the less populated regions of the south — the outback of Australia, the Alps of New Zealand, or South America's barren Altiplano — simply to glimpse a part of astronomical history. But Scotty knew that these pilgrimages could yield even greater prizes. "The comet is going to have an effect on deep-sky observing," he predicted. "Flocks of amateurs are planning trips to the Southern Hemisphere, where many will have their first encounters with the glories of the southern sky . . . I wouldn't be surprised if observers spend more time exploring the southern heavens than looking at the comet." And he was right. What he wanted to remind us, however, was that many celestial sights in the southern sky can be seen from northern locations. Here are a few of them that Scotty wanted readers to enjoy. Their declinations, he reminded us, are no farther south than that of the Scorpion's tail.

One of the great treasures of life is heaven's starry vault on a clear night, when the familiar constellations blaze forth in mystical glory. It's an extra treat if the sky is clear right down to the horizon and we can explore regions normally lost in the haze.

In theory, everyone living south of the 45th parallel can see to the very bottom of Sagittarius. So they can try to hunt out four spiral galaxies in the constellation's southeast corner; their presence was brought to my attention several years ago by New York amateur and author Phil Harrington. All are plotted on *Sky Atlas 2000.0*, and Harrington remarks that they are "challenging finds for even the most accomplished deep-sky observer."

Although they have eluded me from Connecticut, I saw them dimly with a 4-inch rich-field reflector in the Arizona desert. Farther south at Puerto Escondito, Mexico, they were bright and easy in the same telescope. All are between 1' and 2' in diameter. The most obvious member of the group is 12th-magnitude NGC 6902, which shows some detail in a 10-inch aperture.

If these are too far south for your observing site, but you can see brilliant Fomalhaut in Piscis Austrinus, then try looking for another worthwhile quartet of galaxies (**Figure 10.8**). NGC 7172 and the tightly bunched NGC 7173–74 and 7176 are visible in my 4-inch Clark refractor from here in Connecticut. They're all about 12th magnitude, and plotted on *Sky Atlas 2000.0*. NGC 7172 is 2' across, but the other three are only half as large. John Herschel chanced upon them while sweeping the sky from the Southern Hemisphere. When he published a general catalog of deep-sky objects in the mid-1860s, over 90 percent of its 5,000-plus objects were those discovered by him and his father William, and nearly 100

Figure 10.8
If you can see Fomalhaut, you can find the galaxies NGC 7172, 7173, 7174, and 7176, which are clustered about 12° to the west.

of the exceptions were from Messier's famous list. Observers in the Americas had little to contribute.

After the U. S. Civil War, however, Americans went on an observatory-building binge. Funding for many installations came from state legislatures, since the astronomers provided time signals to their local areas. Almost every observatory from that era had a transit instrument for determining time. In return for this service the lawmakers funded a large telescope to keep the astronomers happy. When I was at the University of Wisconsin in the 1930s, Washburn Observatory still had the big brass fittings on the control board that routed time signals to commercial customers.

Most American observatories did not have special programs to search for deep-sky objects. Nevertheless, astronomers found new nebulae in the course of other work and published short lists of these accidentally discovered objects. By the time J. L. E. Dreyer compiled the *NGC* in 1888, the list had grown to nearly 8,000 objects.

About two dozen of these were discovered by Edward S. Holden with the 15½-inch reflector at Washburn Observatory in Wisconsin. One of them, NGC 6912, lies less than 1° west of Omicron (o) Capricorni. When Holden came across it on August 17, 1881, he noted it as very faint. I still look at it when I get a chance — after all, there weren't many deep-sky objects discovered in the state where I grew up.

About 20° east-southeast of NGC 6912 is the globular cluster M30 (**Figure**

Figure 10.9
The globular cluster M30 lies about 40,000 light-years away in the constellation Capricornus. It is best viewed with larger telescopes.

10.9), which Messier discovered in 1764. Although he saw it as a round nebula without stars, William Herschel resolved it into a "brilliant cluster" two decades later. About 8th magnitude and a bit over 10′ in diameter, M30 is not one of the great globulars. Its bright center and easily resolved edges do, however, make it an interesting object for small telescopes. I find this sight rewarding in a 4-inch scope at 40×.

M30 can be a frustrating object for amateurs attempting a Messier marathon in March or April. Because of its position relative to that of the Sun, the cluster is almost impossible to find in either the evening or the morning sky, spoiling the chance to view all the Messier objects during a single night.

Sweeping northward from M30 into Aquarius brings us to several interesting objects. M72 is a globular both smaller and fainter than M30. In the 1930s I viewed it with the 13-inch reflector that belonged to the Milwaukee Astronomical Society and remember the edges of the cluster being well resolved.

Just east of M72 is a tiny group of four stars that Messier perceived as a dim glow and therefore included as entry 73 in his catalog. Some observers continue to debate whether this group belongs on the list with the French comet hunter's other discoveries, but there is no doubt about its correct identity.

Only 2° northeast of M73 is a distinguished planetary nebula that Messier overlooked even though it was certainly within range of his telescopes. NGC 7009 is perhaps better known as the Saturn Nebula (**Figure 10.10**). The name comes from Lord Rosse, who first saw two faint ansae extending from the cen-

Figure 10.10
Nicknamed the Saturn Nebula because of its appearance in larger telescopes, NGC 7009 is one of the brightest planetary nebulae in the sky.

tral nebulosity, but the object itself was discovered by the elder Herschel. What is the smallest telescope that will show the planetary's faint extensions? The nebula's central star is about magnitude 11.5. It stands out like a beacon in my eye that had its lens removed in a cataract operation several years ago and is now sensitive to ultraviolet light.

In the Pasture of Grus

Now we'll have a field day in the sky below Fomalhaut, the brilliant 1st-magnitude beacon in Piscis Austrinus. I like to think of this region as the pasture of Grus the Crane, the remarkably birdlike star group to Fomalhaut's south. A number of galaxies here, while challenging objects from mid-northern latitudes because of their low altitude, are no farther down than the lowest stars in Scorpius's tail.

IC 5271 is a spiral galaxy forming a neat little triangle with the 4th-magnitude stars Delta (δ) and Gamma (γ) Piscis Austrini. It is about 2' long and half as wide, large enough to be identified at 50× (though 100× would be a better magnification with which to search for it). At magnitude 12.6 it can be seen in a 4-inch telescope when well above the horizon. Once, while in Mexico's Sonora desert, I tracked it down with a 4-inch rich-field reflector at 40×. The sky was very good that night, and the naked-eye limit for stars was about magnitude 7.4. (It's always a good idea to note the naked-eye limit in your logbook, since I find this a much better indicator of sky transparency than the usual 1-to-10 scale.)

A little over 2° south of IC 5271 is the similar-looking spiral galaxy IC 5269.

The second *Index Catalogue of Nebulae* (which contains objects found between 1895 and 1907) lists it as very faint, pretty small, and round, to which I can reply, "No, yes, and somewhat." IC 5269 is also the northernmost of five galaxies forming a small curving chain. It is easy to locate since it lies just west of the center of a small triangle of naked-eye stars.

The two stars marking the base of this triangle are on the border of Grus. Almost on a line between them is IC 1459, an elliptical galaxy I estimate to be magnitude 10.0. Roughly ½° farther south is NGC 7418, discovered by John Herschel from Africa's Cape of Good Hope. It's a fat 3′ in diameter and magnitude 11.4. My 4-inch reflector gave a good view of it in the clear desert air.

The next galaxy in the chain is NGC 7421, which lies ⅓° almost due south of NGC 7418. It is about 12th magnitude. Another ½° to the southeast is IC 5273. Although about half a magnitude brighter than NGC 7418, it was missed by the observers whose discoveries were published in 1888 and is listed in the second *Index Catalogue*.

Star-hop to these galaxies by starting at Fomalhaut, moving to Delta (δ) and Gamma (γ) Piscis Austrini, and then to the small triangle of stars. This is a simple task — if the telescope and finder show the same sky orientation. On many commercial telescopes either the finder or the main instrument comes equipped with a star diagonal, which gives a mirror image of the sky. Trying to match this view with a star chart is like reading a newspaper in a mirror.

One solution is to use an Amici prism in place of the diagonal's mirror or right-angle prism. Amici prisms combine the comfort of using a diagonal with a correct-reading image of the sky that can easily be matched to a chart. They are becoming increasingly scarce on the surplus market, but if you can find one I certainly recommend getting it.

A nice group of four galaxies forms a rough square straddling the –40° declination line on *Sky Atlas 2000.0*. While I don't usually receive amateur reports of objects this far south, California observer Tokuo Nakamoto has seen at least two of them.

NGC 7410 is an obvious cigar-shaped galaxy at the northwest corner of the square. It is about 5′ long and 2′ wide. With the 4-inch reflector I estimated it as magnitude 10.0. Using a 6-inch f/8 reflector, Nakamoto found it brighter toward the center with a starlike nucleus. Much fainter is 12.7-magnitude NGC 7462 at the square's southeast corner. It is another thin spindle of light, about 3′ long. Nakamoto noted that it has no apparent nucleus.

At the southwest corner, NGC 7424 is about 7′ in diameter, nearly round, and 10.2 magnitude. At the northeast corner is NGC 7456, some 6′ long, 2′ wide, and magnitude 11.9. Can anyone confirm whether it has a stellar nucleus?

Before leaving this region of the sky, try for a tight group of four galaxies all visible in the same telescope field 5° southeast of the square. Three of them are nice big spirals, all apparently barred, cavorting in a group like white beluga whales. The fourth is just to their west, off by itself. This is NGC 7552, listed as visual magnitude 10.7 and 3′ across. The western galaxy of the triplet, NGC 7582, is cataloged as magnitude 10.6, but it seems to outshine NGC 7552 by even more than the official 0.1-magnitude difference. NGC 7590, about a magnitude fainter, is somewhat smaller than its neighbors. Of similar brightness to NGC 7590 is NGC 7599, though it is nearly as large as NGC 7582. I once marveled at these galaxies from the flanks of a smoking volcano in Guatemala.

OCTOBER OBJECTS

Name	Type	Const.	R. A. h m	Dec. ° '	Millennium Star Atlas	Uranometria 2000.0	Sky Atlas 2000.0
H 20	OC	Sge	19 53.1	+18 20	1219	162	8, 16
Helix Nebula, NGC 7293	PN	Aqr	22 29.6	−20 48	1355, 1379	347	23
IC 1459	Gx	Gru	22 57.2	−36 28	1423	384, 385	23
IC 4997	PN	Sge	20 20.2	+16 45	1217, 1218	163, 208	9, 16
IC 5269	Gx	PsA	22 57.7	−36 02	1423	384, 385	—
IC 5271	Gx	PsA	22 58.0	−33 45	1423	384, 385	23
IC 5273	Gx	Gru	22 59.5	−37 42	1423	384, 385	23
M2, NGC 7089	GC	Aqr	21 33.5	−00 49	1286	255, 256	16, 17
M15, NGC 7078	GC	Peg	21 30.0	+12 10	1238	210	16, 17
M30, NGC 7099	GC	Cap	21 40.4	−23 11	1381, 1382	345, 346	23
M71, NGC 6838	GC	Sge	19 53.8	+18 47	1219	162	8, 16
M72, NGC 6981	GC	Aqr	20 53.5	−12 32	1336	299	16
M73, NGC 6994	OC	Aqr	20 59.0	−12 38	1335, 1336	299	16
NGC 1	Gx	Peg	00 07.3	+27 43	150	89, 125	—
NGC 2	Gx	Peg	00 07.3	+27 41	150, 174	89, 125	—
NGC 16	Gx	Peg	00 09.1	+27 44	150	89, 125	4, 9
NGC 6802	OC	Vul	19 30.6	+20 16	1220	161, 162	—
NGC 6879	PN	Sge	20 10.5	+16 55	1218	163, 208	9, 16
NGC 6886	PN	Sge	20 12.7	+19 59	1218	163	9, 16
NGC 6902	Gx	Sgr	20 24.5	−43 39	1451, 1452	411, 412	23
NGC 6912	Gx	Cap	20 26.9	−18 38	1361	343	—
NGC 7172	Gx	PsA	22 02.0	−31 52	1404	383	—
NGC 7173	Gx	PsA	22 02.0	−31 58	1404	383	—
NGC 7174	Gx	PsA	22 02.1	−31 59	1404	383	—
NGC 7176	Gx	PsA	22 02.1	−31 59	1404	383	—
NGC 7410	Gx	Gru	22 55.0	−39 40	1445	384, 415	23
NGC 7418	Gx	Gru	22 56.6	−37 02	1423, 1424	384, 385	23
NGC 7421	Gx	Gru	22 56.1	−37 21	1423, 1424	384, 385	23
NGC 7424	Gx	Gru	22 57.3	−41 04	1445	415	23
NGC 7448	Gx	Peg	23 00.1	+15 59	1209, 1210	213	17
NGC 7454	Gx	Peg	23 01.1	+16 23	1209, 1210	168, 213	17

Ast = Asterism; BN = Bright Nebula; CGx = Cluster of Galaxies; DN = Dark Nebula; GC = Globular Cluster; Gx = Galaxy; OC = Open Cluster; PN = Planetary Nebula; ✶ = Star; ✶✶ = Double/Multiple Star; Var = Variable Star

OCTOBER OBJECTS (CONTINUED)

Name	Type	Const.	R. A. h m	Dec. ° '	Millennium Star Atlas	Uranometria 2000.0	Sky Atlas 2000.0
NGC 7456	Gx	Gru	23 02.1	–39 35	1445	384, 385, 415	23
NGC 7462	Gx	Gru	23 02.8	–40 50	1445	415	23
NGC 7469	Gx	Peg	23 03.3	+08 52	1233, 1257	213	17
NGC 7479	Gx	Peg	23 04.9	+12 19	1233	213	17
NGC 7552	Gx	Gru	23 16.2	–42 35	1444	415	23
NGC 7582	Gx	Gru	23 18.4	–42 22	1444	415	23
NGC 7590	Gx	Gru	23 18.9	–42 14	1444	415	23
NGC 7599	Gx	Gru	23 19.3	–42 15	1444	415	23
NGC 7619	Gx	Peg	23 20.2	+08 12	1256, 1257	214	17
NGC 7626	Gx	Peg	23 20.7	+08 13	1256, 1257	214	17
NGC 7678	Gx	Peg	23 28.5	+22 25	1184	169	9
NGC 7772	OC	Peg	23 51.8	+16 15	—	125, 170	—
NGC 7819	Gx	Peg	00 04.4	+31 29	150	89	—
NGC 7839	**	Peg	00 07.0	+27 38	—	89, 125	—
Pease 1	PN	Peg	21 30.0	+12 10	1238	210	—
S Sagittae	Var	Sge	19 56.0	+16 38	1219	162, 163, 207, 208	16
Saturn Nebula, NGC 7009	PN	Aqr	21 04.2	–11 22	1335	299, 300	16, 17
Theta (Θ) Sagittae	**	Sge	20 09.9	+20 55	1194, 1218	163	9
Zeta (ζ) Sagittae	**	Sge	19 49.0	+19 09	1219	162	16

Ast = Asterism; BN = Bright Nebula; CGx = Cluster of Galaxies; DN = Dark Nebula; GC = Globular Cluster; Gx = Galaxy; OC = Open Cluster; PN = Planetary Nebula; * = Star; ** = Double/Multiple Star; Var = Variable Star

NOVEMBER

CHAPTER 11

The Cassiopeia Milky Way

On November evenings the Milky Way arches over the pole like a phantom bridge. Here we are looking toward the outer regions of our galaxy where the star fields are less congested; a sweep of the telescope will reveal deep-sky objects peering out from the region's great swarms of stars. The W-shaped asterism of Cassiopeia appears to be a favorite proving ground for budding amateur astronomers scouting out deep-sky sights. Star-rich Cassiopeia does not disappoint, for it contains a fine assortment of open clusters scattered along the galactic equator, which passes through the constellation. "Curiously, amateurs seem to be relatively unfamiliar with most of these objects," Scotty wrote, "except the famous Double Cluster in nearby Perseus. Yet here are many open clusters that are relatively bright and easy to distinguish from the crowded star fields of the Milky Way. Cassiopeia, although usually not considered a hunting ground for clusters and not possessing any of the more interesting ones, is actually full of the smaller galactic clusters."

Novice observers have the universe before them to explore. To begin with, there are many dazzling and delightful objects which can be located almost effortlessly, such as the Orion Nebula, globular star cluster M13, and the Sagittarius star clouds. But what of the dimmer offerings, including many planetary nebulae and most of the *NGC* objects?

A fine harvest awaits the amateur who examines the Milky Way in Cassiopeia with a small telescope. Many readers have commented on the profusion of open clusters in the Cassiopeia Milky Way. Here we are looking toward the outer regions of our galaxy where the star fields are less congested and deep-sky objects are not lost among great swarms of stars. Still, some care is needed to recognize some of them against the starry background. On autumn evenings this region rides high in the north, though it is awkwardly situated for finding objects with equatorially mounted refractors. Plate I of the *Atlas of the Heavens* shows 24 open clusters in the northern portion of Cassiopeia. An amateur who attempts to study them all will be kept busy for many nights.

We'll start this month off with several open star clusters I have selected near

Figure 11.1
Look for NGC 457 near the star Phi Cassiopeiae. This cluster has about 100 stars and is about the same brightness as the cluster M103, which lies about 2½° to the northeast.

naked-eye stars in Cassiopeia. It is usually easier to pick out these clusters with a refractor working at f/15 than with the fast f/4 to f/6 reflectors that are popular today. For extended diffuse objects, faster optical systems give superior views, but when individual stars can be seen the high-power systems darken the sky background without diminishing the conspicuousness of the stars.

Queen Cassiopeia is usually portrayed seated in a chair, with the W as the constellation's best-known part. Third-magnitude Delta (δ) Cassiopeiae is the bottom point of the W's eastern half, and just 2° to its southwest is 5th-magnitude Phi (φ) Cassiopeiae. The open cluster NGC 457 is less than 10′ west-northwest of the star and will be in the same low-power field of view. NGC 457 is a coarse group of distant suns which even shows well in a 2-inch aperture. This open cluster (**Figure 11.1**), 10′ across, is a grand sight even in a 4-inch. The experienced French amateur G. Gauthier saw 50 stars here with a 6-inch reflector. As the light-gathering power of the telescope increases, so does the number of stars that will be seen here, with a 20-inch revealing perhaps 100 stars appearing as bright pinpoints with dark sky between them. Many of this cluster's stars are very faint, and even in the 20-inch refractor at Van Vleck Observatory in Middletown, Connecticut, scores of them dance in and out at the limit of vision.

Look about 40′ northwest in the same field for NGC 436, a smaller cluster of fainter stars. (Be careful when observing the polar regions of the sky, for sky directions here will no longer match the corresponding compass points on the horizon.) When star-hopping from one location to another, it is important to know the actual field diameter of your eyepiece. Remembering how large the Moon (some ½° in diameter) appears in your telescope will enable you to estimate the field with enough accuracy for this purpose. NGC 436 is a better target for larger telescopes. A 3-inch aperture will pick up about half a dozen stars here, while a 10-inch will reveal many more, and possibly the background glow of others just below the limit of visibility.

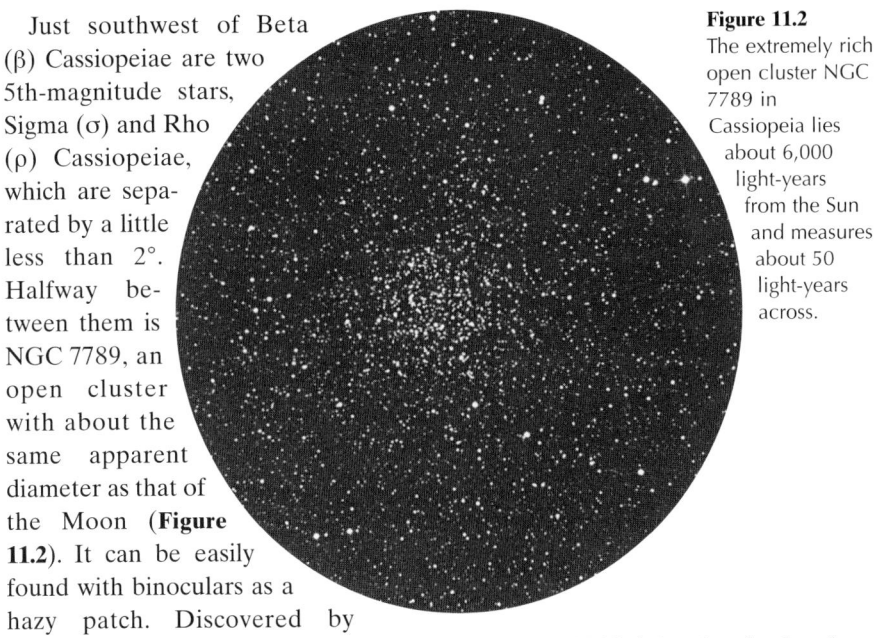

Just southwest of Beta (β) Cassiopeiae are two 5th-magnitude stars, Sigma (σ) and Rho (ρ) Cassiopeiae, which are separated by a little less than 2°. Halfway between them is NGC 7789, an open cluster with about the same apparent diameter as that of the Moon (**Figure 11.2**). It can be easily found with binoculars as a hazy patch. Discovered by William Herschel's sister Caroline in the autumn of 1783, it has inspired authors since William H. Smyth, who in the mid-19th century called NGC 7789 a "mere condensed patch in a vast region of inexpressible splendor." As examined in a 6-inch reflector by G. Gauthier, NGC 7789 is "a splendid swarm of about 200 stars of magnitude 10 to 12," according to the observing manual *Revue des constellations* by Robert Sagot and Jean Texereau. Looking at this cluster, one may gain the impression of a knot of bright stars superimposed on a larger mass of faint points of light.

Figure 11.2
The extremely rich open cluster NGC 7789 in Cassiopeia lies about 6,000 light-years from the Sun and measures about 50 light-years across.

NGC 7789 is one of those rare objects that is impressive in any size instrument. With a 4-inch rich-field telescope the cluster appears as a soft glow nearly ½° across and speckled with tiny, often elusive, individual stars. The 12-inch f/17 Porter turret telescope at Stellafane picks up more than 100 stars. Through a 16-inch aperture the view is spectacular, and the whole field is scattered with diamond dust. And a 22-inch Dobsonian reflector in the clear skies of California gave a most impressive view with countless sparkling points filling an entire 60× field. I particularly like the drawing of NGC 7789 made by Smyth, who observed with a 6-inch refractor. He mentions that the cluster has "rays of stars which give it a remote resemblance to a crab." He imagined the creature's head to be in the northwest with a close double star near where the eyes would be. Does anyone see this pattern today?

Kappa (κ) Cassiopeiae, above the W's center, is the 4th-magnitude star marking the front edge of the Queen's chair. Just 22′ to its north and in the same low-power field is NGC 146. It is a compact cluster containing about 50 stars within a 6′ diameter of sky. However, the Milky Way background is quite rich here, and magnifications between 100× and 200×, especially with 10-inch and larger telescopes, will be necessary to appreciate the cluster fully.

Less than ½° west of NGC 146 is a very similar open cluster, NGC 133. This group shines with a light equal to that of a 9th-magnitude star. This area of the sky is historically interesting because the great supernova of 1572, Tycho's star, appeared some 1½° northwest of these clusters. Attaining the brightness of Venus and remaining visible to the naked eye for 16 months, it was the most spectacular stellar outburst in the past 500 years. Although faint nebulosity believed to be associated with the explosion has been photographed with large telescopes, it is well beyond the visual reach of amateur instruments.

NGC 225 lies 2° northwest of Gamma (γ) Cassiopeiae, the famous irregular variable. The total light of the cluster matches a 7th-magnitude star. It was discovered in 1784 by Caroline Herschel while comet hunting. William Herschel placed this cluster in his class VIII, which is reserved for coarsely scattered star clusters. W. H. Smyth described this as a loose cluster of about 30 9th- and 10th-magnitude stars. About 0.2° in diameter, it contains over a score of stars brighter than visual magnitude 13. I prefer to view this object with my 4-inch Clark refractor, though it is quite evident in a 5-inch Moonwatch Apogee telescope.

NGC 129 is huge, having an apparent diameter as large as the full Moon. It lies midway between Beta and Gamma Cassiopeiae, and the 10 stars brighter than 10th magnitude in it are very scattered, so they do not define the cluster well. But on a very clear night, look with averted vision for a faint background glow caused by the dimmer stars. You'll need a 10-inch or larger instrument to resolve it, for it consists mainly of faint stars. As seen in a 10-inch reflector it is a triangle 11' long. The brightest of its perhaps 50 stars is 9th magnitude

Cassiopeia contains the bright open cluster M52, which lies between Cassiopeia and Cepheus on the edge of the dark lane in the Milky Way that divides these two constellations. The early writers discuss the object under Cepheus, and there one should look for its description in either William Smyth's or Thomas W. Webb's catalogs. It arrived in Cassiopeia only in 1930 after the International Astronomical Union reorganized the official constellation boundaries.

M52 has a total magnitude of 7.2 and is visible in any finder. Recent studies indicate that this object is one of the richer and more compressed open clusters known. It is also relatively young, being some 20 million years old and comparable in age to the Pleiades. Unlike some open clusters, M52 shows increasing richness with larger-aperture telescopes. Smyth, viewing with a 6-inch instrument, was moved to call M52 an object of "singular beauty."

Not large as galactic clusters go, it is 13' in diameter, but packed into that area are more than 150 stars of 11th magnitude and fainter. If it were not for its irregular outlines, M52 might well be mistaken for a globular cluster.

Only a little more than ½° southwest of M52 is the remarkable nebula NGC 7635. Although it is sometimes classed a planetary, it is hardly a typical member of this ill-defined group of objects. I first learned of this bubble of glowing gas in 1957, when our Moonwatch station in Manhattan, Kansas, obtained a copy of the Skalnate Pleso *Atlas of the Heavens* which plotted it. In the transparent skies of the Great Plains my 10-inch reflector easily picked up the 3' nebulous disk surrounding an 8th-magnitude star. My experience was that the nebula was best seen at 150×.

Here in Connecticut I have never been able to find NGC 7635 with a 5-inch Apogee telescope at 20×, and my 4-inch Clark refractor revealed it only on a few nights when the sky conditions were exceptional. Viewing from the Empire Mountains of Arizona, Ron Morales saw the nebula as "two stars close together involved in nebulosity, roundish, faint, and large." He was using an 8-inch at 51×.

The Great Andromeda Galaxy

Amateur astronomers long await the November cold fronts that ferry in clean arctic air over much of the United States. Although these nights are crisp and cold, they offer us some of the best nebula observing of the whole year. During these evenings, M31 — the great Andromeda Galaxy — can be glimpsed with the naked eye regardless of the observer's location in the city or out in the country. Since it is more than two million light-years distant, M31 is one of the most remote objects that can be seen without optical aid. Naturally, because it is bright, so well known, and so easy to locate, M31 is frequently the first deep-sky object sighted by novice telescopic observers. "Initially, many see nothing more than a blob of light," Scotty said. "But with time and patience, beginners gain the ability to perceive greater detail. Seasoned observers find bright and dark lanes, star clouds, and a star-like nucleus to be prominent features of the Andromeda Galaxy." Interestingly, the full visual extent of this galaxy was lost to observers for nearly a century. The mystery has since been solved, and Scotty was present when the tables turned.

Figure 11.3
The great Andromeda Galaxy and its two bright companions — M32 (just below center) and M110 (at upper right).

November is a wonderful month for observers, especially here in New England. If luck is with us the days are mild and the nights are crisp and free of insect marauders. Ever since the summer solstice, the nights have grown longer and deep-sky viewing could begin earlier. The sweet perfume of red and golden foliage is in the air. The beauty of these evenings opens the poetic feelings locked deep inside us all.

November evenings, then, are a time to admire one of the most beautiful deep-sky sights of all — the Andromeda Galaxy, M31.

Well placed for observing near the meridian, M31 (**Figure 11.3**) lies a few degrees above Mirach, Beta (β) Andromedae, about 1° west of Nu (ν) Andromedae, and shines with a total light equivalent to a 4.5-magnitude star. It seems the constellation boundaries of Andromeda were deliberately chosen so that nothing else would compete with the Andromeda Galaxy for the deep-sky observer's attention. And nothing can compare with it. M31 is relatively nearby, being a member of the Local Group. It lies more than 2.2 million light-years distant and has an actual diameter of nearly 180,000 light-years, making it one of the largest galaxies known.

As early as the year 964 the Great Nebula in Andromeda, as M31 was once called, was described as a "little cloud" by the Persian astronomer al-Sūfī. This spiral galaxy can be seen with the naked eye even in a less than perfect sky. Anything from an opera glass to the largest amateur telescope is suitable for viewing the galaxy. It is the chief glory of amateurs and a workhorse at public nights at observatories or star parties.

As with comets, M31 appears to have larger dimensions to the naked eye or through binoculars than when viewed with medium-sized telescopes, and the published values for its size vary widely. Most photographs of M31 show the spiral galaxy to be oval shaped, extending across some 2½° of sky. This dimension is often given in catalogs. However, in 1847 with the 15-inch refractor at Harvard Observatory, George P. Bond traced the long axis to about 4°. His technique for observing the outer portions of the galaxy is still used today by amateurs searching for faint objects. Bond simply rocked the telescope back and forth, since the eye is better able to pick up a faint, low-contrast image when it is moving.

So strong is the desire to believe what astronomical photographs tell us that Bond's observations have largely been overlooked even though they were mentioned in Webb's *Celestial Objects for Common Telescopes*. That M31 is over 4° long was proven in the early 1930s, when Joel Stebbins of Washburn Observatory traced its extent with a photoelectric photometer. (I know, because I was a night assistant in the Washburn Observatory dome that evening!) This stimulated my interest in reproducing the feat with the unaided eye, and I soon learned it was possible to do so on really excellent nights. Here is one of the cases where the eye can surpass photographic film. In 1953, the late French astronomer Robert Jonckheere, using 2-inch binoculars and taking extreme precautions, found the visible length to be 5° 10′. I can see the same with 15 × 75 binoculars. In fact, by moving the bright central region out of the field of view and by using averted vision, many people can do equally well with any low-power telescope. With my 5-inch Moonwatch Apogee telescope, I once seemed to reach 5°.

The apparent length of M31 is a sensitive measure of atmospheric transparency. The slightest haze will drastically reduce what can be seen of the galaxy's faint extensions. I have never been able to see a greater dimension of M31 with optical aid than I can with my naked eye. Indeed, Jonckheere likened the brightness of the galaxy's outer regions to that of the gegenschein. That faint glow is also seen best with the naked eye. A related project for the naked eye is to note how close the edge of the Milky Way comes to M31. On really transparent nights I have seen the luminous band come right up to M31 and then abruptly stop. On other nights a large separation appears between them.

Telescopically, M31 is often a disappointment. Few instruments can fit more than the galaxy's central bulge in their field of view. If those in your group are familiar with photographs of M31, it would be best to tell them about the view through a telescope. Major features of the galaxy that are recorded by the camera are far more subtle in the eyepiece. Its spiral structure, first recognized in the mid-1800s by Ireland's Lord Rosse, probably cannot be seen in amateur instruments despite a few claims to the contrary. Another feature often lost in photographs is M31's nucleus. The telescope shows a much brighter nucleus with the extensions quite faint. At low magnification the core of M31 appears as an even glow. But as the power is increased to several hundred or more, a tiny starlike nucleus becomes visible. High magnifications will help bring out this tiny, starlike part. With my 10-inch reflector in Kansas, this nucleus was very conspicuous in good seeing. Dark lanes in M31 can be identified even in small telescopes if the sky is dark enough.

A challenging project for the owner of a 10-inch or larger telescope is to locate the brighter globular clusters in M31. A description and finder chart for some of them appeared in the November 1979 issue of *Sky & Telescope*, page 490*. That idea came from Doug Welch of Ottawa, Canada. He had used his club's 16-inch reflector to locate several of the brighter globulars, which had been selected from a list of nearly 250 suspect clusters in M31 compiled earlier this century by Edwin Hubble and Walter Baade (and cataloged with HB numbers).

After first mentioning these globulars in the November 1979 column, I received as much mail about them as about any other single topic. A few amateurs sighted some of the globulars easily, while others were just plain skeptical that such observations were possible.

Hank Feijth of Goutum, the Netherlands, held the brightest cluster, HB 12, steadily with his 6.1-inch reflector. He could also glimpse HB 64 and HB 254 with the same instrument. Feijth is an experienced variable-star observer with more than 30,000 magnitude estimates to his credit. On the other end of the experience scale was 15-year-old William Kinney of Whitefish, Montana. With a Celestron 8 he was able to view HB 12 as "fleeting with averted vision." Unfortunately, space does not permit me to list all the amateurs who were able to see this cluster with only a 6-inch aperture. However, from the reports, it seems very likely that this is the minimum-size instrument for such an observation.

HB 254 was the second most observed globular, with several amateurs finding it (with difficulty) in 8-inch telescopes. The most complete report on these clus-

---

* *M31's brightest clusters are plotted on the* Millennium Star Atlas, *too.*

ters came from veteran variable-star observer Thomas W. Wilson of Hunting, West Virginia. He found all 15 known clusters with a 12½-inch f/5.1 homemade Newtonian. HB 90 was his most difficult observation. I find his results with an 8-inch f/5.6 homemade reflector even more impressive. With it he missed HB 90 and only suspected four others, but the remaining 10 clusters were either seen or clearly glimpsed.

M31's Companions

Not all the interesting objects associated with the Andromeda Galaxy are as taxing of the observer's ability as the globulars just mentioned. Familiar to today's amateurs are the companion galaxies M32 and M110. In a small telescope, M32 is well separated south of the disk of M31 and may be mistaken for M110, also known as NGC 205. The latter, however, is much farther away (0.6° from the center of M31) to the northwest. Both of these satellite galaxies are about 9th magnitude. Curiously, its companion galaxy M32 was not noticed until 1749, although at 8th magnitude, it is obvious in very modest instruments. M110, nearly as conspicuous, was discovered as late as 1773 by Charles Messier.

Two more companion galaxies, NGC 147 and 185, lie about 7° north of M31 and are often overlooked by observers. Visually, both these galaxies are several arc minutes in diameter and around 12th magnitude. NGC 147 is often listed in catalogs as being slightly larger and with a total light less than that of NGC 185. Thus, one would reasonably assume that NGC 147's surface brightness would appear noticeably fainter than NGC 185's. However, this is not the case. To me, they usually appear quite equal, and several entries in my notebook even cite NGC 147 as appearing a trace brighter. What is your opinion?

While some writers suggest that a 6-inch telescope is needed to ferret out these companion galaxies from the fringes of the Milky Way in Cassiopeia, I have seen both in a 4-inch instrument. Also, observers with large-aperture telescopes might want to check out a report by Tokuo Nakamoto. Using the 30-inch reflector at Stony Ridge Observatory in California, he suspected NGC 185 as being "granular" and thus on the verge of being resolved.

Sizing up the Fish

Pisces, a long, dim, and usually neglected region of sky, is nearly a ghost town of deep-sky wonders. "Few ever probe this area," Scotty wrote, "for here are found no planetaries, clusters, or nebulosities." Rather, he pointed out, "Pisces is all galaxies. In the region between right ascension 0^h and 2^h and declination 0° to +20°, we list 17 of these outposts of perception." One reason amateurs neglect Pisces is that its weak cast of galaxies must compete with the nearby presence of "magnificent M31," and the somewhat elusive but "marvelous M33." Only one of Pisces' galaxies is bright enough to be a Messier object — M74 — and it is the fabled thorn in the side of amateurs participating in Messier marathons. Therefore few observers even bother to turn their telescopes to it, or to Pisces. But, as Scotty explains here, there are many galaxies in the 12th- to 13th-magnitude range suitable for an 8-inch telescope.

South of the Great Square of Pegasus is the vast emptiness of Pisces, wherein lies M74, one of the more frustrating Messier objects (**Figure 11.4**). This face-on spiral appears some 6′ in diameter and is about 10th magnitude. It seems an easy galaxy from its catalog description, yet it is considered by many to be the most difficult object in Messier's list.

Many famous observers have commented on the difficulty of viewing M74. Pierre Méchain, who discovered the object in September 1780, wrote: "This nebula contains no star; it is fairly large, very obscure, and extremely difficult to observe. One can make it out with more certainty in fine, frosty conditions!" His friend Charles Messier agreed that the object was obscure. Both Smyth and Webb omitted M74 from their observing guides, as have a chain of succeeding writers. Nevertheless, I have seen it in my 4-inch Clark refractor with the aperture stopped down to 2.8 inches. The delicate, sharply defined arms of this galaxy are very regular, which together with the broadside presentation, makes this a perfect object of its type. The arms cannot be seen in small telescopes.

Figure 11.4 M74 is a two-armed spiral galaxy tilted 56° from our line of sight. It has a mass of some 58 billion Suns.

John Herschel mistakenly classified M74 as a globular cluster, and this was not the only time it was misidentified. M74 has a rather bright, almost starlike nucleus, and when the famous *Bonner Durchmusterung* charts were compiled from observations made with a 3-inch refractor, the galaxy was included as the "star" BD +15°238.

One of the reasons amateurs have difficulty locating M74 is that its light is rather evenly distributed across the disk, thus yielding a low surface brightness. Your chances of picking it up are improved if you prepare to see a large, dim mass. M74 is easy to find, 1¼° east and 27′ north of Eta (η) Piscium. As a guide to this spiral galaxy, there is an easily identified chain of stars about a degree to the northeast and visible in the same low-power field. If the observer sets Eta on the southern edge of a low-power field and lets the stars drift across, M74 will enter the field in a few minutes. This technique has a good deal to recommend it, for the eye becomes much more sensitive by staring at a relatively blank sky for a while. Also, try viewing this galaxy with averted vision, which should give a further gain of up to a stellar magnitude. Indeed, in most observing locations eye sensitivity is decreased by stray light, overall sky light, and reflections from the telescope tube. By shielding the head and the eyepiece with a black cloth, as a portrait photographer does, the observer will surely increase his ability to see faint objects, such as M74.

Paradoxically, M74 can sometimes be seen more readily in binoculars or a finder than in a 6-inch telescope. Its feeble contrast with the surrounding sky makes this a tricky object in any aperture. In the mid-1920s I caught a glimpse of M74 with a 1-inch 40× refractor. This was quite possibly due to the sensitivity of my young eyes since I was still in grade school at the time. At 60×, the 4-inch Clark stopped to 3 inches shows M74 as a galaxy. However, at 2 inches a quick glance reveals just the stellar appearance that gave rise to the mistake on the *BD* charts. A 10-inch shows that the central portion is brightest. No spiral arms can be seen, but it is obviously not a "nebulous star," as some older accounts suggest. There is no question about it being a galaxy when viewed with a 20-inch aperture, but even then the spiral structure is not apparent.

NGC 524, a 1.5'-diameter elliptical object listed at 11th magnitude, is easier to see than M74. Because NGC 524 has roughly a twentieth the surface area of M74, but is only about 2½ times fainter in total light, its average surface brightness is some eight times greater. Once again, we find that the relative visibilities of objects should not be judged on the basis of catalog magnitudes alone.

NGC 128 is another interesting galaxy in Pisces. It is a tiny torpedo-shaped object about 2' long and a quarter as wide, and I estimate it to be magnitude 12.2. It can be seen in my 5-inch Apogee telescope at 20×. This telescope works well here because the star background does not overwhelm the deep-sky objects the way it does in places like the Milky Way in Aquila. NGC 128 has a challenging feature for observers: its central core is rectangular in appearance. Years ago in Kansas I could see this shape quite well with a 10-inch reflector, but it is beyond the grasp of the 4-inch Clark.

Smyth mentioned another difficult Pisces object, near the western end of the constellation: NGC 7541. It lies 1.4° north-northwest of the 4th-magnitude star Gamma (γ) Piscium. This 3' long oval is similar in magnitude to NGC 128. As viewed with his 6-inch refractor, this galaxy appeared "so dim as to be only perceptible under settled gazing, and clock-work motion, when it faintly gleams among the telescopic stars in the field." Yet my 5-inch Apogee telescope on a good night in Connecticut distinctly shows this 3' by 1' spindle, which I estimate as of visual magnitude 11.8.

Pisces also contains the interesting but elusive set of twin galaxies NGC 470 and NGC 474, which are separated by just 6'. While catalogs differ as to which of them is brighter, I favor NGC 470 with a magnitude of 11.2. They were found quickly one night when I observed at my former home in Kansas, but there have been many nights here in Connecticut with my 4-inch Clark refractor when this instrument would not reach them. I wonder if they would be seen more in a sky of better transparency. This galaxy's light is concentrated within a 2' patch, and it is seen in the field of my refractor's 10× eyepiece without any trouble.

Nearby is another pair of galaxies, this time more suitable for a 16-inch aperture. NGC 7619 and NGC 7626, separated by 7', are located on the border of Pisces and Pegasus. They are both about 1' in diameter and have cataloged magnitudes of around 12.7. There are over a dozen *NGC* objects within a 2° radius of this pair, and I suspect that many of them would be visible with a 10-inch telescope used in good skies.

As a challenge, look for NGC 7534, which is a part of a small chain of very faint galaxies a little more than 1° long running roughly east to west just below the celestial equator. A 16-inch telescope may be needed to show all of them. Work through this group carefully.

The Splendors of Sculptor

Although Northern Hemisphere observers cannot view the south celestial pole, they can see the South Galactic Pole and the splendors surrounding it in Sculptor. This dim constellation remains low in the southern sky for observers in mid-northern latitudes. However, its placement — as Scotty reminds us — is as good as the Milk Dipper of Sagittarius and the tail of Scorpius, which amateurs probe every summer night. "Sculptor contains many objects that would be the talk of every late-summer star party," he imparted, "if it were not for their relatively poor placement near the southern horizon." It's true. Without question, the deep-sky objects in Sculptor are some of the best in the entire sky, so they are well worth hunting down on those exceptional nights when cold fronts scrub the haze from the horizon and allow us to slant our telescope tubes ever deeper toward the south.

Amateurs find it difficult to obtain visual descriptions of clusters and nebulae with large southern declinations. There is the extensive 1847 catalog of John Herschel, but the Southern Hemisphere seems never to have had a Smyth or a Webb.

Observers who have seen the whole sky often contend that some southern objects surpass the best the north can offer. When a clear, dark night is at hand, amateurs in mid-northern latitudes should attempt to view some of these southern wonders. Though low in the sky, their image quality is occasionally surprising. In November one may easily explore to the South Galactic Pole (declination –27°) in Sculptor.

Less than a degree from the South Galactic Pole lies NGC 288. This interesting globular cluster (**Figure 11.5**) is often overlooked by amateurs, though I readily find it whenever the sky is especially clear. One August, after a cold Canadian

Figure 11.5
Look for the loose globular cluster NGC 288 in the constellation Sculptor. Larger telescopes reveal a bright core of stars.

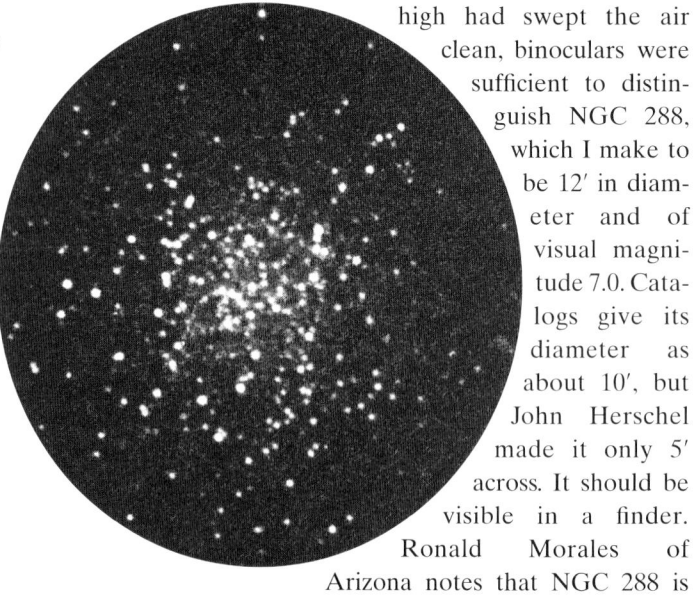

high had swept the air clean, binoculars were sufficient to distinguish NGC 288, which I make to be 12' in diameter and of visual magnitude 7.0. Catalogs give its diameter as about 10', but John Herschel made it only 5' across. It should be visible in a finder. Ronald Morales of Arizona notes that NGC 288 is larger than M15 in Pegasus, but otherwise the two globulars appear about the same. In a pair of 5-inch 20× binoculars, it was possible to see individual stars around the edge of the cluster. And with a borrowed 12½-inch f/8 Newtonian reflector, the cluster could be resolved almost to its center. Only the regularity of the star pattern indicated that NGC 288 was a globular and not a compact open cluster. Furthermore, open clusters are concentrated along the Milky Way.

About 1.5° to the northwest is NGC 253, one of the largest galaxies in the southern sky (**Figure 11.6**). It is about 25' long (almost as long as the Moon is wide) and one-third as thick. At 7th magnitude it is a spectacular object for visual

Figure 11.6
NGC 253 is a member of the Sculptor Group of galaxies, and a known radio source. It can be found with a 6-inch telescope, but a larger instrument is required to see the details of its dusty arms.

observers and astrophotographers alike. It is one of the 14 nebulae and clusters discovered by Caroline Herschel in England during her sweeps for comets in the years 1782–83. Observing from the Cape of Good Hope a half-century later, her nephew John Herschel described its appearance in his 18¾-inch speculum-metal reflector: "Very, very bright and large; a superb object . . . 24′ in length; breadth about 3′; position angle of long diameter 143.8°. Its light is somewhat streaky, but I see no stars in it except 4 large and 1 very small one, and these seem not to belong to it, there being many near."

Figure 11.7 NGC 300 is among the largest galaxies in the sky. It resembles M33, with a compact nucleus and beautiful spiral arms full of nebulae and stars.

The galaxy was easily seen in Vermont with a 6-inch rich-field at about 20×. Long-exposure photographs reveal a complicated pattern of dust lanes and patches across its surface. Morales notes the galaxy's internal mottling as "very impressive" seen with a 6-inch f/8 reflector. In the Wright telescope it was quite bright, and dark lanes were just visible. The sight reminded me a little of M31 in a 1.5-inch refractor in Connecticut.

There is always the possibility of discovering a supernova while viewing galaxies, but the chances are slim. One difficulty is that amateur telescopes usually reveal only the bright inner core of a galaxy, and an outlying supernova will look like any other field star. This was my experience in 1972 when viewing the bright supernova that erupted in NGC 5253 in Centaurus. Of course, if an unfamiliar star is seen superimposed on the galaxy's image, it is a much more likely supernova candidate.

Southward from the galactic pole and near the 5th-magnitude star Eta (η) Sculptoris is NGC 134. Though 11th-magnitude and only 5′ × 1′ in size, this object was held steadily in my 4-inch Clark refractor. The view was not as good as one I had from Louisiana in 1943. But despite its low altitude NGC 134 is no more difficult than NGC 6207, a well-known spiral galaxy near M13.

If you have an unobstructed horizon, try adding two even more southerly galaxies to your Sculptor list. NGC 300 is another large spiral (**Figure 11.7**) seen face-on that spans 20′ by 10′. Steve Coe of Arizona found it to be very much like the well-known M33 in Triangulum. With a 13-inch telescope he can just discern a dim S-shaped structure. To my eye its total light is equivalent to an 8.5-magnitude star, but since it is spread out over so large an area, the surface brightness is low. Care is needed to locate this object, but it can be seen in a 5-inch Moonwatch Apogee telescope. With a 10-inch reflector in Kansas, I found this galaxy to be fascinating, with abundant detail that challenges anyone who attempts to make a drawing.

Figure 11.8
While scanning the Sculptor Group, look for the large spiral galaxy NGC 55. It lies within 15° of the South Galactic Pole.

Some deep-sky objects offer beautiful, breathtaking visual experiences. NGC 55 is one such object (**Figure 11.8**). It is the largest galaxy in the nearby Sculptor Group and by itself is one of the giants of the southern sky. In fact, these objects form the nearest cluster to our Local Group, being only some 13 million light-years away. From observations made in Mexico, where NGC 55 rises much higher than as seen from most of the United States, I estimated its total magnitude to be 7.1. When I first saw NGC 55 in binoculars from Chiapas, I was surprised that such low power would reveal this galaxy so clearly.

Most amateur telescopes can trace NGC 55's slender, spindle-shaped form across 20' of sky. Long-exposure photographs record more than a ½° length, and densitometer measurements of these negatives reveal a disk that approaches 1° long. This makes it an excellent candidate for viewing in a black sky with binoculars. Remember that the Andromeda Galaxy, M31, can be visually traced out to almost 5° with binoculars, whereas photographs usually stop at 2½°. Has anyone tried viewing NGC 55 with a light pollution filter? The results might be surprising.

Houston, Texas, amateur Barbara Wilson, whose prowess as a deep-sky observer is legendary, sent some of her notes regarding NGC 55. "August 9, 1986, 3 a.m., seeing good, very transparent night at Columbus, Texas. Using the 13.1-inch [at 115×], NGC 55 appears very large, extremely elongated, very mottled, slightly tilted, very bright, however east end fades rapidly."

A year later from the same site she viewed NGC 55 again, this time with a

17.5-inch telescope: "A long streak narrower at one end with mottling in the central region visible at 97×. It fills the 0.52° field of a 21-mm eyepiece. Incredible! Spiral almost edge-on, extremely bright, dimmer [east] end seen easily as it fades into the sky background, it is wider, but of lower brightness, than west end."

Makes you want to go out and take a look tonight, doesn't it?

The Sculptor Dwarf Challenge

Can the famous Sculptor Dwarf Galaxy be detected visually? I still remember the excitement in 1938 when Harvard astronomers announced a hitherto unknown stellar system at only about an eighth the distance of the great Andromeda Galaxy, M31.

Although it has a total magnitude of about 8 and is more than a degree in diameter, the Sculptor system is most elusive because of its very low surface brightness. It consists of a loose swarm of thousands of stars of magnitude 18 and fainter. The discovery was made on a photograph taken with a 24-inch telescope, with a long enough exposure for the individual stars to show.

But the Sculptor system was also detected on a plate exposed for 23.3 hours with a 1-inch f/13 patrol camera, which revealed it as a very dim, unresolved smudge. Harvard's medium-sized photographic instruments recorded nothing.

The best hope of seeing the Sculptor Dwarf Galaxy (**Figure 11.9**) is probably with a small, low-power, wide-field telescope or big binoculars, on an exceptionally clear night. Success may require considerable observing skill and first-class vision. Amateurs with fast telephoto lenses of medium focal length (300- to 500-mm) can record this system.

Figure 11.9 The loosely organized Sculptor Dwarf Galaxy measures a mere 8,000 light-years in diameter and has a mass of only about 2 million Suns.

On November evenings Sculptor is near the meridian about the time evening twilight ends. The elusive galaxy is about 4° south of 4th-magnitude Alpha (α) Sculptoris. When I first presented readers with a Sculptor challenge in 1971, no one answered. This wasn't surprising since the galaxy consists of very dim stars spread over a very large area. Some people have compared the galaxy to a huge globular cluster with 99 percent of the stars removed. Recently, however, I received a letter from Steve Coe in Arizona. From a dark-sky site he had searched without success for the Sculptor system with a 17½-inch reflector. He then tried a 4½-inch f/4 reflector at 16×. After carefully dark-adapting his eyes and covering his head with a dark cloth to prevent stray light from interfering with the view, he swept the sky again. There it was, a blob of light without any detail, but visible nonetheless. Why don't you give it a try?

The Naked-Eye Milky Way

It's ironic, but we amateur astronomers often miss the forest for the trees. How many nights have we spent squinting behind an eyepiece just to glimpse the dim and fuzzy glows of distant galaxies? How many nights have we tried to seek out hyperfine details — a dust lane, a stellar nucleus, an H II region, a knot — within these ghostly extragalactic bodies? And while we strain our eyes, the most detailed galaxy of all goes virtually unnoticed. It's called the Milky Way, and we hardly know what it looks like to the naked eye. In the November 1990 Deep-Sky Wonders column, Scotty awakened himself — and us — to that obvious fact. While attending the Texas Star Party one spring, he sat in a chair under the stars and rediscovered the Milky Way. "I just observed," he said, "with that marvelous and often ignored optical instrument, the human eye."

One of the best observing experiences of my life came unexpectedly in May 1990 at the Texas Star Party (TSP). From 11 p.m. to 3 a.m. I simply sat in a chair and watched the Milky Way.

Held in the shadow of Mount Locke, home of McDonald Observatory, the TSP usually has the best observing conditions of the big three annual conventions in the United States — the other two being the Riverside Telescope Makers Conference in California and Stellafane in Vermont. The TSP has virtually no light pollution, and this year the air was very dry and transparent. The stars actually looked like sharp points.

For my Milky Way survey I used no telescope, though several dozen were close at hand and probably mine for the asking. The little 8 × 15 binoculars never left my shirt pocket, either. Even my Lumicon nebula filters, which I'm never without these days, stayed safely in a box. I just observed with that marvelous and often ignored optical instrument, the human eye.

Furthermore, I just *looked* at the summer Milky Way (**Figure 11.10a and b**), not searching for anything in particular but alert for anything unusual. I wanted

Figure 11.10a The glorious summer Milky Way and the Sagittarius star cloud.

Figure 11.10b The sparkling winter Milky Way, accompanied by the constellation Orion and the bright star Sirius.

purely naked-eye impressions, not photographic, not photoelectric, only what I could see.

It was a wonderful experience. The turbulent star clouds appeared so crisp and well-defined. The most luminous part of the strip, running from Cygnus at the zenith to below Scorpius on the mountainous southern horizon, was remarkably narrow. I was suddenly struck by how thin the disk of our galaxy really is, and how closely packed the stars are within it. I was also surprised by the band's straightness. The concept of the Milky Way as an ill-defined, irregular, pythonlike stellar monster had given way to the image of a slender shaft following a purposeful straight line across the sky. It was not a sight you glean from looking at star maps. I now have a better feel for the legends that call the Milky Way the backbone of the sky.

I was most surprised by how isolated Antares and the globular cluster M4 are from the galactic disk. Except for a faint haze, the sky east of Antares is vacant — rather like the sky west of M4, which all star charts place outside the edge of the Milky Way. Now, you can reason that the void east of Antares is due to dark dust clouds, and this is fine scientifically. But the eye is not aware of what is happening here, and thus Antares *appears* outside the Milky Way.

In getting this column together I searched through my library for amateur oriented books that devote solid space to the visual Milky Way. I found many brief accounts of localized areas such as the star clouds of Cygnus and Sagittarius, but little on the naked-eye Milky Way as a whole. Then I turned to Garrett P. Serviss's classic *Astronomy with the Naked Eye*. It contains an entire chapter on the subject. Serviss relates a folk tale he picked up from Lafcadio Hearn, a turn-of-the-century writer who used words to paint full-color pictures in three dimensions.

According to Hearn, several Eastern cultures had variations on a tale of a herdsman falling in love with a weaver named Orihimé. In anger, Orihimé's father banishes the lovers to the sky where they become the constellations Aquila and Lyra. Once a year they are allowed to meet if they can cross the Celestial River of the Milky Way. Serviss concludes his chapter by quoting Hearn:

> In the silence of transparent nights, before the rising of the moon, the charm of the ancient tale sometimes descends upon me out of the scintillant sky, to make me forget the monstrous facts of science and the stupendous horror of Space. Then I no longer behold the Milky Way as that awful Ring of the Cosmos . . . but as the very Amanogawa itself — the River Celestial. I see the thrill of its shining stream, the mists that hover along its verge, and the water-grasses that bend in the winds of autumn. White Orihimé I see at her starry loom, and the Ox that grazes on the farther shore — and I know that the falling dew is the spray of the Herdsman's oar.

Although the Milky Way has forever been so conspicuous in the night sky that it is woven into folk tales everywhere around the world, it is slowly attaining the status of a "test object" because of the ever-spreading threat of light pollution.

NOVEMBER OBJECTS

Name	Type	Const.	R. A. h m	Dec. ° ′	Millennium Star Atlas	Uranometria 2000.0	Sky Atlas 2000.0
Andromeda Galaxy, M31, NGC 224	Gx	And	00 42.7	+41 16	104, 105	60	4, 9
M32, NGC 221	Gx	And	00 42.7	+40 52	105	60	4, 9
M52, NGC 7654	OC	Cas	23 24.2	+61 35	1069, 1070	15, 34, 58	3
M74, NGC 628	Gx	Psc	01 36.7	+15 47	194	173	10
M110, NGC 205	Gx	And	00 40.4	+41 41	105	60	4, 9
NGC 55	Gx	Scl	00 14.9	−39 11	410, 430	350, 351, 386	18, 23
NGC 128	Gx	Psc	00 29.2	+02 51	245, 269	216	10
NGC 129	OC	Cas	00 29.9	+60 14	49, 50	15, 35, 36	1, 3
NGC 133	OC	Cas	00 31.2	+63 22	36, 49, 50	15	—
NGC 134	Gx	Scl	00 30.4	−33 15	387, 409	351	18
NGC 146	OC	Cas	00 33.1	+63 18	36, 49, 50	15, 16	—
NGC 147	Gx	Cas	00 33.2	+48 30	85	60	4, 9
NGC 185	Gx	Cas	00 39.0	+48 20	85	60	4, 9
NGC 225	OC	Cas	00 43.4	+61 47	49	16, 36	1, 3
NGC 253	Gx	Scl	00 47.6	−25 17	364	306, 307	18
NGC 288	GC	Scl	00 52.8	−26 35	364, 386	307	18
NGC 300	Gx	Scl	00 54.9	−37 41	408	351, 352	18
NGC 436	OC	Cas	01 15.6	+58 49	48	36	—
NGC 457	OC	Cas	01 19.1	+58 20	48	36	1
NGC 470	Gx	Psc	01 19.7	+03 25	242, 243, 266, 267	217, 218	10
NGC 474	Gx	Psc	01 20.1	+03 25	242, 243, 266, 267	217, 218	10
NGC 524	Gx	Psc	01 24.8	+09 32	218, 242	173	10
NGC 7534	Gx	Psc	23 14.5	−02 39	1281, 1305	258, 259	—
NGC 7541	Gx	Psc	23 14.7	+04 32	1257	258, 259	17
NGC 7619	Gx	Peg	23 20.2	+08 12	1256, 1257	214	17
NGC 7626	Gx	Peg	23 20.7	+08 13	1256, 1257	214	17
NGC 7635	BN	Cas	23 20.7	+61 12	1069, 1070	15, 34, 58	3

Ast = Asterism; BN = Bright Nebula; CGx = Cluster of Galaxies; DN = Dark Nebula; GC = Globular Cluster; Gx = Galaxy;
OC = Open Cluster; PN = Planetary Nebula; ∗ = Star; ∗∗ = Double/Multiple Star; Var = Variable Star

NOVEMBER OBJECTS (CONTINUED)

Name	Type	Const.	R. A. h m	Dec. ° ′	Millennium Star Atlas	Uranometria 2000.0	Sky Atlas 2000.0
NGC 7789	OC	Cas	23 57.0	+56 44	50, 66, 1069, 1083	35	3
Sculptor Dwarf Galaxy, ESO 351-G30	Gx	Scl	00 59.9	−33 42	408	351, 352	18

Ast = Asterism; BN = Bright Nebula; CGx = Cluster of Galaxies; DN = Dark Nebula; GC = Globular Cluster; Gx = Galaxy; OC = Open Cluster; PN = Planetary Nebula; *= Star; ** = Double/Multiple Star; Var = Variable Star

DECEMBER

CHAPTER 12

The Challenge of the Seven Sisters

On December evenings the most remarkable star cluster in the heavens, the Pleiades (often called the Seven Sisters), rises to prominence. "Even on the coldest winter nights, when time spent at the telescope is better measured in minutes than hours, it would be a very *un*-amateur act not to look in on the Pleiades," Scotty quipped. As far as he was concerned, the Pleiades were worth looking at every clear night, and for good reason. Seasoned amateur astronomers know that the state of Earth's atmosphere can change from night to night or minute to minute. And the clarity of our atmosphere makes the difference in seeing not only the fainter members of the star cluster with the naked eye — the exact number of which has long been disputed in amateur circles — but also any telescopic traces of the delicate nebulosity that swaddles the entire cluster, like jewels in a bed of vapors. "On an ordinary night," Scotty pointed out, "there are usually a few wisps of nebulosity seen around the Pleiad Merope. But on really exceptional nights (or more likely, half-hours), the glow swells out to encompass the entire cluster in a big cocoon."

December brings winter, and with it many cold but often clear nights. On such evenings, when the stars sparkle like diamonds, there is no sight as spectacular as M45, the Pleiades (**Figure 12.1**). Currently, this open star cluster rides high in the eastern sky at the end of astronomical twilight. It is delightful in any instrument, from the naked eye to the largest amateur telescope, although I find large binoculars give the most impressive view. Almost every culture, past and present, mentions in its folklore the dazzling stars of this nearby cluster. They have enhanced the imaginations of gifted poet and commoner alike as far as we can remember. They are the starry seven of Keats, the fireflies tangled in a silver braid of Tennyson, the fire god's flame of the old Hindus, and the ceremonial razor of old Japan. No other celestial configuration appears so often on the pages of the poet.

Astronomers have studied the Pleiades in great detail. The 19th-century author Agnes Clerke wrote that the cluster is "the meeting-place in the sky of mythology and science." Our present knowledge of stellar evolution suggests that the

Figure 12.1
The incomparable Pleiades and associated nebulosity. North is to the left.

Pleiades are only 20 million years old. Thus, mammals were well established on Earth when the Pleiades' light first fell upon it. I am always taken by the fact the dinosaurs never saw the Pleiades, for even as the last of those beasts roamed the land 65 million years ago, it would be another 45 million years before the Pleiades would begin to shine.

M45 is a little less than 2° in diameter — too large for the Moon to occult all of its stars at once. Partial occultations of the cluster recur in an 18.6-year cycle, as the Moon's nodes regress along its orbit.

Have you ever tried to count the Pleiades with the naked eye? Experiments show that light entering the eye from the side reduces sensitivity and contrast. Thus, I made a cardboard tube about a foot long and a foot in diameter that I painted flat black on the inside. Unfortunately it didn't help me see more stars. In fact, so much light bounced off the sides of the tube that I saw fewer stars with it than without it. Do not consult a chart while you are trying to count Pleiads. Instead, make a careful drawing of what you see and compare it with a chart later.

Depending on light pollution and sky conditions, most persons can see between four and six naked-eye Pleiads. Traditionally, the average eye can see six stars here, the exceptional eye seven, and 10 bear names or Flamsteed numbers. However, during the 1800s the noted British amateurs Richard Carrington and William Denning both counted 14 stars. The late dean of visual observers, Leslie Peltier, told me he could always see 12 to 14 stars there on any good moonless night. Perhaps because it has been repeated so many times that the number of naked-eye Pleiads is six or seven, too many observers quit counting before really reaching their limit. Many observers can reach magnitude 7.5 with the naked eye. Were it not for the bright Pleiads, these observers should be able to count upward of 30 stars in the group.

In 1935 at Tucson, Arizona, I was able to make out 14 Pleiads with ease, despite my "class 2 eyes," and as many as 18 under exceptional skies. Forty years later from the same location, deteriorating observing conditions had reduced this number to five. The reason was apparent even before the sky darkened, for heavy clouds of smog from copper smelters had settled into the natural bowl where Tucson is located. In the sooty skies of our populated areas, it is now not uncommon that no stars can be distinguished; the eye sees just a shimmering patch. So the number of Pleiades stars visible is really an index of the transparency of the atmosphere, and the cluster does not make a valid eyesight test. If you look at this cluster only infrequently, one glance will not tell you much about the sky conditions. But with practice you'll know by the cluster's appearance whether the night is a particularly good one.

A Nebula Challenge

The stars of the Pleiades were born so recently (by astronomical standards) that some of the lingering dust and gas of their birth still surrounds the stars*, especially Merope (**Figure 12.2**). As early as 1859, the German astronomer Wilhelm Tempel observed this nebulosity surrounding the Pleiad Merope with a 4-inch refractor in Italy. Others using larger instruments failed to see it, and some doubted its existence. Photography, however, proved that the nebulosity is real, and the quest was on to see as much of it as possible visually.

From Tucson my 4-inch showed it readily. In Connecticut, a 10-inch reflector failed, but in Vermont a 5-inch Moonwatch Apogee telescope succeeded. At the August convention of the Astronomical League in Tennessee, I was surprised to find several observers who had seen the Merope Nebula more than once. It was readily visible in a 6-inch reflector made by Fred Lossing of Ottawa. Once its position southwest of the star Merope was pointed out, others saw the dim glow too. In the 16-inch, the nebula seemed much more obvious, and averted vision was not required.

---

* *Recent research suggests that the Pleiades may simply be passing through unrelated nebulosity.*

Figure 12.2
The most obvious part of the visual nebulosity surrounding the Pleiades lies near Merope, the cluster's southernmost bright member.

In addition to the well-known glowing "thumbprint" of material south of Merope, the cluster is surrounded by a faint cloud of nebulosity. As with seeing naked-eye Pleiads, its visibility is more a test of atmospheric transparency than observer skill. The 19th-century comet observer Heinrich d'Arrest wrote of the Pleiades, "Here are nebulae, invisible or barely seen in great telescopes, which can easily be seen in finders." Most amateurs are content to glimpse the wisp near Alcyone. But I have seen nebulosity curdling and weaving well beyond the brightest stars of the cluster. My 6-inch f/4 Cave mirror in Connecticut occasionally gives fine views of this.

With low-power telescopes and the excellent dark skies of Arizona, California, and Kansas, I have easily seen the nebulosity as a bright cotton ball encompassing the cluster. But even the slightest traces of dew on the optics will give the same impression, and I constantly check for this by looking back at the nearby Hyades cluster. Today observers have little trouble seeing the entire Pleiades cluster immersed in nebulosity when observing conditions are right.

A number of amateurs claim to have observed this nebulosity with the naked eye, and I tend to believe them. You might investigate this by carefully comparing the appearances of the Pleiades and the Hyades, to see if there is any difference in what might first be perceived as sky background brightness.

Another feature of the Pleiades is mentioned in a letter by skilled variable-star observer Stephen Knight of East Waterford, Maine. He writes, "There is a dimly visible dark ring that encircles the outer boundary of the diffuse nebula. It is irregular in width and darkness. One of the darker and more easily seen parts is just north of Asterope and between it and 18 Tauri. I first spotted the dark ring with my 6-inch reflector and low power, and I followed it all the way around the bright nebulosity. It appears as an area with an absence of faint background stars or glow from the outskirts of the nearby Milky Way. The most difficult part to see is south of Merope and Electra." Later Knight reported seeing weakly luminous material outside the ring.

At the turn of the century Edward Emerson Barnard photographed this bright outer nebulosity, so its reality is not questioned. But how many people can see it visually? It is certainly more difficult than the California Nebula, and probably on a par with the Sculptor system. I suspect that this will be a test object that will be around for a while.

Taurus: The Observer's Paradise

When December rolled around, Scotty said he often first turned his attention to the famous Double Cluster in Perseus, high overhead. Then, looking to the southeast, he paused for a long look at the brilliant V of the Hyades cluster — the face of Taurus the Bull. "The 7° field of view in standard 7 × 50 binoculars," he said, "will hold the cluster comfortably along with 1st-magnitude rose-tinted Aldebaran." The area of sky containing the head and horns of Taurus is a veritable playground for amateurs using binoculars and small, rich-field telescopes. Especially noteworthy are its bright open clusters, and of course, the famous supernova remnant of A.D. 1054, more commonly known as M1 or the Crab Nebula. Taurus also contains another object of historical significance: the planetary nebula NGC 1514. As Scotty explains here, Herschel's discovery of this object "marked a turning point in the thinking of this great astronomer."

The constellation Taurus is well placed in the evening sky this month. Situated along the western edge of the Milky Way, Taurus might be expected to contain the swarms of open star clusters that pepper the constellations of Cassiopeia, Perseus, and Auriga. But this is not the case, and observing guides rarely list more than half a dozen clusters in Taurus. Furthermore, despite being the 17th-largest constellation, it contains no galaxies and only one planetary nebula within the reach of amateur telescopes.

Yet, Taurus is still an observer's paradise. Objects within its borders range from the magnificent naked-eye Pleiades and Hyades star clusters to a giant 3° bubble of glowing gas similar to the Veil Nebula in Cygnus (called S147). The Hyades star cluster is rather close to us, and its stars appear spread over quite a large area. Unlike the Pleiades, there is no nebulosity associated with the group. In fact, whenever I think I've sighted the wispy glow between the Pleiades, I quickly turn to the Hyades to check for a glow there, too. If I see any, then I know to blame a slight dewing of the optics, even if their surfaces look clear.

Another cluster is NGC 1807. I estimate it to be about 15' in diameter with a total magnitude a little brighter than 8. However, the 30 or so stars that a 10-inch telescope shows here are so scattered that the total magnitude has little bearing on the cluster's actual visibility. NGC 1807 is seen quite well in my 4-inch Clark refractor, and the view in 60-mm binoculars is also pleasing.

Just 0.4° to the northeast, and in the same field of a low-power eyepiece, is NGC 1817. This open cluster is about 20' across, and some astronomers suspect it may be physically associated with NGC 1807. English observer Kenneth Glyn Jones reports that, with an 8-inch reflector at 40×, NGC 1817 looks like two clusters in contact with each other. While a 10-inch telescope may reveal 20 stars here, the number may more than double in a 12-inch. Amateurs with access to large-aperture telescopes should record their counts.

The open cluster NGC 1746 presents a classic problem for visual observers — what is its apparent diameter? Modern catalogs often list its size as 45'. However estimates range from 25' to 1°, with the larger telescopes usually giving the smaller values.

Figure 12.3 Look for the planetary nebula NGC 1514 near the Perseus/Taurus border. Its central star easily outshines the surrounding nebula.

NGC 1514 is the sole planetary nebula that amateurs can find easily in Taurus (**Figure 12.3**). It was discovered by William Herschel in 1790, and it marked a turning point in the thinking of this great astronomer. Until then it was widely accepted that nebulae were just clusters of stars either too faint or too distant to be resolved. But Herschel saw NGC 1514 differently:

> A most singular phenomenon! A star of about 8th magnitude with a faint luminous atmosphere, of circular form, and about 3 minutes in diameter. The star is in the centre, and the atmosphere is so faint and delicate and equal throughout that there can be no surmise of its consisting of stars; nor can there be a doubt of the evident connection between the atmosphere and the star.

Thus, for the first time it occurred to Herschel that there existed a "shining fluid of a nature totally unknown to us." By the mid-19th century, spectroscopy proved beyond a doubt that glowing nebular gas was an astronomical reality, but when you look at NGC 1514 bear in mind that the concept began here.

To me NGC 1514 appears more like a nebulous star, for, unlike other planetaries where the central star is often a challenge, NGC 1514's luminary almost dominates the view. Years ago in Kansas I saw the star (listed as 9th magnitude) in a 4-inch telescope stopped down to 2 inches. The surrounding nebula is slightly oval and about 2' in diameter. Some observers consider this object difficult for an 8-inch telescope.

NGC 1555, Hind's Variable Nebula, deserves more attention from amateurs. It is associated with the variable star T Tauri, which fluctuates irregularly between about magnitude 9 and 13. There is no clear relationship between the star's brightness and that of the nebula. Both objects were discovered in 1852 with a 7-inch refractor by the Englishman John Russell Hind.

He reported NGC 1555 to be 4" southwest of the star. By 1868 the nebula had faded from view. It was rediscovered by Barnard and Sherburne W. Burnham in 1890 with the 36-inch refractor at Lick Observatory. Five years later the nebula had again vanished, but it was recovered photographically in 1899 and has been followed ever since.

NGC 1555 has brightened significantly since the early '30s. Not only has its brightness changed, but apparently so have its shape and position. Currently it is due west of T Tauri, and photographs suggest that NGC 1555 is part of a shell of material surrounding the star. My last observation of it was in 1977 with the 4-inch Clark at 150×. At lower magnifications it could have been easily overlooked.

Dissecting the Crab

It's a short hop from the heart of the Hyades to M1, the Crab Nebula (**Figure 12.4**). During the 20th century M1 was discovered to be a source of strong radio emission. In the 1950s, Cliff Simpson and I operated several radio telescopes at Manhattan, Kansas. Although they were built from odds and ends, they were carefully assembled and quite sensitive. The signal from M1 was among the half-dozen strongest on our list. M1 has been carefully studied by radio and X-ray astronomers. It is known to be the remains of a supernova which exploded in A.D. 1054, and the pulsar at its center is suspected to be a rotating neutron star. The supernova was visible even in the daytime, but it wasn't until 1731 that the nebula was first seen, by the English amateur astronomer John Bevis. Messier independently discovered it 27 years later in 1758. Traditionally, it was this object that induced Messier to compile his famous catalog of nebulae and clusters, so these deep-sky features would not again be mistaken for comets in his small telescopes; it was also in the search region for the 1835 return of Halley's Comet.

In December 1852, the English telescope maker and observer William Lassell used a 24-inch speculum-metal reflector to view M1 from Malta. He noted,

"With 160× it is a very bright nebula, with two or three stars in it, but with 565× . . . long filaments run out from all sides and there seems to be a number of minute and faint stars scattered over it." On January 6th of the following year Lassell reobserved the Crab with 565×: "The brightest parts are about 2' in length, while the outlying claws are only just circumscribed by the edge of the field of 6' in diameter."

The Crab can be seen in 2-inch finders. Small telescopes reveal only a shapeless 8th-magnitude blur variously sketched as oval, rectangular, or more often something in between. The Crab Nebula usually shows in small telescopes as a featureless gray ghost. My 4-inch Clark refractor has revealed hints of the nebula's ragged edge that appears so prominently in photographs. These edge serrations are usually apparent in a 12-inch telescope and easy in a 17-inch. Increased magnification does not seem to change the appearance much. Telescopes of 12-inch aperture and larger often reveal delicate filamentary structure in the nebula.

My impression is that large amateur reflectors do not show much more of the Crab Nebula than a 6-inch does, though of course they show it better. Amateurs with access to a 16-inch or larger telescope, perhaps a club telescope, can perform some interesting experiments on how the appearance of M1 varies with different magnifications and telescope f/ratios. The latter can be changed by making aperture masks of varying diameters. Mike Mattei of Littleton, Massachusetts, has suggested using a nebula filter in the "flicker" mode. By rapidly moving it in and out of the space between the eyepiece and your eye, it is easy to note the effect of the filter. I tried this technique using a 15 × 65 monocular at my home in East Haddam. The Crab was easily detected with the flicker method, but I was unable to hold the nebula steadily when viewing either without the filter or with it fixed in place.

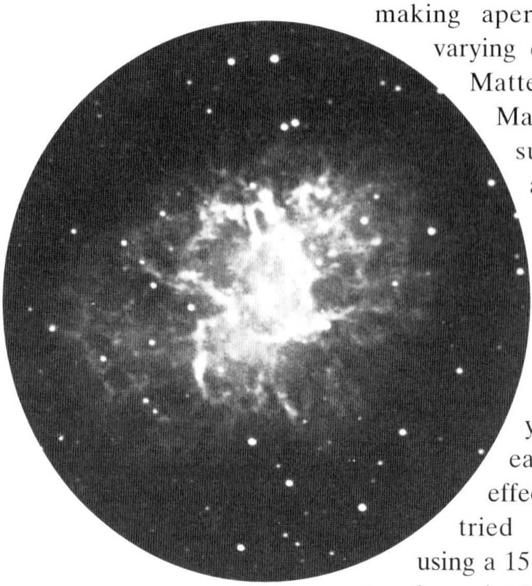

Figure 12.4
M1, the Crab Nebula in Taurus, is a supernova remnant that is expanding at a rate of 600 miles per second.

Light pollution does affect the visibility of M1, and many amateurs, especially from the East Coast, report that they have despaired of seeing this nebula, even though it is 9th magnitude and about 4' in diameter. My mail is divided on how well nebula filters work on this object. What do you find?

Targeting the Cetus Seyfert

The key to successful galaxy hunting is being able to read and interpret a star chart. Of course, such a skill is indispensable for finding virtually all deep-sky objects. There is no escaping that fundamental necessity. Fortunately, the sky has its celestial coincidences — such as a bright galaxy lying near a naked-eye star — which benefits beginners looking for a good place to practice their hunting skills. "The job of locating faint deep-sky objects is always made easier if there are nearby finder stars," Scotty once instructed his readers. The constellation Cetus is home to one of these celestial coincidences, for near the 4th-magnitude star Delta Ceti lies the bright Messier galaxy, M77, which is also one of the sky's weirdest. As Scotty explains, the galaxy not only perplexed early telescopic observers but also modern-day astronomers.

As I gather my thoughts and begin writing this column, it is a musty, rain-soaked August here in Connecticut. But I look at "December" typed at the top of the page and think to myself, "Ah, in December the rains will be long gone and the cold arctic air overhead will let the Milky Way erupt, trailing clouds of glory across the heavens. Then, with the help of a hefty snow shovel, those of us in the Northeast may occasionally dig out our telescopes enough to do some useful work."

I can only hope that this will be the case, for 1990 has been a particularly poor year for clear skies in New England. Then again, treasured in my old-letter file is a 1930s missive from Leon Campbell, the first recorder for the American Association of Variable Star Observers. In it he complains that during the previous month the weather was so bad that the AAVSO received not a single observation from Massachusetts or Connecticut. So if the weather confounds your observing program, the remedy is simple: live long.

December is a fine month for viewing Cetus, the Whale. This constellation, which swims mostly just under the celestial equator, is presently near the meridian during evening hours. If you work with 8- to 12-inch telescopes, there are many fascinating galaxies in Cetus worth hunting down. In this part of the heavens our gaze is away from the plane of the Milky Way. Indeed, just over the Cetus border in Sculptor is the South Galactic Pole. Here we are looking into the depths of the universe, beyond our own galaxy's veil.

Those who use 17-inch and larger telescopes will no doubt be able to find many more galaxies than shown on charts like Tirion's *Sky Atlas 2000.0*. While many of them can be found in the *New General Catalogue of Galaxies and Clusters of Stars* (*NGC*), others will be listed only in specialized catalogs. Correctly identifying these galaxies can be a real challenge, even for the amateur who has a good reference library.

Learning to use a good atlas, preferably one showing 9th-magnitude stars, is another important project. Frequent glances back and forth between eyepiece and atlas, the latter illuminated by a dim red light, will soon tell you how much sky appears in the eyepiece at one time. Recent psychological studies of pattern recognition have shown the importance of orienting the chart to match the visual field. This is easy when a region near the meridian is viewed in a refractor, but some care may be needed with a reflector, or if the region is far from the meridian.

The job of locating faint deep-sky objects is always made easier if there are nearby finder stars. Since most telescopes have a field of view less than 1° in diameter when used at 100× (a good magnification for searching out many small objects), it would be nice to have charts with at least one star in each 1°-diameter field. You can test your own charts by cutting the correct size circle in a piece of paper and moving it around the charts to see how often there are no stars showing in the 1° opening.

Figure 12.5
The energetic nucleus of the Seyfert galaxy M77 in Cetus is spewing out clouds of gas (each with the mass of about 10 million Suns) at speeds of nearly 400 miles per second.

The galaxy M77 in Cetus is a good practice object near a naked-eye star (**Figure 12.5**). Put on a low-power eyepiece and center Delta (δ) Ceti in the field (perhaps after sighting along the side of the telescope tube). Then, by remembering how large the ½° Moon looks in the same ocular, slide the telescope an estimated 1° southeast. A small, steady glow seen in this region will probably be M77.

If this attempt fails, put Delta back in the center of the field, then wait 3.2 minutes — the difference in right ascension between Delta and M77 — and scan along a north-south line. You should spot the galaxy south of where the star had been by some 21′, the difference in their declinations.

M77 is an old friend from my days in Kansas. Even under 400× in my 10-inch reflector, I could never quite distinguish the starlike nucleus that some observers

270   DEEP-SKY WONDERS

have reported. In viewing extended deep-sky objects, maximum image contrast is essential to discern small differences of tone. But any stray light in the optical system (from dirty or imperfect lenses), sky brightness, or artificial light will compress the scale of recognizable shades by eliminating black and lightening all grays. Light that does not go into the image goes elsewhere to brighten the field. The result is a washed-out image, and near the limit of visibility it may mean the difference between seeing or not seeing the subject.

This fact was emphasized one night when I was observing M77 with the 4-inch Clark refractor. The night was good and the Milky Way especially vivid. Normally, M77 appears as a round patch of light 2′ in diameter, fading irregularly near its edge. This night, however, the edge of the galaxy was sharply seen against the background. Also, for the first time in so small an instrument, I could see mottling in its core. The excellent optics and the fine night changed this minor galaxy into a marvelous sight, rivaling much more popular objects.

The Skalnate Pleso *Atlas Catalogue* puts M77's visual magnitude at 8.9 and photographic at 9.6. But the *Revised New General Catalogue* (*RNGC*) gives 10.5 photographic for it. Such discrepancies between catalogs are not at all unusual, emphasizing the fact that magnitudes assigned to nebular objects depend strongly on the size and type of instrument and the method of observing. Galaxies are roughly one magnitude fainter in blue light (photographic) than in yellow-green (visual). Also an estimate of a galaxy's magnitude is apt to be brighter with a small telescope than a large one. This effect was noted long ago by comet observers. It is particularly marked for very extended objects of low surface brightness.

M77 inspired a remarkable set of seesaw descriptions by early observers. The galaxy was discovered in October 1780 by Pierre Méchain. Although Méchain initially described it as a nebula, that December Messier called it "a cluster of small stars which contains some nebulosity." The observation that really surprised me, however, was William Herschel's in the early 19th century. This skilled observer called M77 "a cluster of stars . . . [with a] stellar appearance when it is viewed in a very good common telescope." (Keep in mind that a common telescope in Herschel's day was not very good when compared with the average equipment owned by amateurs today.) His son John called M77 "partly resolved."

At this point one might rightly ask, what gives? Does M77 look like stars or nebulosity? The pioneering English astrophotographer Isaac Roberts indirectly offers a possible answer to the discordant visual descriptions. In the late 1800s his photographs revealed M77 to be a stellar nucleus surrounded by nebulosity "studded with strong condensations resembling stars." Could others have seen these condensations as stars? I had one such view, but it was with a 36-inch reflector back in the 1930s, and other visual descriptions made with similar-size instruments make no mention of the "stars." Thus, with proper reverence for Messier and his contemporaries, we must erase any idea that M77 can be resolved like a star cluster.

There's no question, on the other hand, that the galaxy has a bright nucleus. M77 is in fact among the brightest examples of a Seyfert galaxy, one having a miniature quasar at its core.

Galaxy Chains in Cetus

Just as open clusters dominate our view of the starry band of the Milky Way, galaxies tend to be most numerous in those parts of the sky farthest from this region, where dense star clouds and obscuring dust can block our view of deep space. Such is true for the constellation Cetus, now in the evening sky. "It is a vast stellar desert," Scotty wrote, "but a rich area for galaxy hunters." Amateurs equipped with only a 3-inch or 4-inch telescope can pick out many galaxies, and a 12½-inch will show many more of them. Here Scotty describes a series of galaxy chains in Cetus and a mysterious planetary nebula that should keep novices and seasoned observers busy during the long December nights.

For those who want to sweep the field around M77, several other galaxies are within the range of amateur telescopes. This is an especially good field for beginners, since the 4th-magnitude star Delta (δ) Ceti provides an easy starting point just under 1° northwest of M77. The star and the moderately bright Messier object serve as landmarks to return to if you get lost while hunting for the other sights.

Some ¾° due north of Delta Ceti lies the faint galaxy NGC 1032. *Sky Catalogue 2000.0* lists it as a spiral with photographic magnitude 13.2. A good rule of thumb is that such objects will appear a magnitude brighter visually. I've estimated NGC 1032's brightness as 12.1 with the 4-inch Clark. Herschel called it "pretty bright." This may seem strange in light of his description of nearby 10.5-magnitude NGC 1055 as "pretty faint," but it may be due to NGC 1032's smaller size (about 3' × 1'), which increases the surface brightness of the galaxy.

This effect is also demonstrated by two other galaxies in the area. Both NGC 1073 and NGC 1087 are listed as magnitude 11.0 in Roger Sinnott's *NGC 2000.0*. Herschel, however, called the smaller galaxy (NGC 1087, 3.5') "pretty bright" and the larger one (NGC 1073, 4.9') "very faint." I have examined a photograph of the field, and at first glance it would seem that more than a magnitude separates these galaxies. Trying to convert from one system of brightness determinations to another can be an amateur's nightmare.

Eight degrees south of Delta Ceti, the stars 80 and 77 Ceti mark another clump of galaxies. The three brightest are seen in my 4-inch Clark with averted vision, and once so found they can be glimpsed with only a 3-inch. First try NGC 1022, a barred spiral, 1⅓° northeast of the two stars. From Louisiana, back in 1945, my 10-inch reflector showed traces of structure in this 1.4' × 1.8' oval; the night was superb but the dewing was so heavy that I had to use heaters on the main mirror to observe at all!

At my Joseph Meek Observatory, the galaxy was featureless in the 4-inch. I made its magnitude 10.8, while the *RNGC* gives its value as 12.5 photographically.

NGC 1052's tiny 1′ diameter bears magnification well. In a 10-inch with 200× at Milwaukee, Wisconsin, I once estimated it to be 11th magnitude; the *RNGC* makes it 12.0 in blue light. About equally dim is NGC 1084, a distinct oval 2′ long.

Those equipped with large amateur instruments may want to fill out the grouping with a trio of very faint galaxies.

Even a 10-inch will pick up NGC 991 with little trouble. This is a roundish galaxy less than 2′ across and ⅔° north of 80 Ceti. According to the catalog, its photographic magnitude is 12.5.

NGC 1048 lies ⅓° south of NGC 1052 and slightly west. Decidedly cigar-shaped, it is 4′ long and looks about 12th magnitude. The *RNGC* lists it as a close pair of galaxies, each of photographic magnitude 14.0.

Using 100× or more, try your luck with NGC 1035, a sliver 2′ long (between 80 Ceti and NGC 1052). Here, the photographic magnitude and my visual impression are both 13.

The last three are atmospheric test objects for us in the United States, and even more so for Canadians, where these galaxies culminate somewhat lower in the southern heavens.

Northeast of 4th-magnitude Theta (θ) Ceti is a chain of four galaxies just right for 6- to 8-inch telescopes. The chain extends northwest to southeast. The northernmost galaxy is NGC 584, an elliptical whose slightly oval disk is a bit less than 2′ across with a visual magnitude of 10.8. It has been seen with a 2½-inch aperture and shows well in my 4-inch Clark refractor. If you are using a large instrument, try looking for 14th-magnitude NGC 586 just 5′ to the southeast.

The next galaxy in the chain, 25′ southeast of NGC 584, is NGC 596, which lies just west of a faint naked-eye star. The best view of this 11.5-magnitude elliptical will be had if the relatively bright star is kept out of the field. It may also help to rock the telescope gently, as slight motion sometimes improves the visibility of faint objects. I have found that the rocking technique works well when I am viewing extended objects like galaxies, but just the opposite is true for stars — the least field motion will wipe out faint stars.

About ½° southeast of NGC 596 is NGC 615. It is a spiral galaxy about 2′ long with roughly the same brightness as NGC 596. NGC 615 has a distinctly bright nucleus that seems displaced from the center of the galaxy. This object was easy for my old 10-inch reflector in Kansas.

The last galaxy in the chain is NGC 636. It is about 1° from NGC 615 and displaced slightly to the east of a line connecting the other three. I estimate this faint elliptical to be magnitude 12.0 and just under 1′ across. NGC 636 looks very much like a small planetary nebula. When I placed an ultrahigh-contrast (UHC) filter between the eyepiece and my eye, however, the object dimmed along with the stars, thus ending any thought that it was a nebula.

It was quite another story when the filter was used on the nearby large plan-

Figure 12.6
The planetary nebula NGC 246 (with four stars superimposed on it) appears bright and round in large telescopes.

etary NGC 246. Flipping the filter in and out of the view caused the stars to blink while the nebula remained conspicuous. In my opinion filter flipping is a much better way to locate small planetaries than the older method of using a prism or spectroscope.

NGC 246 was discovered by William Herschel, who called it large and very faint. I question his description of it as faint, since I can see it with the 4-inch Clark. Modern estimates placed the 4'-diameter planetary at magnitude 8.5. Ronald Morales saw NGC 246 "well" in a 6-inch reflector. With an 8-inch f/5 reflector, he reports that the nebula was "easily seen as a round, diffuse glow behind three stars of similar brightness. A fourth was glimpsed, but the central star was not seen." Other reports suggest that the planetary is a complete ring in apertures larger than 10 inches, while smaller instruments show the ring as broken.

If you are interested in hunting out some of Cetus's fainter galaxies, begin with NGC 309, located about 3° northeast of NGC 246. It is a 12.5-magnitude spiral about 2' in diameter. Objects like this show better in long-focus instruments since they tend to scatter less light in the field. My old 10-inch reflector was f/8.6, and it gave fine views of faint deep-sky objects. Some of my best deep-sky views have been with the Stellafane 12-inch Porter turret telescope, which is f/17. I also suggest using a Barlow lens to obtain higher magnification. A low-power eyepiece and Barlow seem to scatter less light and produce a higher-contrast view than a high-power eyepiece alone.

Less than 1° northeast of NGC 246 is NGC 255, another spiral that is about the same magnitude and slightly larger than NGC 309. I have glimpsed it with the 4-inch Clark, but a 10-inch would be a more practical instrument to fish it out.

Lastly, there are two faint galaxies about 3° southwest of NGC 246. The brighter is the spiral NGC 210. Its oval disk is about 4' long and catalogs list it as magnitude 11.8. While in Louisiana during World War II, I had a fine view of this object with the 10-inch reflector. Less than 1° to the southwest is NGC 178, which was also visible in the 10-inch. However, at 13th magnitude and about 1' in diameter it is a more challenging object.

Another Cetus galaxy, NGC 578, seems never to have been mentioned in this column before, nor is it plotted in *Norton's Star Atlas*, but over the years I have seen it several times with apertures ranging from 13 inches down to 4. With averted vision, look a for a dim oval glow, about 4' by 2' in extent. It is best to prepare a finder chart by plotting this position on a map showing stars as faint as magnitude 9, for this object is very difficult to acquire by simply sweeping. If you don't find NGC 578, you have the consolation that William Herschel missed it too. My estimate of the total magnitude is about 11.2, as seen in the 4-inch.

By the way, I would like to thank all the people at the 1990 Stellafane convention in Vermont who participated in the "Hello, Scotty!" greeting on Saturday evening. That shout may have made it all the way to Connecticut (I've learned of unusual happenings in the area that were glibly attributed to weather conditions), but at the time I was enjoying a meal of delicious mussels at a sidewalk cafe in Belgium following the AAVSO's European meeting. It was the first Stellafane I've missed in many years.

A Romp in the December Wonderland

"It is December again and the stars have an extra snap and sparkle. The humidity is down, objectives do not dew up, and the warmly dressed observer is as happy as on June nights." Or so Scotty thought when he wrote from Kansas in 1955. Nearly four decades later he noticed that the skies had steadily deteriorated and wondered what our skies would be like in the future. Despite this growing nemesis, which we now commonly refer to as light pollution, the night sky with all its splendors continues to inspire passionate souls. In this last installment, Scotty takes us on a visual romp to some of December's most astonishing sights, all of which have been covered in more detail throughout this book. Still, they are, as Scotty has called them, the old favorites of our youth. The romp ends with M42, the Orion Nebula. Thus the book comes full circle. In the end, when called upon to reflect on the importance of amateur astronomy, Scotty looked into his crystal ball and saw the promise of progress, mainly because in his heart he cherished a belief in the positive impact of the night sky on the affairs of humanity.

Fifty years ago, any kid in a big city like my old hometown of Milwaukee could set up a telescope in his or her backyard and enjoy really first-rate views of the night sky. This is far from the case today, and the problem facing most urban observers is to find a dark-sky site that can be reached quickly with a telescope that can be stuffed into the back of an automobile. Long-time author and telescope maker Robert E. Cox once noted that amateur telescopes in the 1930s

Figure 12.7
The open cluster M38 (top) and its smaller companion NGC 1907 (bottom) both lie in Auriga.

stayed collimated better than those of today. One reason may be that years ago we didn't drag our equipment over so many back roads.

As many of us know, the telescope is a wondrous invention, and the heavens contain all manner of marvels that can still astound the imaginative mind, no matter what the smog density may be. Some of the better sights await us in the December evening sky. The Northern Cross is erect in the northwest: Albireo has already set. Pegasus is now a great diamond-shape sloping slowly to the west, as Orion mounts closer to the meridian. This is no time for routine or difficult objects: it is better that we sweep again the old favorites of our youth — the sights that enthralled us with our first homemade reflector.

The Pleiades must come first, that marvelous cluster which in opera glasses is more splendid than most galactic clusters in a 10-inch. But if we have a 10-inch the splendor may keep us at the eyepiece for long moments. Perhaps we can faintly see the Merope Nebula — it is not impossible. Next we pause for a brief glance at the Hyades, bright with piercing sparkle.

After this comes the great Double Cluster in Perseus. To the naked eye it shines with a steady glow between Cassiopeia and Perseus, and in the telescope this tremendous blaze of scintillating suns makes a commanding entrance into the eyepiece field. One can look for a long time at the many doubles, the colors, the winding patterns as the dense cores of the cluster thin out slowly to merge finally into the star-rich background of the galaxy itself.

Moving down the Milky Way we run into such variegated star fields and clusters in Auriga that it is almost impossible to know where to halt, but this might very well be at M38. Evenly compressed into a glowing ball two thirds the diameter of the full Moon are over 100 softly blazing stars (**Figure 12.7**). Nearby is

M36, a rich cluster of fainter stars, somewhat smaller but also impressive. It is well to trail the telescope slowly over the whole length of the Milky Way in Auriga, for objects unmentioned in these regions would be major sights in most other parts of the sky.

Farther to the south, we pause for a moment at M1, the famous Crab Nebula, although our telescopes will show little more than an oval glow, with little trace of wispy filaments. Less well-known, but still one of my favorites, is the diffuse nebula M78 in Orion, with a few stars apparently superimposed. It has a curious "ink-blob" symmetry.

Next in numerical order is M79 in Lepus, an 8th-magnitude globular cluster eight arcminutes in diameter. Finally, saving the most impressive for last, is the incomparable M42, the great Orion Nebula, about which words fail. No amount of intensive gazing ever encompasses all its vivid splendor.

What Is an Amateur Astronomer?

Fortunately most amateurs don't bother with the semantics of their title. But it seems appropriate to consider what the name means. For starters I turned to the *New English Dictionary*. This scholarly work informed me that the word "amateur" is rather new to the English language, having appeared in print sometime after 1700. It is taken from a French word meaning "lover." In the 1700s, to be an amateur simply meant loving a subject. The word was used in bird-watching, lichen counting, painting, and all such sorts of human devotion. The connotations were always favorable.

Shortly after 1800 a derisive use began to appear. And in astronomy the modern division began to form between admiration for the stars and earning a living. The separation between amateur and professional astronomers became wide and deep. In the 1880s, when New York amateur astronomer and newspaper writer Garrett P. Serviss formed an organization called the American Astronomical Society, Simon Newcomb of the Nautical Almanac Office loudly protested that no amateur group should be allowed to carry so lofty a name.

Other professionals were less disapproving. Edward C. Pickering of Harvard recruited amateurs to monitor variable stars, and his efforts led to the formation of the American Association of Variable Star Observers. This and other organizations helped raise the status of amateurs.

Today the observations of amateurs can redirect efforts at professional observatories around the world. Australia's Robert Evans, observing from his backyard, has alerted professionals to many supernovae, allowing them to gather data on some very unusual stars in a timely fashion.

In 1931 Harvard astronomer Harlow Shapley gave a talk to the Milwaukee Astronomical Society (all 18 members). Afterward, during a discussion that went into the wee small hours, we talked of the role amateur astronomers play. Shapley did not see amateurs as volunteers who only did chores for the professionals. He saw them as a vital link between professionals and the public, a link that must exist if observatories hope to survive.

Lewis Epstein had a slightly different assessment of amateurs, one that he outlined in a talk before the Astronomical Society of the Pacific. He sees amateurs as the ones who plant and cultivate the seeds for the next generation of professional physicists, mathematicians, and engineers, as well as astronomers. So it's with a great sense of pride that we, as amateurs, go outside and enjoy the night sky.

Walter Scott Houston
1912 — 1993

Whether listening to lectures, participating in swap meets, sharing tips on telescopes, or showing off their equipment — as shown in these scenes from the July 1987 Stellafane convention in Springfield, Vermont — amateur astronomers spread their passion for astronomy. It all comes down to the love of the night sky, which Scotty helped to foster. In 1994, the year after Scotty died, Stellafaners showed their love and affection for him with a moment of silence by candlelight. It was a fitting tribute to a man whose words helped so many grow in their own love of the sky.

DECEMBER 279

280   DEEP-SKY WONDERS

DECEMBER OBJECTS

Name	Type	Const.	R. A. h m	Dec. ° '	Millennium Star Atlas	Uranometria 2000.0	Sky Atlas 2000.0
Crab Nebula, M1, NGC 1952	BN	Tau	05 34.5	+22 01	158	135, 136	5
Double Cluster, (east), NGC 884	OC	Per	02 22.4	+57 07	46, 47, 62	37	1
Double Cluster, (west), NGC 869	OC	Per	02 19.0	+57 09	46, 62	37	1
Hind's Variable Nebula, NGC 1555	BN	Tau	04 21.8	+19 32	185	133	5, 11
Hyades	OC	Tau	04 27.1	+16 55	185, 186, 209, 210	133, 134, 178, 179	5, 11
M36, NGC 1960	OC	Aur	05 36.1	+34 08	113	97, 98	5
M38, NGC 1912	OC	Aur	05 28.7	+35 50	113, 114	97	5
M77, NGC 1068	Gx	Cet	02 42.7	−00 01	262	220	10
M78, NGC 2068	BN	Ori	05 46.7	+00 03	253	226	11, B2
M79, NGC 1904	GC	Lep	05 24.5	−24 33	350	315	19
NGC 178	Gx	Cet	00 39.1	−14 10	316, 317	261	—
NGC 210	Gx	Cet	00 40.6	−13 52	316, 317	261	10
NGC 246	PN	Cet	00 47.0	−11 53	316	261, 262	10
NGC 255	Gx	Cet	00 47.8	−11 28	316	261, 262	10
NGC 309	Gx	Cet	00 56.7	−09 55	316	262	10
NGC 578	Gx	Cet	01 30.5	−22 40	362	308	18
NGC 584	Gx	Cet	01 31.3	−06 52	290	263	10
NGC 586	Gx	Cet	01 31.6	−06 54	290	263	—
NGC 596	Gx	Cet	01 32.9	−07 02	290	263	10
NGC 615	Gx	Cet	01 35.1	−07 20	290	263	10
NGC 636	Gx	Cet	01 39.1	−07 31	290	263	10
NGC 991	Gx	Cet	02 35.5	−07 09	287	265	10
NGC 1022	Gx	Cet	02 38.5	−06 40	287	265	10
NGC 1032	Gx	Cet	02 39.4	+01 06	262, 263	220	10
NGC 1035	Gx	Cet	02 39.5	−08 08	286, 287	265	10
NGC 1048	Gx	Cet	02 40.6	−08 33	286, 287	265	—
NGC 1052	Gx	Cet	02 41.1	−08 15	286	265	10
NGC 1073	Gx	Cet	02 43.7	+01 23	262	220	10

Ast = Asterism; BN = Bright Nebula; CGx = Cluster of Galaxies; DN = Dark Nebula; GC = Globular Cluster; Gx = Galaxy;
OC = Open Cluster; PN = Planetary Nebula; ∗ = Star; ∗∗ = Double/Multiple Star; Var = Variable Star

DECEMBER OBJECTS

Name	Type	Const.	R. A. h m	Dec. ° ′	Millennium Star Atlas	Uranometria 2000.0	Sky Atlas 2000.0
NGC 1084	Gx	Eri	02 46.0	−07 35	286	265	10
NGC 1087	Gx	Cet	02 46.4	−00 30	262	220	10
NGC 1514	PN	Tau	04 09.2	+30 47	139	95	5
NGC 1746	OC	Tau	05 03.6	+23 49	159	134, 135	5
NGC 1807	OC	Tau	05 10.7	+16 32	183	135, 180	5, 11
NGC 1817	OC	Tau	05 12.1	+16 42	183	135, 180	5, 11
Orion Nebula, M42, NGC 1976	BN	Ori	05 35.4	−05 27	278	225, 226, 270, 271	11, B2
Pleiades, M45	OC	Tau	03 47.0	+24 07	163	132	4, A2
T Tauri	Var	Tau	04 22.0	+19 32	185	133	5, 11

Ast = Asterism; BN = Bright Nebula; CGx = Cluster of Galaxies; DN = Dark Nebula; GC = Globular Cluster; Gx = Galaxy; OC = Open Cluster; PN = Planetary Nebula; ✷ = Star; ✷✷ = Double/Multiple Star; Var = Variable Star

SOURCES

Listed below by month and year of issue are the Deep-Sky Wonders columns from *Sky & Telescope* that were mined to create this book. Also listed are the sources who generously provided the images that accompany the text, and whose permission is gratefully acknowledged.

CHAPTER 1 — JANUARY

The Glory of a Thousand Stars in a Thousand Hues
January 1991.

The Nebulous Wonder of Orion
February 1950, January 1952, December 1955, February 1971, September 1980, January 1987, January 1991.

The Quest For Barnard's Loop and the Horsehead
Barnard's Loop: February 1984, January 1987, January 1991.

Horsehead: January 1952, December 1957, January 1969, January 1970, January 1979, January 1987, February 1990, January 1991.

The Great Triangulum Spiral
July 1949, January 1955, April 1955, October 1958, October 1959, January 1962, November 1962, December 1965, January 1975.

The Mystery of Nonexistent Star Clusters
December 1975, January 1976.

Jewel of the Night
December 1953, December 1955, December 1956, January 1962, December 1964, November 1979, October 1980, December 1982, December 1983, November 1984, December 1985, December 1988, November 1990.

Winter's Furnace
January 1957, December 1963, January 1972, December 1972, January 1986, December 1986, February 1992.

Images: 1.1, 1.2, 1.3, 1.5, 1.6, 1.10, 1.11, 1.12: Akira Fujii. 1.4: Chuck Vaughn. 1.7, 1.8, 1.9: *Sky & Telescope.* 1.13: NGS–POSS.* 1.14: DSS–S.**

CHAPTER 2 — FEBRUARY

Wonders in the Void
April 1954, March 1962, August 1963, February 1968, May 1968, December 1970, December 1973, December 1980, March 1985, October 1986, February 1991, November 1991, November 1993, February 1994.

Kemble's Cascade and Pazmino's Cluster
February 1968, December 1973, December 1980, March 1985, February 1991, November 1991, November 1993, February 1994.

Going to California
December 1981, March 1982, December 1982, November 1983, November 1985, January 1987, October 1988, December 1988, September 1989, January 1993.

The Little Dumbbell
December 1947, November 1954, December 1963, May 1969, December 1976, January 1979, December 1981,

SOURCES   283

December 1982, January 1984, December 1988, November 1993.

Probing the Depths of Perseus
December 1947, November 1954, December 1956, November 1960, January 1964, December 1976, December 1982, November 1983, December 1984, December 1988, January 1989.

Navigating the Celestial River
April 1954, February 1962, December 1966, November 1967, January 1971, January 1972, January 1973, December 1978, January 1980, January 1984, December 1986, January 1992, March 1994.

Averted Vision and the Celestial Jellyfish
March 1949, February 1962, April 1964, May 1966, November 1967, May 1979, January 1980, November 1981.

Images: 2.1: Preston Scott Justis. 2.2, 2.7: Martin C. Germano. 2.3, 2.9: NGS–POSS. 2.4, 2.5: DSS–N.*** 2.6: George Greaney. 2.8: George R. Viscome. 2.10: Akira Fujii. 2.11: Mt. Wilson and Palomar Observatories, courtesy the California Institute of Technology. 2.12: Dennis di Cicco.

CHAPTER 3 — MARCH

The Elusive Winter Wreath
March 1962, February 1974, November 1983, February 1987, January 1988, February 1990, January 1993.

The Gem of Gemini
April 1957, March 1960, March 1964, January 1966, March 1977, February 1980, March 1982, December 1983, February 1984, January 1985, December 1985, February 1986, March 1989, March 1992.

M35's "Comet" Companion
April 1957, March 1960, March 1964, March 1977, February 1980, March 1982, February 1984, January 1985, February 1986, March 1989.

The Domain of Castor
February 1961, February 1970, February 1971, February 1972, February 1980, March 1981, February 1983, February 1986, March 1989, March 1992, December 1993.

The Great Corridor of Open Clusters
January 1947, February 1954, December 1955, February 1960, January 1961, January 1974, February 1975, March 1977, December 1979, December 1985, January 1990.

The "Leader of the Host of Heaven" and Its Neglected Entourage
March 1947, February 1955, January 1960, February 1973, February 1978, February 1982, February 1988, February 1993, December 1993, January 1994.

Columba and the March Hare
January 1954, February 1956, February 1969, January 1973, January 1980.

Images: 3.1: Chuck Vaughn. 3.2, 3.7, 3.9: Akira Fujii. 3.3: Preston Scott Justis. 3.4: John Chumack. 3.5: *Sky & Telescope*. 3.6: NGS–POSS. 3.8: DSS–N. 3.10: DSS–S. 3.11: Martin C. Germano.

CHAPTER 4 — APRIL

Intergalactic Wanderer and Extragalactic Wonders
March 1949, March 1956, March 1971, March 1975, April 1978, February 1980, February 1981, April 1984, April 1985, February 1989.

The Dynamic Duo
April 1947, March 1955, May 1987, June 1987, May 1988, May 1992.

Seeing Double, and a Mysterious Planetary in Lynx
February 1981.

The Beehive Challenge
March 1951, February 1959, April 1960, March 1961, March 1963, March 1965, March 1976, March 1981, February 1983, March 1984, February 1985, March 1988, April 1989, March 1990, January 1993.

Hydra Hysteria
March 1963, February 1967, March 1970, April 1971, March 1972, April 1972, March 1978, March 1981, May 1981, February

1983, March 1984, April 1984, May 1984, May 1991, May 1994.

The Ghost of Jupiter
May 1947, June 1947, May 1969, March 1970, April 1971, March 1978, March 1984, April 1984.

Images: 4.1: Preston Scott Justis. 4.2: NGS–POSS. 4.3, 4.5, 4.6: Akira Fujii. 4.4, 4.8: Martin C. Germano. 4.7: DSS–N.

CHAPTER 5 — MAY

The Grandeur of Omega Centauri
June 1956, June 1961, May 1965, May 1970, May 1973, June 1978, May 1979, May 1981, May 1984, May 1987, May 1988, February 1992, May 1994.

Galactic Visibility
May 1951, March 1962, April 1968, April 1969, March 1974, April 1976, March 1981, April 1982, March 1983, April 1984, April 1985, March 1987, March 1988.

The Dwarfs that Dwell in Leo
March 1980, April 1990.

Hunting Galaxies in Leo
April 1948, March 1954, February 1958, April 1961, April 1963, March 1973, April 1977, March 1980, April 1982, March 1983, May 1983, April 1986, April 1987, April 1988, April 1989.

Lure of the Little Lion
May 1951, February 1952, March 1962, April 1968, April 1969, March 1971, March 1974, April 1976, March 1981, April 1982, March 1983, April 1984, April 1985, March 1987, March 1988.

Navigating Sextans
April 1964, April 1970, April 1971, March 1981, April 1982, April 1984, April 1985, March 1991, April 1992.

Images: 5.1a, 5.1b, 5.2, 5.8, 5.10: Akira Fujii. 5.3, 5.6: NGS–POSS. 5.4, 5.5: DSS–N. 5.7, 5.9: Chuck Vaughn. 5.11: Jeffrey Jones.

CHAPTER 6 — JUNE

The Bowl of Night
April 1959, January 1963, May 1974, April 1980, May 1987, May 1992.

The Mystery of M102
February 1948, March 1952, February 1967, July 1980, July 1986, May 1991.

The Northern Deep-Sky Triangle
June 1977, May 1986, June 1986, April 1991.

The Wonder of M106
February 1948, April 1991.

Forgotten Corridors
May 1953, June 1964, May 1967, May 1972, June 1982, May 1985.

Cup and Crow
May 1948, April 1962, May 1966, April 1979, April 1981, May 1981, April 1983, May 1985, May 1987, May 1988, June 1992, April 1994, June 1994.

Images: 6.1, 6.2, 6.3, 6.4, 6.5, 6.6: Akira Fujii. 6.7: Kim Zussman. 6.8: National Maritime Museum, London. 6.9, 6.10: NGS–POSS. 6.11: DSS–N. 6.12: Harvey Freed. 6.13: Martin C. Germano. 6.14: George R. Viscome.

CHAPTER 7 — JULY

Peering into the Cat's Eye
August 1949, July 1967, May 1969, June 1969, June 1979, March 1982, July 1985, July 1986, May 1991.

The Crown Jewels
July 1988, July 1991, July 1993.

Dueling Globulars
M13: June 1958, June 1963, June 1976, September 1982, June 1983, January 1985, August 1986, June 1987, September 1989, July 1992, July 1993.

M5: June 1950, July 1958, June 1965, July 1973, July 1974, June 1980, June 1986.

The Orphans of Ophiuchus
August 1961, July 1979, July 1980, June 1981, August 1982, July 1983, June 1985, July 1985.

More Surprises in the Serpent Bearer
July 1967, August 1970, July 1971, Septem-

SOURCES   285

ber 1978, July 1980, June 1984, August 1984.

Naked-Eye Globular Clusters
July 1994.

Images: 7.1, 7.7: Martin C. Germano. 7.2, 7.4, 7.8: Akira Fujii. 7.3: Mt. Wilson and Palomar Observatories, courtesy the California Institute of Technology. 7.5: From *The Scientific Papers of William Parsons, Third Earl of Rosse, 1800–1867.* Collected and republished by the Hon. Sir Charles Parsons, K.C.B., F. R. S., 1926. 7.6: Roger Sliva.

CHAPTER 8 — AUGUST

Scanning the August Pole and More Sights in Cepheus
September 1970, November 1972, December 1973, October 1974, November 1975, July 1985, December 1985, November 1986, November 1988, June 1992.

The Great Planetaries of Summer
August 1953, August 1962, August 1976, August 1980, July 1984, August 1987, July 1988, August 1989, August 1990.

The Dumbbell's Many Faces
September 1957, September 1963, August 1978, July 1984, July 1986, September 1986, September 1987, August 1990.

Houston's Uncertainty Principle
August 1989.

Telescopic Delights in Delphinus
October 1956, October 1961, August 1963, August 1967, October 1980, September 1991, August 1993.

Images: 8.1: NGS–POSS. 8.2: Robert Bickel. 8.3, 8.10: Preston Scott Justis. 8.4, 8.6: Martin C. Germano. 8.5, 8.7, 8.8: Akira Fujii. 8.9: DSS–N.

CHAPTER 9 — SEPTEMBER

Wandering Through Lacerta, the Lizard
October 1971, September 1972, January 1976, October 1977, October 1981, November 1982, August 1983, September 1989, November 1991, October 1992.

Cruising Through Cygnus
August 1948, September 1948, October 1948, September 1956, August 1965, September 1968, August 1972, September 1973, October 1973, August 1980, September 1980, October 1980, September 1981, September 1982, August 1983, October 1984, December 1985, September 1989, November 1991.

Unveiling the Veil
September 1966, December 1967, November 1969, September 1980, August 1983, July 1984, October 1984, September 1987, December 1987.

The Mystery of NGC 6811
September 1968, September 1973, October 1986, November 1988, December 1991.

Hunting Cosmic Pearls in Aquila
September 1979, September 1982, September 1993.

Images: 9.1, 9.2, 9.5, 9.7, 9.8: Martin C. Germano. 9.3: DSS–N. 9.4: Paul Lind. 9.6: Chuck Vaughn.

CHAPTER 10 — OCTOBER

The Great Square of Pegasus
October 1961, October 1962, November 1977, October 1978, December 1978, October 1983, November 1985, October 1988, January 1994.

Two Spectacular Autumn Globulars
October 1952, October 1955, September 1962, November 1974, October 1978, October 1979, October 1980, July 1982, October 1982, October 1983, November 1985, December 1985, October 1988, November 1992.

Sweeping Through Sagitta
September, 1956, September 1958, August 1966, July 1979, September 1985, July 1986, September 1986.

Unraveling the Helix
October 1968, October 1969, October 1979, October 1984, December 1987,

November 1992.

A "Field Day" in the South
October 1985, October 1987.

Images: 10.1, 10.5: Akira Fujii. 10.2, 10.6, 10.10: Martin C. Germano. 10.3: Lee C. Coombs. 10.4: DSS–S. 10.7: Mt. Wilson and Palomar Observatories, courtesy the California Institute of Technology. 10.8: NGS–POSS. 10.9: George R. Viscome.

CHAPTER 11 — NOVEMBER

The Cassiopeia Milky Way
January 1948, October 1953, October 1954, September 1967, December 1968, December 1969, November 1976, November 1981, September 1982, October 1983.

The Great Andromeda Galaxy
November 1946, November 1955, October 1958, January 1962, December 1976, December 1977, September 1980, November 1980, September 1981, November 1981, December 1981, November 1985, January 1986.

Sizing Up the Fish
November 1954, September 1959, December 1959, November 1973, November 1977.

The Splendors of Sculptor
October 1963, November 1970, August 1971, November 1972, October 1976, November 1988, December 1991.

The Naked-Eye Milky Way
November 1990.

Images: 11.1: Gerald Manley. 11.2, 11.5, 11.7: Martin C. Germano. 11.3, 11.4, 11.10a, 11.10b: Akira Fujii. 11.6: Preston Scott Justis. 11.8: Andrew Peters. 11.9: Royal Observatory, Edinburgh.

CHAPTER 12 — DECEMBER

The Challenge of the Seven Sisters
October 1950, January 1967, December 1971, October 1972, January 1975, December 1976, December 1977, December 1979, November 1980, December 1980, December 1982, November 1983, December 1983, December 1984, January 1985, December 1985, January 1988, November 1990, January 1994.

Taurus: The Observer's Paradise
January 1979, December 1979, January 1981, December 1983, October 1992.

Targeting the Cetus Seyfert
January 1966, November 1974, December 1974, December 1984, December 1990.

Galaxy Chains in Cetus
November 1971, December 1974, December 1984, December 1990.

A Romp in the December Wonderland
December 1955, December 1983, November 1988.

Images: 12.1, 12.2, 12.3, 12.4: Chuck Vaughn. 12.5: Dale E. Mais. 12.6: Martin C. Germano. 12.7: Preston Scott Justis. 12.8a–h: Dennis di Cicco.

* From the *National Geographic Society–Palomar Observatory Sky Survey,* courtesy the California Institute of Technology.

** From the Digitized Sky Survey, Southern Hemisphere, courtesy U. K. Schmidt Telescope Unit and NASA/AURA/STScI.

*** From the Digitized Sky Survey, Northern Hemisphere, courtesy the Palomar Observatory and NASA/AURA/STScI.

BIBLIOGRAPHY

This bibliography lists the books, atlases, journals, and other published materials cited in the text. If a work is currently available in a new edition, the most recent bibliographical information is given.

Allen, R. H., *Star Names: Their Lore and Meaning,* New York, 1963: Dover Publications.

Alfonsine Tables. A set of astronomical tables widely used during the Middle Ages. Named after Alfonso X, King of Castille and Leon (1252–1284).

Archinal, B. *The "Non-Existent" Star Clusters of the RNGC,* Typeset and printed in Great Britain by Don Miles, Portsmouth: The Webb Society, 1993.

Aristotle, *Meteorologica.* In Greek, with English translation by H. D. P. Lee, Cambridge, 1952: Cambridge University Press.

Barns, C. E., *1001 Celestial Wonders, as observed with home-built instruments,* Morgan Hill, CA, 1927: Science Service Press; released in 1929 by Pacific Science Press.

Bečvár, A., *Atlas Coeli II Katalog 1950.0 (Atlas of the Heavens Catalogue 1950.0),* Prague, 1960: Ceskoslovenske Academie.

Bečvár, A., *Skalnate Pleso Atlas of the Heavens,* Cambridge, MA, 1964: Sky Publishing Corporation.

Bonner Durchmusterung Catalog, Bonn, Germany, 1855: Universitats-Sternwarte zu Bonn.

Brocchi, D. F., *AAVSO Star Atlas,* Cambridge, MA, 1936: American Association of Variable Star Observers.

Burnham, R., Jr., *Burnham's Celestial Handbook,* (3 vols.) New York, 1978: Dover Publications.

Burritt, E. H., *Geography of the Heavens,* New York, 1873: Sheldon and Company. (First published in 1833.)

Chambers, G. F., *Descriptive Astronomy,* Oxford, 1867: Clarendon Press.

Cragin, M., J. Lucyk, and B. Rappaport, *The Deep-Sky Field Guide to Uranometria 2000.0,* Richmond, VA, 1993: Willmann-Bell.

The Deep-Sky Observer, Journal of the Webb Society. Typeset and printed in Southsea, Hampshire, Great Britain for The Webb Society.

Dreyer, J. L. E., *New General Catalogue of Nebulae and Clusters of Stars* (1888), *Index Catalogue* (1895), *Second Index Catalogue* (1908), London, 1962: Royal Astronomical Society.

Encke, J. F., ed., *Astronomisches Jahrbuch für 1834,* Berlin, 1832: Druckerei der Königl, Akademie der Wissenschaften.

Galilei, G., *The Sidereal Messenger (Sidereus nuncius).* Translated, with introduction, conclusion, and notes by

Albert van Helden. Chicago, IL, 1989: University of Chicago Press.

Harrington, P., *Touring the Universe Through Binoculars,* New York, 1990: John Wiley & Sons.

Hartung, E. J., *Astronomical Objects for Southern Telescopes,* Cambridge, England, 1968: Cambridge University Press. (Republished in 1995 as *Hartung's Astronomical Objects for Southern Telescopes,* 2nd ed., rev. and illus. by David Malin and David J. Frew.)

Hirshfeld, A., R.W. Sinnott, and F. Ochsenbein, eds., *Sky Catalogue 2000.0, Volume 1: Stars to Magnitude 8.0,* 2nd ed., Cambridge, MA, 1991: Sky Publishing Corporation and Cambridge University Press.

Hirshfeld, A., and R. W. Sinnott, eds., *Sky Catalogue 2000.0, Volume 2: Double Stars, Variable Stars, and Nonstellar Objects,* Cambridge, MA, 1985: Sky Publishing Corporation and Cambridge University Press.

Hogg, H. S., *Bibliography of Individual Globular Clusters,* Publications of the David Dunlap Observatory, Vol. 1, No. 4, (1939), Vol. II, No. 2 (1955), Vol. III, No. 6 (1973), Toronto, Canada: University of Toronto Press.

Jones, K. G., *Messier's Nebulae and Star Clusters,* Cambridge, England, 1991: Cambridge University Press.

Jones, K. G., ed., *Webb Society Deep-Sky Observer's Handbook,* Vol. 1, Double Stars, Short Hills, NJ, 1979: Enslow Publishers.

Jones, K. G., ed., *Webb Society Deep-Sky Observer's Handbook,* Vol. 2, Planetary and Gaseous Nebulae, Hillside, NJ, 1979: Enslow Publishers.

Jones, K. G., ed., *Webb Society Deep-Sky Observer's Handbook,* Vol. 3, Open and Globular Clusters, Hillside, NJ, 1980: Enslow Publishers.

Jones, K. G., ed., *Webb Society Deep-Sky Observer's Handbook, Vol. 4,* Galaxies, Hillside, NJ, 1981: Enslow Publishers.

Jones, K. G., ed., *Webb Society Deep-Sky Observer's Handbook,* Vol. 5, Clusters of Galaxies, Hillside, NJ, 1982: Enslow Publishers.

Jones, K. G., ed., *Webb Society Deep-Sky Observer's Handbook,* Vols. 6–8, Anonymous Galaxies, Hillside, NJ, 1975: Enslow Publishers

Journal of the Royal Astronomical Society of Canada. Richmond Hill, Ontario: David Dunlap Observatory.

Kukarkin, B. V., *General Catalog of Variable Stars,* Leiden, The Netherlands, 1949: Sternberg Astronomical Institute and International Astronomical Union.

Luginbuhl, C. B. and B. A. Skiff, *Observing Handbook and Catalogue of Deep-Sky Objects,* 2nd ed., Cambridge, England 1998: Cambridge University Press.

Mallas, J. H., and E. Kreimer, *The Messier Album,* Cambridge, MA, 1978: Sky Publishing Corporation.

McKready, K., *A Beginner's Star-Book,* New York, 1923: G. P. Putnam's Sons.

Mitchell, L., *Mitchell's Anonymous Catalog,* self-published.

National Geographic Society, *National Geographic Society–Palomar Observatory Sky Survey,* Pasadena, CA, 1954, 1958: California Institute of Technology.

Olcott, W. T., *Field Book of the Stars,* New York, 1907: G. P. Putnam's Sons.

Perek, L., and L. Kohoutek, *Catalogue of Galactic Planetary Nebulae,* Prague, 1967: Academia Publishing House of the Czechoslovak Academy of Sciences.

Ridpath, I. A. *Norton's Star Atlas and Reference Handbook,* 19th ed., Essex, England, 1998: Addison Wesley Longman. (First edition published in 1910.)

Ross, F. E., and M. R. Calvert, *Atlas of the Northern Milky Way,* Chicago, 1934: University of Chicago Press.

Sagot, R., and J. Texereau, *Revue des constellations,* Paris, 1964: Société astronomique de France.

Scovil, C. A., AAVSO *Star Atlas,* Cambridge, MA, 1990: American Association of Variable Star Observers.

Serviss, G. P., *Astronomy with the Naked Eye,* New York, 1908: Harper and Brothers.

Serviss, G. P., *Astronomy with an Opera-Glass,* New York, 1888: D. Appleton and Company.

Serviss, G. P., *Pleasures of the Telescope,* New York, 1901: D. Appleton and Company.

Shapley, H., *Galaxies,* Harvard Books on Astronomy, revised ed., Cambridge, MA, 1961: Harvard University Press.

Shapley, H. and A. Ames, *A Survey of the External Galaxies Brighter than the 13th Magnitude,* Annals of the Astronomical Observatory of Harvard College, Vol. 88, No. 2., Cambridge, MA,1932: Harvard College. (Note: this was revised in 1981. See Sandage, A., and G. A. Tammann, *A Revised Shapley-Ames Catalog of Bright Galaxies,* Washington, DC, 1981: Carnegie Institution, and again in 1987.)

Sinnott, R. W., ed., *NGC 2000.0: The Complete New General Catalogue and Index Catalogues of Nebulae and Star Clusters by J. L. E. Dreyer,* Cambridge, MA, 1988: Sky Publishing Corporation and Cambridge University Press.

Sinnott, R. W., and M. A. C. Perryman, *Millennium Star Atlas,* Cambridge, MA, and Noordwijk, The Netherlands, 1997: Sky Publishing Corporation and European Space Agency.

Smyth, W. H., *The Bedford Catalogue,* from *A Cycle of Celestial Objects,* Vol. 2., Richmond, VA, 1986: Willmann-Bell.

Smyth, W. H., *A Cycle of Celestial Objects,* Vol. 1, London, 1844: J. L. Parker; Oxford, 1881, Clarendon Press.

Star Atlas (Workbook of the Heavens), Middletown, CT, 1968: American Education Publications.

Sulentic, J. W., and W. G. Tifft, *The Revised New General Catalogue of Nonstellar Astronomical Objects,* Tucson, AZ, 1980: University of Arizona Press.

Tirion, W., and R. W. Sinnott, *Sky Atlas 2000.0,* 2nd ed., Cambridge, MA, 1998: Sky Publishing Corporation and Cambridge University Press.

Tirion, W., B. Rappaport, and G. Lovi, *Uranometria 2000.0,* 2 vols., Richmond, VA, 1987: Willmann-Bell.

Vehrenberg, H., *Atlas of Deep-Sky Splendors,* 4th ed., Cambridge, MA, 1983: Sky Publishing Corporation and Cambridge University Press.

Webb, T. W., *Celestial Objects for Common Telescopes,* Vol. 2, New York, 1962: Dover Publications.

INDEX

Scotty loved to set challenges for observers, so it is appropriate that this distillation of his *Sky & Telescope* columns proved something of a challenge to index.

Scotty peppered his writing with references to past observers; how such deep-sky greats as Charles Messier; William, John and Caroline Herschel; Lord Rosse; William H. Smyth; and Thomas W. Webb described this or that object was an essential ingredient of his mix. To record every mention of the names that appear so often would have generated an unwieldy batch of index entries, so for these only a representative selection has been chosen. (The most frequently quoted of Scotty's own correspondents have not been trimmed in this way.)

Included in this index are the use of various accessories (indexed individually) and important general tips and techniques (collected under "observing tips and techniques"). Telescope types are not indexed. Many specific instruments that Scotty and others used are indexed under the names of the observatories where they are located. Not included are Scotty's own telescopes — he mentions his trusty 4-inch Clark refractor far too often.

In a sense, every Deep-Sky Wonders column threw out tests to its readers. A selection of these have been indexed under "challenges," along with a variety of targets historically regarded by amateur astronomers as tests of either their observing skills or equipment. (Many of these objects have been rendered less daunting over the years by advances in amateur telescope technology.) The "challenges" entry, along with some others that contain a dozen or more page references, has not been divided into subentries.

Every deep-sky object Scotty describes in the book is indexed here under its most familiar label. Thus, for example, the main entry for the Orion Nebula is under that name, rather than M42 or NGC 1976; such catalog numbers will also be found as cross-references. Stars mentioned in the text simply as guideposts are not indexed. There are no generic entries for types of deep-sky objects, such as spiral galaxies or dark nebulae. Again, this would have made for a number of unwieldy entries.

Object designations such as Theta (θ) Orionis, FI Hydrae, and 19 Lyncis, which contain the genitive form of a constellation name, are indexed under the constellation, in these cases under Orion, Hydra, and Lynx. Each constellation entry starts with the numbers of pages on which these and differently designated objects in that constellation are described.

Page numbers in **boldface** refer to photographs or illustrations and their captions. Numbers in *italics* refer to the tables of objects at the end of each chapter. The suffix *n* indicates a footnote.

"4-H cluster" *see* NGC 1664
AAVSO *see* American Association of Variable Star Observers
Abbey, Leonard B., Jr., 57
Abell, George, 113
Abell 426 *see* Perseus Cluster
Abell 1367, 113, *121*
Abell 2065 *see* Corona Borealis cluster of galaxies
Acamar (Theta Eridani) *see* Eridanus
Achernar (Alpha Eridani) *see* Eridanus
Adams, R., 11
Alfonsine Tables, 16*n*
Allen, Richard H., 60, 85, 188
al-Sūfī, 204, 246
Altair (Alpha Aquilae) *see* Aquila
Alvan Clark & Sons, 6*n*
amateur astronomy, 277–8
    development of, 39–40, 49, 50–52, 105–7, 114, 155–6, 211, 219
Ambrosi, Dave, 11
American Association of Variable Star Observers (AAVSO), 27, 45, 269, 277
American Astronomical Society, 277
Ames, Adelaide, 81*n*
Amici prism, 237
Andromeda, 95, 245–8
Andromeda Galaxy (M31), 45, 204, 245–8, **245**, *259*

Antares (Alpha Scorpii) *see* Scorpius
Antennae *see* Ring-tail Galaxy
Antlia, 92
Aquarius, 224–6, 230–33, 235
Aquila, 210–13, **227**
    Altair (Alpha [α] Aquilae), **185**, **187**, 211, **227**
Aratus, 85, 118
Archinal, Brent, 12
Aristotle, 63
Arizona, University of, 36-inch reflector, 131
Aselli, 85*n*
Ashbrook, Joseph, 82
Astrofest (convention), 210
Astronomical Society of the Pacific, 278
astrophotography, 6–7, 111, 226
    photographs vs. visual appearance, 3, 5–6, 8, 38, 43, 110, 111, 125, 134, 142, 153, 158, 160, 163, 180–81, 209, 227, 246, 263
atmosphere, quality of *see* sky conditions
atmospheric absorption, 75, 79, 101
Auriga, 64–8, **65**
averted vision, 44–6

Baade, Walter, 247
Barker, Ed, 167, 229–30
Barlow lens, 69, 88, 126, 153–4, 176, 183, 223, 274
Barnard, E. E., 5, 7, 34, 60, 165, 182, 228, 264, 267
Barnard's Loop (Sh2-276), 5–6, **6**, *25*
Barnard 33 *see* Horsehead Nebula
Barnard 64, 164, *171*
Barnard 168, 203, *214*
Barnard 201, 19, *25*
Barnard 259, 165, *171*
Barnhart, Stephen, 7
Barns, Charles E., 16, 124, 206

294   DEEP-SKY WONDERS

Bartek, Mr., 11
Bartels, John F., 6, 51, 207, 219
Bayer, Johann, 85, 100
Beçvár, Antonín, 12*n*
Beehive (Praesepe, M44), 84–7, **85**, **87**, *96*
Belt of Orion, 8
Beowulf, 84–5
Berkeley 10, 32, **32**, *47*
Berlin Observatory, 81
Bevis, John, 267
Big Dipper, 108, **123**, 123–7
"Big Four" of Perseus (NGC 869 and 884, M34, M76, NGC 1499), 36
Bigourdan, Guillaume, 105
Birr Castle Observatory (of Lord Rosse)
    36-inch reflector, 4
    72-inch reflector ("Leviathan of Parsonstown"), 37*n*, 180
Boardman, L. J., 45
Bochart de Saron, Jean-Baptiste-Gaspard, 134
Bode, Johann, 38, 81, 195
Bond, George P., 246
Bortle, John, 8, 35, 159
"Box Nebula" *see* NGC 6309
Bradley, James, 59
Branchett, Brenda, 116, 117
Branchett, D., 164
Brashear, John, 142
Brennan, Pat, 11–15, 31, 165, 199, 229
Brocchi, Dalmiro F., 27, 76
Brocchi's cluster, **229**
Brooks, William R., 28
Brown, Douglas, 29
Brown, James P., 195
Buffham, W., 224
Burnham, Robert, Jr., 19, 28, 176
Burnham, Sherburne W., 61, 77, 83, 119, 142, 188, 267
Burnham 576, 82, *96*
Burritt, Elijah H., 75
Burton, Tom, 231
Buta, Nancy, 83
Buta, Ronald, 81, 83
Caelum, 69
Cain, Lee, 207
California Nebula (NGC 1499), 33–6, **34**, *47*
Camelopardalis, 14, 27–33, 177
    SZ Camelopardalis, 31
Campbell, Leon, 269
Cancer, 84–9
Canes Venatici, 125, 131–5, 179
    Cor Coroli (Alpha [α] Canum Venaticorum), 132; *see also* "Deep-Sky Triangle"
Canis Major, 61–5
    Sirius (Alpha [α] Canis Majoris), 61–4, **62**, *73*, 83, **257**
Cannon 3-1, 167, *171*
Capricornus, 234–5
Carpenter, Dr., 11
Carrington, Richard, 263
Cassiopeia, **16**, 17, 241–5
Castor (Alpha Geminorum) *see* Gemini
catalogs, descriptions in vs. visual appearance, 12–15, 29, 36, 55–6, 80, 103, 116–17, 173, 174, 229, 246, 251, 266, 271, 272
cataracts *see under* visual acuity
Cat's Eye Nebula (NGC 6543), 151–4, **152**, *171*
Centaurus, 93, 99–102
    Alpha (α) Centauri, 62
    Omega (ω) Cen (NGC 5139), 99–102, **100**, **101**, *122*
Cepheus, 87, 173–9
Cetus, 46, 269–75
challenges, 5–11, 23, 33–6, 49, 51–2, 57–60, 102–5, 131–2, 180–81, 198, 204–8, 209–10, 213, 229, 255, 263–5
Chambers, George F., 160
Chapin, Bruce, 95

Chaple, Glen, Jr., 163
Chéseaux, Phillippe de, 52
Christensen, Tommy, 208–9
circumpolar, 79
Clark, Alvan G., 6*n*, 61
Clark, Jeannie, 37
Clark, Tom, 37, 99
Clerke, Agnes, 261
Cochran, Harry, 231
Cocoon Nebula (IC 5146), 202–3, *214*
Coe, Steve, 226, 253, 256
Colfax Observatory, 8
Collinder 256, 136, *147*
Columba, 68–70
Coma Berenices, 136–7
Coma-Virgo cluster of galaxies, 136
Combs, Christine, 151
common proper motion, 177
constellations
    (changes to) boundaries, 75, 82, 160, 175, 226
    history, 114–15, 117–18, 195
    as signposts, 218
Cooke, S. R. B., 11
Copeland, Leland S., 77
Cor Coroli (Alpha Canum Venaticorum) *see* Canes Venatici
Corn, James, 11, 29, 46, 103, 204
Corona Borealis cluster of galaxies (Abell 2065), 156, **156**, *171*
Corona Borealis, 154–6
    R Coronae Borealis, 154–5, **155**, *172*
    T CrB, 127, 155, **155**, *172*
Corvus, 140, 142–5
Cox, Robert E., 275
Crab Nebula (M1), 267–8, **268**, 277, *281*
Crater, 142–5
craters, terrestrial, 105
Cuffey, J., 66
Curtis, Heber D., 218, 229
Cygnus, 175, 200–210
    52 Cygni, 206, 207, *214*

Deneb (Alpha [α] Cyg), **185**
SS Cyg, 127
Cyrus, Charles, 82
Cysat, Johann, 3

Darquier, Antoine, 179
d'Arrest, Heinrich, 29, 40, 88, 224, 264
Davey, William, 55
Dawes, William, 177
Dawes limit, 82
De Chéseaux's comet, 222
Dearborn Observatory, 61
deep-sky filters *see* nebula filters
"Deep-Sky Triangle" (Eta [η] Ursae Majoris, Alpha [α] Canum Venaticorum, Gamma [γ] Boötis), 129, **130**
Delphinus (Job's Coffin), 186–91, **187**
    Alpha (α) Delphini, 189, *192*
    Beta (β) Del, 82, *96*, 188–9, *192*
    Gamma (γ) Del, 188, *192*
    Theta (θ) Del, 191, *193*
Delporte, Eugene, 175
Deneb (Alpha Cygni) *see* Cygnus
Denning, William, 29, 263
di Cicco, Dennis, 18, 159
Dorpat Observatory, 31*n*
Double Cluster (h and Chi [χ] Persei, NGC 869 and 884), 15–19, **16**, **17**, *25*, 276, *281*
Double Double (Epsilon [ε] Lyrae) *see* Lyra
Draco, 151–4
Draper, Henry, 4
Dreyer, J. L. E., 11, 234
Dumbbell Nebula (M27), 182–4, **183**, *192*, 204–5
Dunlop, James, 23

ecliptic, 151–2
Epstein, Lewis, 278

Eratosthenes, 226
Eridanus, 41–6
    Acamar (Theta [θ] Eridani), 42
    Achernar (Alpha [α] Eri), 41–2, **42**, *47*
ESO 351-G30 *see* Sculptor Dwarf Galaxy
ESO 356-4 *see* Fornax System
Espin, Thomas E., 202
Evans, Robert, 167, 277
Everhart, Edgar, 202, 207
extinction *see* atmospheric absorption
eyepieces *see also* Barlow lens
    Erfle, 67, 207
    König, 131
    monocular, 35
    Nagler, 113, 224
    occulting bar in, 114
    orthoscopic, 35
    Plössl, 54, 153, 207
    wide-field, 19, 54
eyesight *see* visual acuity

Farrar, Leonard P., 8, 231
Feijth, Hank, 247
filters *see* nebula filters
Finkelstein, Jan, 232
Flammarion, Camille, 88, 92, 146
Flamsteed, John, 63
Flynn, E. D., 101
Fornax Group of galaxies, 20–23, **21**
Fornax System (ESO 356-4), 22–3, *25*
Fornax, 20–24, **20**
    Alpha (α) Fornacis, 21, *25*
Frederici Honores, 195
Galbraith, William, 103
Galileo Galilei, 3, 16, 85
Gardner, Michael, 153
Gauthier, G., 242, 243
gegenschein, 52
Gemini, 52–61
    Castor (Alpha [α] Geminorum), 57–61, **58**, **59**, *72*
    Pollux (Beta [β] Gem), 57
    YY Gem, 60;
Ghost of Jupiter (NGC 3242), 93, 94, **94**, *98*
Gingerich, Owen, 90, 111–12, 125
Gleason, Geoffrey, 59
Gottlieb, Steve, 232
Grabenhorst, Fred, 103
Grunwald, Mark, 7, 231, 232
Grus, 237–8

H20, 229, *239*
h3752, 71, *72*
Hahn, Friedrich von, 180
Halbach, Ed, 40
Hale Telescope *see* Palomar Mountain Observatory
Halley, Edmond, 100, 132, 157
Halley's Comet, 53, 185, 233, 267
Hansen, Todd, 23, 229
Harrington, Philip, 119, 191, 233
Hartung, Ernst J., 23, 166
Harvard Observatory, 15-inch refractor, 4, 246
Hastings, Charles, 142
HB 12, 247
HB 64, 247
HB 90, 248
HB 254, 247
H-beta (Hβ) filters *see* nebula filters
Heartwell, Bryce, 207
Heidelberg Observatory, 16-inch refractor, 114
Heintz, Wulff D., 177
Helix Nebula (NGC 7293), 230–33, *239*
Henzl filters, 207
Hercules, 157–60
Herring, Alika, 231
Herschel, Caroline, 90, 243, 244, 253
Herschel, John, 4, 15, 22, 23, 58, 70–71, 83–4, 94, 100, 133, 161, 173, 176,

179, 253, 271
Herschel, William, 3–4, 15, 49, 50, 52, 58, 67, 76, 83–4, 92, 111, 137, **137**, 205–6, 266–7, 271
Hevelius, Johann, 75, 114, 118
Hind, John Russell, 164, 267
Hind's Variable Nebula (NGC 1555), 267, *281*
Hipparchus, 16, 85
Hogg, Helen Sawyer, 129, 134
Holden, Edward S., 181, 234
Holden, William, 92
Holetschek, Johann, 111–12, 113
Horsehead Nebula (B33), 5, 6–9, **7**, *25*, 51
"Houston's Uncertainty Principle," 186
Hubble, Edwin, 11, 247
Hubble's Variable Nebula (NGC 2261), *25*
Huggins, William, 4, 46, 152
Hurst, Guy, 66, 67, 198
Huygens, Christiaan, 3
Hyades (Melotte 25), 265, 276, *281*
Hydra, 89–95: FI Hydrae, 92, *96*
"Hydra Hysteria," 89–93

IAU *see* International Astronomical Union
IC 342, 29, *47*
IC 410, 66, *72*
IC 418, 71, *72*
IC 434, 7, **7**, *25*
IC 591, 107, *121*
IC 1296, 182, *192*
IC 1318, 202, *214*
IC 1434, 198, *214*
IC 1459, 237, *239*
IC 1470, 178, *192*
IC 2156, 57, *72*
IC 2157, 57, *72*
IC 2196, 61, *72*
IC 3568, 29, *47*

IC 4665, 163, *171*
IC 4756, 163, *171*
IC 4997, 229, *239*
IC 5070 *see* Pelican Nebula
IC 5146 *see* Cocoon Nebula
IC 5217, 199, **199**, *214*
IC 5269, 237, *239*
IC 5271, 236, *239*
IC 5273, 237, *239*
Ihle, Abraham, 170
International Astronomical Union (IAU), and constellation boundaries, 75, 82
IRAS-Araki-Alcock, Comet, 184

James, Thomas, 86
Job's Coffin *see* Delphinus
Johnson, Gus E., 168
Johnson, Richard, 6
Jonckheere, Robert, 246
Jones, Kenneth Glyn, 57, 90, 161, 226, 266

Kaler, James B., 23
Karnes, Stephan, 232
Kemble, Lucian J., 30, 33, 35
Kemble's Cascade, 30–32, **31**, *47*
Keystone of Hercules, 158
Kinney, William, 247
Kirch, Gottfried, 160, 168
Klein, Fred, 23
Knight, Stephen, 264
Kobayashi-Berger-Milon, Comet, 225
Koken, Harry, 101
Komorowski, Ted, 7, 231
Konecny, Marton, 210
Konst, Joanne, 208, 232
Krumenaker, Larry, 7

Lacaille, Nicolas-Louis de, 92
Lacerta, 15, 195–9
    8 Lacertae, 196, *214*
Lassell, William, 46, 50, 182, 222, 267

Le Gentil, Jean-Baptiste, 202
Leo, 105–14
    Regulus (Alpha [α] Leonis), **106**, 107
Leo I, 105, **106**, 107, *121*
Leo II, **107**, 107–8, *121*
Leo Minor, 78, 102–5, 114–17
Leonard, Arthur, 19, 54, 87
Lepus, 68, 70–71
"Leviathan of Parsonstown" (72-inch reflector) *see* Birr Castle Observatory
Lick Observatory, 36-inch refractor, 267
Liddell, J. P., 18
light-pollution reduction filters *see* nebula filters
Lima, Ray, 231
Ling, Alister, 35, 201
Little Dumbbell (M76), 36–8, **37**, *47*
Local Group of galaxies, 29, 107
Lossing, Fred, 51, 232, 263
Lovi, George, 75
low surface brightness (LSB), 34–5, 249, 271
LSB *see* low surface brightness
Luginbuhl, Christian, 146
Lumicon filters, 6, 8, 49, 51, 95, 169, 199, 208, 213, 224
Lundmark, Knut, 10
Lynx, 75–9, 82–4
    19 Lyncis, 82, *96*
    Alpha (α) Lyn, **78**
Lyra, 179–82, 186
    Delta (δ) Lyrae, 186, *192*
    Double Double (Epsilon [ε] Lyr), 186, *192*
    Vega (Alpha [α] Lyr), **185**, 186, *193*
    Zeta (ζ) Lyr, 186, *193*

M1 *see* Crab Nebula
M2, 170, *171*, 224–6, **225**, *239*
M3, 170, *171*
M4, 157, 169, **169**, *171*, 258
M5, 139, **139**, *147*, 160–61, **161**, 168, *171*
M9, 162, 164, **164**, *171*
M10, 157, 162
M12, 157, 162
M13, 157–60, **158**, **159**, 168, *171*, 224
M14, 157
M15, 170, *171*, 222–4, **223**, *239*
M19, 157, 162, 165, *171*
M22, 169, *171*
M27 *see* Dumbbell Nebula
M29, 203, **203**, *214*
M30, 235, **235**, *239*
M31 *see* Andromeda Galaxy
M32, **245**, 248, *259*
M33 *see* Triangulum Spiral
M34, 38–9, **39**, *47*
M35, 52–5, **53**, 57, *72*
M36, 64, 66, *72*, 277, *281*
M37, 64, **65**, 65–6, *72*
M38, 64, 66, *72*, **276**, 277, *281*
M39, 202, *214*
M41, **62**, 62–3, **64**, *72*
M42 *see* Orion Nebula
M44 *see* Beehive
M45 *see* Pleiades
M48, 90, *96*
M51 *see* Whirlpool Galaxy
M52, 79, 244, *259*
M53, 136, *147*
M57 *see* Ring Nebula
M62, 157, 162
M63 *see* Sunflower Galaxy
M65, 111–12, **112**, *121*
M66, 111–12, **112**, *121*
M67, **87**, 87–8, *96*
M68, 92, *96*
M70, 93, *96*
M71, 227–9, **227**, *239*
M72, 235, *239*
M73, 235, *239*
M74, 249, **249**, 250, *259*
M76 *see* Little Dumbbell

M77, 270–72, **272**, *281*
M78, 277, *281*
M79, 70–71, **70**, *72*, 277, *281*
M80, 157
M81, 79, 80, **80**, 81, 82, *96*
M82, 79, 80, **80**, 81, *96*, 125
M83, 92, 93, *96*
M92, 157, 170, *171*
M95, 108–9, **109**, *121*
M96, 108–9, **109**, *121*
M97 *see* Owl Nebula
M100, 168
M101, 127, **128**, 129, *147*
M102, 127
M103, 79
M104 *see* Sombrero Galaxy
M105, 109–10, **109**, *121*
M106, 125, 134–5, **135**, *147*
M107, 157, 162
M108, 124, **124**, 125, *147*
M109, 125, *147*
M110, **245**, 248, *259*
Machholz, Don, 89, 126
magnitudes of deep-sky objects, reasons for differences in, 271
Mallas, John, 5, 92, 93, 112, 131, 146, 160, 165, 183, 224, 226, 228
Maraldi, Jean-Dominique, 222, 225
Marseilles Observatory, 31-inch reflector, 182
Marth, Albert, 49–50, 182, 222
Mattei, Mike, 95
McDonald, Lee, 105, 107
McDonald Observatory, 30-inch telescope, 81
McMahon, James H., 231
Méchain, Pierre, 70, 92, 108, 109–10, 111, 128–9, 227, 249–50, 271
Meek, Joseph, 27, 127
Meier, Rolf, 51
Meketa, Jim, 232
Melotte 25 *see* Hyades
Melotte 111 *see* Collinder 256

Merope (23 Tauri) *see* Taurus
Merope Nebula (NGC 1435), 263–5, **264**, 276
Messier catalog
   additions to, 125, 127–9, 145–6
   origin of, 267
Messier marathon, 235
Messier, Charles, 16, 38, 52, 90, 92–3, 108, 111, 127–8, 134–5, 225, 227, 267, 271
micrometer, 59–60
Milky Way, 36, 44–5, **65**, 66, 139, **185**, 217, 241, 247, **257**, 277
   boundaries, 184–5, 186
   in Eastern legend, 258
   individual stars, visibility of, 86
   as naked-eye object, 256
Milton, Russell, 102
Milwaukee Astronomical Society, 277
   13-inch reflector, 39, 40, 55, 235
"missing" objects *see* "nonexistent" objects
Mitchell, Larry, 38
Monnig, Oscar, 127
Monoceros, 12, 49–51
   12 Monocerotis, 50
Morales, Ronald, 44, 78, 91, 95, 111, 115, 116, 131, 135, 143, 145, 156–7, 166, 167, 191, 245, 252, 253, 274
Moseley, Robert, 213
Mount Wilson Observatory, 5, 127
   60-inch reflector, 153
   100-inch reflector, 224

Nakamoto, Tokuo, 95, 111, 115, 237, 248
naked eye, faint objects visible to, 10–11, 49, 52–3, 63, 80, 84–5, 86, 139, 161, 181, 218–19, 223, 224–5, 256, 264
nebula filters (light-pollution reduc-

tion filters), 34, 38, 49, 52, 84, 230, 232, 268
   deep-sky, 169
   H-beta (Hβ), 8–9
   O III, 6, 34, 84, 199, 213, 224
   ultrahigh-contrast (UHC), 6, 49, 51–2, 95, 201, 207, 208, 232, 273
nebulium, 153
Newcomb, Simon, 18, 277
NGC 1, 221, *239*
NGC 2, 221, *239*
NGC 16, 219, *239*
NGC 40, 178, **178**, *192*
NGC 55, 254–5, **254**, *259*
NGC 128, 250, *259*
NGC 129, 244, *259*
NGC 133, 244, *259*
NGC 134, 253, *259*
NGC 146, 243, *259*
NGC 147, 248, *259*
NGC 178, 275, *281*
NGC 185, 248, *259*
NGC 188, 87, 175–6, *192*
NGC 205 *see* M110
NGC 210, 274, *281*
NGC 221 *see* M32
NGC 224 *see* Andromeda Galaxy
NGC 225, 244, *259*
NGC 246, 46, *47*, 274, *281*
NGC 253, 252, **252**, *259*
NGC 255, 274, *281*
NGC 288, 252, **252**, *259*
NGC 300, 253, **253**, *259*
NGC 309, 274, *281*
NGC 362, 224
NGC 436, 242, *259*
NGC 457, 242, **242**, *259*
NGC 470, 250–51, *259*
NGC 474, 250, *259*
NGC 524, 250, *259*
NGC 578, 275, *281*
NGC 584, 273, *281*

NGC 586, 273, *281*
NGC 596, 273, *281*
NGC 598 *see* Triangulum Spiral
NGC 604, 11, *25*
NGC 615, 273, *281*
NGC 628 *see* M74
NGC 636, 273, *281*
NGC 650-51 *see* Little Dumbbell
NGC 782, 42, *47*
NGC 869 *see* Double Cluster
NGC 884 *see* Double Cluster
NGC 991, 273, *281*
NGC 1022, 272, *281*
NGC 1032, 272, *281*
NGC 1035, 273, *281*
NGC 1039 *see* M34
NGC 1048, 273, *281*
NGC 1049, **22**, 23, *25*
NGC 1052, 273, *281*
NGC 1068 *see* M77
NGC 1073, 272, *281*
NGC 1084, 273, *282*
NGC 1087, 272, *282*
NGC 1129, 40, *47*
NGC 1130, 40, *47*
NGC 1131, 40, *47*
NGC 1232, 43–4, *47*
NGC 1275, 40–41, **41**, *47*
NGC 1297, 43, *47*
NGC 1300, 43, **43**, *47*
NGC 1316, 22, *25*
NGC 1317, 22, *25*
NGC 1318, 22, *25*
NGC 1325, 44
NGC 1325A, 44, *47*
NGC 1332, 44, *47*
NGC 1360, 23–4, *25*
NGC 1365, **21**, 22, *25*
NGC 1435 *see* Merope Nebula
NGC 1440, 44, *47*
NGC 1499 *see* California Nebula
NGC 1501 *see* Oyster Nebula
NGC 1502, 30–31, **31**, *47*

NGC 1514, 266, **266**, *282*
NGC 1531, 43, *47*
NGC 1532, 43, *47*
NGC 1535, 45–6, **46**, *47*
NGC 1555 *see* Hind's Variable Nebula
NGC 1569, 29, *47*
NGC 1637, 44, *47*
NGC 1662, 12, *25*
NGC 1664 ("4-H cluster"), 67–8, *72*
NGC 1708, 14–15, **14**, *25*
NGC 1746, 266, *282*
NGC 1792, 69, **69**, *72*
NGC 1807, 265–6, *282*
NGC 1808, 69, **69**, *72*
NGC 1817, 266, *282*
NGC 1851, 69, *72*
NGC 1857, 67, *72*
NGC 1883, 68, *72*
NGC 1893, 66, *72*
NGC 1904 *see* M79
NGC 1907, 68, *72*, **276**
NGC 1912 *see* M38
NGC 1931, 66, *72*
NGC 1952 *see* Crab Nebula
NGC 1960 *see* M36
NGC 1976 *see* Orion Nebula
NGC 2024, 7, *25*
NGC 2063, 14, *25*
NGC 2068 *see* M78
NGC 2090, 69, *72*
NGC 2099 *see* M37
NGC 2126, 68, *72*
NGC 2158, **53**, 54, 55–7, **56**, *72*
NGC 2168 *see* M35
NGC 2169, 12, *25*
NGC 2180, **13**, 14, *25*
NGC 2184, 13–14, **13**, *25*
NGC 2237-39 *see* Rosette Nebula
NGC 2244, 49–51, **50**, *72*
NGC 2251, 12, *25*
NGC 2261 *see* Hubble's Variable Nebula
NGC 2276, 173–4, *192*

NGC 2281, 68, *72*
NGC 2287 *see* M41
NGC 2292, 64, **64**, *72*
NGC 2293, 64, **64**, *72*
NGC 2295, 64, **64**, *72*
NGC 2300, 173–4, *192*
NGC 2403, 27–9, **28**, *47*
NGC 2410, 61, *72*
NGC 2419, **76**, 76–7, *96*
NGC 2469, 84, *96*
NGC 2474, 83, *96*
NGC 2475, 83, *96*
NGC 2500, 78, *96*
NGC 2537, 78, *96*
NGC 2541, 78, *96*
NGC 2548 *see* M48
NGC 2549, 78, *96*
NGC 2552, 79, *96*
NGC 2610, 90, 91, *96*
NGC 2632 *see* Beehive
NGC 2642, 90, *96*
NGC 2667A, 88, *96*
NGC 2667B, 89, *96*
NGC 2672, 88, **88**, *96*
NGC 2673, 88, **88**, *96*
NGC 2677, 89, *97*
NGC 2682 *see* M67
NGC 2683, 77, *97*
NGC 2713, 90, *97*
NGC 2749, 88, *97*
NGC 2763, 91, *97*
NGC 2764, 88, *97*
NGC 2781, 91, *97*
NGC 2782, 79, *97*
NGC 2787, 81, *97*
NGC 2793, 78, **78**, *97*
NGC 2811, 91, *97*
NGC 2831, 77–8, *97*
NGC 2832, 77, **78**, *97*
NGC 2844, 79, *97*
NGC 2848, 91, *97*
NGC 2851, 91, *97*
NGC 2855, 91, *97*

NGC 2859, 78, **78**, 97, 116, *121*
NGC 2889, 91, *97*
NGC 2903, **110**, 110–11, *121*
NGC 2905, 110–11, *121*
NGC 2962, 90, *97*
NGC 2967, 90, *97*
NGC 2976, 81, *97*
NGC 2985, 81, *97*
NGC 2992, 91, *97*
NGC 2993, 91, *97*
NGC 2997, 92, *97*
NGC 3003, 116–17, *121*
NGC 3021, 116–17, *121*
NGC 3031 *see* M81
NGC 3034 *see* M82
NGC 3077, 81, *97*
NGC 3109, 91, *97*
NGC 3115 *see* Spindle Galaxy
NGC 3130, 113–14, *121*
NGC 3145, 91, *98*
NGC 3158, 105, *121*
NGC 3162, 113, *121*
NGC 3163, 105, *121*
NGC 3172 *see* Polarissima
NGC 3177, 113, *121*
NGC 3184, 115, *121*
NGC 3185, 113, *121*
NGC 3190, 113, *121*
NGC 3193, 113, *121*
NGC 3200, 92, *98*
NGC 3242 *see* Ghost of Jupiter
NGC 3245, 116, *121*
NGC 3294, 117, *121*
NGC 3344, **115**, 115–16, *121*
NGC 3351 *see* M95
NGC 3368 *see* M96
NGC 3379 *see* M105
NGC 3384, 110, *121*
NGC 3389, 110, *121*
NGC 3395, 117, *121*
NGC 3396, 117, *121*
NGC 3414, 103, **104**, *122*
NGC 3486, 103, **104**, *122*

NGC 3504, 103, **104**, *122*
NGC 3556 *see* M108
NGC 3587 *see* Owl Nebula
NGC 3588, 113, *122*
NGC 3593, 112, *122*
NGC 3610, 126, *147*
NGC 3613, 126, *147*
NGC 3619, 126, *147*
NGC 3623 *see* M65
NGC 3627 *see* M66
NGC 3628, **112**, 113, *122*
NGC 3634, 142, *147*
NGC 3635, 142, *147*
NGC 3637, 142, *147*
NGC 3642, 126, *147*
NGC 3672, 142, *147*
NGC 3683, 126, *147*
NGC 3690, 126, *147*
NGC 3732, 142, *147*
NGC 3738, 126, *147*
NGC 3756, 126, *147*
NGC 3842, 113, *122*
NGC 3865, 142, *147*
NGC 3887, 141, **141**, *147*
NGC 3894, 126, *147*
NGC 3898, 126, *147*
NGC 3945, 126, *147*
NGC 3982, 126, *147*
NGC 3992 *see* M109
NGC 4027, 145, *148*
NGC 4036, 126, *148*
NGC 4038-39 *see* Ring-tail Galaxy
NGC 4041, 126, *148*
NGC 4147, 137, *148*
NGC 4153, 137, *148*
NGC 4217, 125, *148*
NGC 4258 *see* M106
NGC 4321 *see* M100
NGC 4361, 142–4, **143**, *148*
NGC 4590 *see* M68
NGC 4594 *see* Sombrero Galaxy
NGC 4605, 126, *148*
NGC 4782, 145, *148*

NGC 4783, 145, *148*
NGC 4792, 145, *148*
NGC 4794, 145, *148*
NGC 4814, 126, *148*
NGC 4868, 132, *148*
NGC 4914, 132, *148*
NGC 5005, 132, *148*
NGC 5024 *see* M53
NGC 5033, 132, *148*
NGC 5053, 136, *148*
NGC 5055 *see* Sunflower Galaxy
NGC 5085, 92, *98*
NGC 5101, 92, *98*
NGC 5139 *see* Omega (ω) Centauri
NGC 5150, 92, *98*
NGC 5194 *see* Whirlpool Galaxy
NGC 5195, **130**, *148*
NGC 5198, 132, *148*
NGC 5236 *see* M83
NGC 5253, 93, *98*, 253
NGC 5272 *see* M3
NGC 5296, 179
NGC 5350, 133, *148*
NGC 5353, 133, *148*
NGC 5354, 133, *148*
NGC 5371, 133, *148*
NGC 5390, 133
NGC 5457 *see* M101
NGC 5694, 93, *98*
NGC 5740, 138, *148*
NGC 5746, 138, **138**, *148*
NGC 5846, 139, *148*
NGC 5850, 139, *148*
NGC 5866, 128–9, *148*
NGC 5904 *see* M5
NGC 6093 *see* M80
NGC 6121 *see* M4
NGC 6144, 169, *171*
NGC 6205 *see* M13
NGC 6217, 174, *192*
NGC 6218 *see* M12
NGC 6254 *see* M10
NGC 6266 *see* M62

NGC 6273 *see* M19
NGC 6284, 165, *171*
NGC 6293, 165, *171*
NGC 6309, 167, *171*
NGC 6333 *see* M9
NGC 6341 *see* M92
NGC 6342, 165, *171*
NGC 6356, 165, *171*
NGC 6366, 164, *172*
NGC 6369, 166, *172*
NGC 6371 *see* M107
NGC 6402 *see* M14
NGC 6517, 163, *172*
NGC 6535, 163–4, *172*
NGC 6539, 163, *172*
NGC 6543 *see* Cat's Eye Nebula
NGC 6572, 167, *172*
NGC 6633, 167, *172*
NGC 6656 *see* M22
NGC 6681 *see* M70
NGC 6700, 182, *192*
NGC 6713, 182, *192*
NGC 6720 *see* Ring Nebula
NGC 6802, 229, *239*
NGC 6803, 213, *214*
NGC 6804, **212**, 212–13, *214*
NGC 6811, 208–210, **209**, *214*
NGC 6819, 203, *214*
NGC 6822, 107, *122*
NGC 6826, 204, *214*
NGC 6838 *see* M71
NGC 6853 *see* Dumbbell Nebula
NGC 6866, 203, *214*
NGC 6871, 201, *214*
NGC 6879, 229, *239*
NGC 6886, 229, *239*
NGC 6891, 191, *192*
NGC 6894, 204, *214*
NGC 6902, 233, *239*
NGC 6905, 190, *192*
NGC 6910, 201, *214*
NGC 6912, 234, *239*
NGC 6913 *see* M29

NGC 6928, 191, *192*
NGC 6934, 190, **190**, *192*
NGC 6939, 174, *192*
NGC 6946, 175, *192*, 204, *214*
NGC 6956, 191, *192*
NGC 6960 *see* Veil Nebula
NGC 6979 *see* Veil Nebula
NGC 6981 *see* M72
NGC 6992-5 *see* Veil Nebula
NGC 6994 *see* M73
NGC 6997, 201, *214*
NGC 7000 *see* North America Nebula
NGC 7006, **189**, 189–90, *192*
NGC 7009 *see* Saturn Nebula
NGC 7023, **177**, 178, *192*
NGC 7078 *see* M15
NGC 7089 *see* M2
NGC 7092 *see* M39
NGC 7099 *see* M30
NGC 7129, 179, *192*
NGC 7142, 176, *192*
NGC 7172, 233, **234**, *239*
NGC 7173, 233, **234**, *239*
NGC 7174, 233, **234**, *239*
NGC 7176, 233, **234**, *239*
NGC 7209, 196, **197**, 197–8, *214*
NGC 7243, **196**, 196–7, *214*
NGC 7245, 198, *214*
NGC 7293 *see* Helix Nebula
NGC 7296, 198, *214*
NGC 7317 *see* Stephan's Quintet
NGC 7318A *see* Stephan's Quintet
NGC 7318B *see* Stephan's Quintet
NGC 7319 *see* Stephan's Quintet
NGC 7320 *see* Stephan's Quintet
NGC 7331, 199, *214*
NGC 7380, 175, **175**, *192*
NGC 7394, 15, **15**, *26*, 199, *215*
NGC 7410, 237, *239*
NGC 7418, 237, *239*
NGC 7421, 237
NGC 7424, 237, *239*

NGC 7448, 220, *239*
NGC 7454, 220, *239*
NGC 7456, 237, *240*
NGC 7462, 237, *240*
NGC 7469, 220, *240*
NGC 7479, 219–20, **220**, *240*
NGC 7510, 176, *192*
NGC 7534, 251, *259*
NGC 7541, 250, *259*
NGC 7552, 237–8, *240*
NGC 7582, 238, *240*
NGC 7590, 238, *240*
NGC 7599, 238, *240*
NGC 7619, 221, *240*, 251, *259*
NGC 7626, 221, *240*, 251, *259*
NGC 7635, 244–5, *259*
NGC 7654 *see* M52
NGC 7662, 95, *98*
NGC 7678, 221, *240*
NGC 7772, 221, *240*
NGC 7789, 243, **243**, *260*
NGC 7819, 221, *240*
NGC 7839, 221, *240*
NGC 7840, 221
"nonexistent" objects
    Messier objects, 90, 127–9, 235
    NGC objects, 11–15, 133, 221, 226–7
North America Nebula (NGC 7000), **201**, 201–2, *214*
Northern Cross *see* Cygnus
Nutley, Anthony, 8

O III filters *see* nebula filters
O'Brian, William, 231
observing tips and techniques
    aperture mask, 83
    for beginners, 45, 129
    binocular vision, 208
    charts and their limiting magnitude, 140, 211, 269–70
    clusters, drawing, 33
    defocusing, 56

double stars, 83, 189
faint and diffuse objects, locating, 45
filter "flicking" or "flipping," 51, 95, 213, 268, 274
image contrast, maximizing, 271
iris diaphragm, 8
low-surface brightness objects, 34–5, 36
occulting mask, 67
planetary nebulae, 153–4, 166
rocking telescope to see faint objects, 29, 246
scattered or stray light, reducing, 86, 146, 219, 250
spectroscopic, 166–7
star-hopping, 118–19
strip method, 137–9
sweeping, 35
suboptimal skies, locating objects in, 17
telescopes, choice of, 242
see also averted vision
occulting bar *see under* eyepieces
Olcott, William Tyler, 114, 140
O'Meara, Donna Donovan, 186
O'Meara, Stephen J., 86
Omega Centauri, 99, **100**, **101**, 102, *122*
Ophiuchus, 162–7
70 Ophiuchi, 166, *171*
BF Oph, 165, *171*
Orion, 1–9, **2**, 12, 13–14, **257**
Betelgeuse (Alpha [α] Orionis), pronunciation, 188
T Ori, 4, *26*
Theta (θ) Ori *see* Trapezium
Zeta (ζ) Ori, 7, 8
Orion Nebula (M42), 2–5, **3**, *26*, 277, *282*
Orion's Belt, 8
Orion's Sword, 3, 8
Otto Struve 44 (OΣ 44), 39, *48*
Owl Nebula, 124, **124**, *148*
Oyster Nebula (NGC 1501), 29, **30**, *47*
Palomar Mountain Observatory

48-inch Schmidt telescope, 12*n*, 77, 105, 107, 133
200-inch Hale Telescope, 134
Pannekoek, Antonie, 185
Paris Observatory, 12-inch refractor, 105
Parsons, William *see* Rosse, 3rd Earl of
Pasterfield, Dunstan, 167, 176
Pazmino, John, 32
Pazmino's Cluster (Stock 23), 32–3, **33**, *48*
Pease 1, 224, *240*
Pegasus, 199–200, 217–24
Peiresc, Nicolas, 3
Pelican Nebula (IC 5070), 201, **201**, *215*
Peltier, Leslie C., 7, 28, 127, 225, 263
Pennington, Harvard, 89
Perkins, Billy, 232
Perseus, 15–19, 33–41
h and Chi (χ) Persei *see* Double Cluster
Xi (ξ) Per, 35
Perseus A , 41
Perseus Cluster (Abell 426), 41, **41**, *48*
Peterson, A. D., 11
Peterson, Harold, 10
photography *see* astrophotography
Pickering, Edward C., 277
Pickering, William H., 5
Pisces, 248–51
Piscis Austrinus, 233, 236
PK 38+12.1 *see* Cannon 3-1
PK 45–2.2, 213, *215*
PK 52–2.2, 213, *215*
PK 52–4.1, 213, *215*
PK 164+31.1, 83, **83**, *98*
PK 171–25.1, 46, *48*
Pleiades (Seven Sisters, M45), 1, 1*n*, *26*, 261–6, **262**, **264**, 276, *282*
Pleinis, Michael, 231

Polaris (Alpha Ursae Minoris) *see* Ursa Minor
Polarissima (NGC 3172), 173, **174**, *193*
Pollux (Beta Geminorum) *see* Gemini
Porter turret telescope *see* Stellafane
Praesepe *see* Beehive
precession, 2, 152
prisms, 166, 229, 274
    Amici prism, 237
Proctor, Richard A., 185
Ptolemy, 16, 16*n*, 100, 118
Pulkovo Observatory, 31*n*

Rachal, Darian, 8
radio observations, amateur, 40–41, 267
Rattley, Gerry, 105, 107, 126
Regulus (Alpha Leonis) *see* Leo
Reiland, Tom, 178
Riddle, David, 232
Ring Nebula (M57), 179, **189**, *193*
Ring-tail Galaxy (Antennae, NGC 4038-39), 140, **144**, 144–5, *149*
Roberts, Isaac, 271
Rolwicz, Frank, 145
Römer, Jan, 159–60
Rosette Nebula (NGC 2237-39), 49–51, **50**, *73*
Rosse, 3rd Earl of (Lord Rosse), 4, 37–8, 37*n*, 76, 83–4, 105, 124, 131, 158, **159**, 179, 180, 236, 247
Royer, Augustin, 195

S147, 205
Sagitta, 226–30, **227**
    S Sagittae, 230, *240*
    Theta (θ) Sge, 230, *240*
    Zeta (ζ) Sge, 230, *240*
Sagittarius, 233
Sagot, Robert, 23, 54, 95, 174, 197, 243
Saturn Nebula (NGC 7009), 235–6, **236**, *240*

Schlyter, Paul, 86
Schmidt, Joseph, 117
Schmidt, Robert, 77
Scorpius
    Antares (Alpha [α] Scorpii), 83, 169, *171*, 258
Scotten, George, 37
Sculptor Dwarf Galaxy (ESO 351-G30), 255, **255**, *260*
Sculptor group of galaxies, 254
Sculptor, 251–6
seeing *see* sky conditions
Serpens, 160–61, 163
Serviss, Garrett P., 4, 18, 52, 136, 160, 258, 277
Sextans, 117–20
    Alpha (α) Sextantis, 117
Sh2-276 *see* Barnard's Loop
Shapley, Harlow, 81*n*, 277
Simmons, Harold, 8
Simpson, Cliff, 40, 267
Sirius (Alpha Canis Majoris) *see* Canis Major
Skalnate Pleso Observatory, 12*n*
Skiff, Brian, 17, 51, 52, 56, 63–4, 67, 88, 146, 170
sky conditions, 17, 35, 82, 84, 185–6
    measures of, 19, 38, 53, 84–5, 174, 218, 236, 246, 261, 263, 273
    see also atmospheric absorption
Smyth, William H., 3, 18, 54, 65, 88, 90, 94, 110, 111, 128, 153, 158, 160–61, 200, 206, 222, 243
Snow, Margaret, 8, 9, 51
Sombrero Galaxy (M104), 145–6, **146**, *149*
South Galactic Pole, 251
southern objects visible to northern observers, 19–24, 41–4, 68–71, 93, 99–102, 233–8, 251–6
spectra
    of meteors, 127
    of NGC 1535, 46

of Orion Nebula 4
of planetary nebulae, 152–3, 166, 213
spectroscope, direct-vision *see* prisms
Spindle Galaxy (NGC 3115), 118–20, **119**, **120**, *122*
Square of Pegasus, 217–221, **218**
star-hopping, 118–19
Stebbins, Joel, 246
Stein, Mark K., 56, 160
Stellafane (convention), 11, 19, 37, 133, 275, **279–80**
    Porter turret telescope, 12-inch, 19, 117, 125, 159, 181, 183, 203, 219–20, 226, 274, 243
Stephan, Édouard, 182, 186
Stephan's Quintet (NGC 7317, 7318A, 7318B, 7319, 7320), 199–200
Steward Observatory, 36-inch reflector, 4, 11, 224
Stock 23 *see* Pazmino's Cluster
Stockton, Edward, 232
Stoney, Bindon, 4, 158–9
Stony Ridge Observatory, 30-inch reflector, 95, 248
Struve, Otto, 31*n*
Struve, Wilhelm, 31*n*, 167, 177, 188
Struve 484 and 485 (Σ484 and Σ485), 31, *48*
Struve 1694 (Σ1694), 177, *193*
Struve 2894 (Σ2894), 196, *215*
Struve 2923 (Σ2923), 177, *193*
Struve 2924 (Σ2924), 177, *193*
Sulentic, Jack W., 11, 14
Summer Triangle, **185**, 210, 211, **257**
Sunflower Galaxy (M63), 132, **132**, *147*
Swift, Lewis, 23, 50, 91, 113
Sword of Orion, 3, 8

Taurus, 46, 205, 261–8
    Merope (23 Tauri), 263–8, **264**
    T Tau, 267, *282*
Taurus Poniatovii, 166

Taylor, N., 231
Telescopium Herschelii, 76
Tempel, Wilhelm, 263
Tempest, Buddy, 231
Tennyson, Alfred, Lord, 1*n*
test objects *see* challenges
tests *see* sky conditions, visual acuity
Texas Star Party, 34, 151, 170, 208, 256
Texereau, Jean, 9, 23, 54, 95, 174, 197, 243
Thomson, Malcolm J., 179
Thoreau, Henry David, 30
Tifft, William, 11
Tirion, Wil, 162
Tombaugh, Clyde, 63
Tombaugh 1, 63–4, *73*
Tombaugh 2, 64, *73*
Trapezium (Theta [θ] Orionis), 3, *26*
Triangulum, 9–11, **9**
Triangulum Spiral (M33), 9–11, **9**, **10**, 25
Trouvelot, Leopold, 4
Tycho's star, 244

U. S. Naval Observatory, 26-inch refractor, 181
U. S. Naval Research Laboratory, 45
ultrahigh-contrast (UHC) filters *see* nebula filters
universe, age of, 175*n*
Ursa Major, 79–82, 115, 123–7, 173
    10 Ursae Majoris, 82, *96*
Ursa Minor, 173–4
    Polaris (Alpha [α] Ursae Minoris), 173, 177, *193*

Van Vleck Observatory *see* Wesleyan University
Vanderbilt University, 6-inch refractor, 34
Vega (Alpha Lyrae) *see* Lyra
Vehrenberg, Hans, 164, 232

Veil Nebula (NGC 6960, 6979, 6992-5), 204–8, **205**, *215*
Vince, Samuel, 225
Virgo, 137–9, 145–6
visual acuity, 10, 45, 52, 63, 206
    binocular vision, 232
    effect of cataract surgery, 94–5, 143, 153
    tests for, 34
    variation in, 223
Vulpecula, 182–4

Wallenquist, Åke, 39
Warner Observatory, 16-inch refractor, 91
Washburn Observatory, 234, 246
    6-inch refractor, 77, 188
    15½-inch refractor, 92
Webb, Thomas W., 18, 38, 39, 55, 59, 63, 159, 169, 183, 191, 200, 222, 224
Welch, Doug, 247
Wesleyan University (Van Vleck Observatory), 20-inch refractor, 91, 111, 116, 142, 161, 164, 169, 181, 225, 228, 242
Whirlpool Galaxy (M51), **130**, 131–2, *147*
Wilds, Richard, 206
Wilson, Barbara, 190, 254
Wilson, Thomas W., 248
Winnecke, August, 23
Winter Star Party, 37, 42, 99, 169
Wolf, Max, 114
Wolfe, Agnes, 204
Wooten, Wayne, 8

zodiacal band/light, 52, 211

Return the attached card for a FREE ISSUE of *Sky & Telescope* magazine, the Essential Magazine of Astronomy.

FREE issue for readers of *Deep Sky Wonders!*

If observing is your passion, we can help you make the most of it. Whether you're looking for a Messier object, an obscure lunar crater, or a challenging double star, the editors of *Sky & Telescope* are there to guide you every step of the way. We pack every issue with helpful observing tips, in-depth astronomical news coverage, thought-provoking science articles, stunning maps and star charts, and awe-inspiring astrophotography. Also inside are our renowned S&T Test Reports, in which we subject telescopes, astronomical gear, and software to intense scrutiny. **Simply fill out the card below and drop in the mail.**

YES! I accept your invitation to sample a FREE issue of *Sky & Telescope*. If I like what I see, I'll pay $42.95 for 11 more issues (12 issues total). If I decide not to subscribe, I'll mark "cancel" on your invoice. Either way, the FREE issue is mine to keep without further obligation.

NAME

ADDRESS

CITY   STATE   ZIP

PHONE

E-MAIL

Offer good in US and possessions. Canadian rate: one year for $49.95 (includes GST). International rate: one year for $61.95. All rates payable in US funds. Offer subject to change.

VISIT US AT *SkyandTelescope.com*

BUSINESS REPLY MAIL
FIRST-CLASS MAIL   PERMIT NO. 10565   CAMBRIDGE MA

POSTAGE WILL BE PAID BY ADDRESSEE

SKY & TELESCOPE

49 BAY STATE ROAD
CAMBRIDGE, MA 02138-9700
USA

NO POSTAGE
NECESSARY
IF MAILED
IN THE
UNITED STATES